NITRATES *in* GROUNDWATER

NITRATES *in* GROUNDWATER

Larry W. Canter
University of Oklahoma
Norman, Oklahoma

Library of Congress Cataloging-in-Publication Data

Canter, Larry W.
Nitrates in groundwater / Larry W. Canter.
p. cm.
Includes bibliographical references and index.
ISBN 0-87371-569-1 (alk. paper)
1. Nitrates--Environmental aspects. 2. Agricultural pollution.
3. Groundwater--Pollution. I. Title.
TD427.N5C36 1996
628.1′684--dc20 96-5855
CIP

Lewis Publishers is an imprint of CRC Press

International Standard Book Number 0-87371-569-1
Library of Congress Card Number 96-5855
Printed in the United States of America 2 3 4 5 6 7 8 9 0
Printed on acid-free paper

PREFACE

This book summarizes current information on nitrogen in the subsurface environment, with emphasis given to agricultural sources of nitrates and appropriate prevention and treatment measures. Efforts have been made to include both historical and current reference materials; however, it is realized that such efforts frequently miss major works. Accordingly, this book should be viewed as a "snapshot" in time, related to the subject of nitrates in groundwater. The importance of the subject is best reflected by the realization that nitrates are probably the most ubiquitous of all chemical contaminants in groundwater.

Nitrate contamination of groundwater is typically a result of nitrogen movement into the unsaturated (vadose) zone, the occurrence of transformation processes in either the unsaturated or saturated zones, and nitrate movement with groundwater. Following an introductory chapter (1) that highlights the nitrogen cycle, sources of nitrates, and health concerns related to nitrates in groundwater, Chapter 2 addresses subsurface processes. A number of hydrogeological factors and agricultural practices influence the concentration of nitrates in groundwater. Examples of such factors and practices include precipitation/runoff, irrigation, soil type, soil depth, geological features (e.g., karst areas, fertilizing intensity, and crop types), and land usage. Chapter 3 highlights these factors and practices through a series of case studies from the United States and Europe.

Vulnerability mapping of groundwater resources represents an emerging tool useful in management decisions. Numerous techniques for vulnerability mapping have been developed, with these techniques generally involving the composite consideration of selected factors descriptive of the hydrogeology in given geographical areas. Vulnerability mapping is the subject of Chapter 4. Chapter 5 presents a nitrate pollution index, which can be used in evaluating actual or potential nitrate pollution from agricultural areas. Included in Chapter 5 is information on the development and testing of the index. There are numerous mathematical models and submodels that address the transport and fate of nitrogen and nitrates in the subsurface environment; Chapter 6 provides a summary. Categories addressed include source characterization models, nitrate transport models, and management models for developing pollution prevention and control programs. Detailed information is included on several nitrate transport models, particularly those involving mass balances associated with the unsaturated and saturated zones.

There are several fertilizer and management measures potentially useful in minimizing nitrate pollution of groundwater that results from agricultural practices. Included in such measures is "Best Management Practice." This measure and other measures are the subject of Chapter 7. Examples include matching the timing of fertilizer application to meet crop needs; matching fertilizer application rate to crop needs, and limiting fertilizer application in well recharge areas that are vulnerable to pollution. These measures represent a compendium of approaches that could be used to control nitrate pollution of groundwater from agricultural practices. The approaches can be used either singly or in various combinations. Finally, there are several treatment measures that can be used to remove excessive concentrations of nitrate from groundwater systems. The measures are primarily for application to pumped groundwater, and they are addressed in Chapter 8. Cost-effective treatment processes include ion exchange involving anionic resins, and biological denitrification. *In situ* denitrification is also addressed in Chapter 8.

This book could be used as a specialized reading component of an upper division or graduate course addressing groundwater quality protection. The course could be offered through the auspices of an engineering, geology, environmental science, or geography department, or it could be offered as an interdisciplinary effort of several departments. The book could also be used by practicing hydrogeologists; agricultural, chemical, and environmental engineers; environment scientists; chemists and microbiologists; soil scientists; geographers; and agronomists working on groundwater quality issues. In addition, the book would be useful to policy specialists working in groundwater protection programs at all governmental levels.

The author wishes to express his appreciation to several persons associated indirectly with the development of this book. First, Luciano Mulazzani of C. Lotti and Associates in Rome, Italy and Eric Giroult of the World Health Organization in Copenhagen, Denmark provided an opportunity for the author to work on a groundwater pollution study involving nitrates from agricultural sources. The study was conducted in Tunisia, Algeria, and Morocco in North Africa; it was sponsored by the United Nations Development Program.

The author wishes to express his gratitude to a number of individuals who have participated directly or indirectly in the assemblage of information related to this book. These include former students, such as Drs. Robert Knox, Mohammed Lahlou, and Nadim Farajalla, and also Ken Herrington, Satish Kamath, and Laetitia Ramolino. These students have conducted research or participated in various related projects as part of their graduate work. In particular, Chapter 5 herein is largely based on the thesis research of Ms. Ramolino. The support and assistance of current and former colleagues at the University of Oklahoma is also gratefully acknowledged. Included are Drs. Robert Knox, James Robertson, David Sabatini, and Leale Streebin, and Professor George Reid.

Of major importance to the author is the positive attitude and helpfulness of Ms. Mittie Durham and Ms. Ginger Geis of the Environmental and Groundwater Institute, University of Oklahoma, in conjunction with typing and retyping original parts of this manuscript. Their technical abilities and pleasant attitude have made this book possible. The excellent typing skills of Ms. Geis is particularly acknowledged for the final copy.

This author also expresses his gratitude to the College of Engineering, University of Oklahoma for its support during the preparation of this book. Included are Dr. Ronald Sack of the School of Civil Engineering and Environmental Science and Dean Billy Crynes of the College of Engineering. Finally, the author thanks his wife for her encouragement and continued support in the process of developing this manuscript.

Larry W. Canter

AUTHOR

Larry W. Canter, P.E., Ph.D., is the Sun Company Chair of Ground Water Hydrology, George Lynn Cross Research Professor, and Director of Environmental and Ground Water Institute, University of Oklahoma, Norman, Oklahoma. Dr. Canter received his Ph.D. in environmental health engineering from the University of Texas in 1967, M.S. in sanitary engineering from the University of Illinois in 1962, and B.E. in civil engineering from Vanderbilt University in 1961. Before joining the faculty of the University of Oklahoma in 1969, he was on the faculty at Tulane University and was a sanitary engineer in the U.S. Public Health Service. He served as Director of the School of Civil Engineering and Environmental Science at the University of Oklahoma from 1971 to 1979 and as Co-Director of the National Center for Ground Water Research (a consortium of the University of Oklahoma, Oklahoma State University, and Rice University) from 1979 to 1992.

Dr. Canter's research interests include environmental impact assessment (EIA) methodologies, groundwater pollution source evaluation and groundwater protection, soil and groundwater remediation technologies and market-based approaches for air quality management and impact mitigation. Currently, he is conducting research on cumulative impact assessment and valuation methods for groundwater resources. In 1982 he received the Outstanding Faculty Achievement in Research Award from the College of Engineering at the University of Oklahoma and in 1983 the Regent's Award for Superior Accomplishment in Research.

Dr. Canter has co-authored five books related to groundwater; examples include *Groundwater Pollution Control* (Lewis, 1985), *Ground Water Quality Protection* (Lewis, 1987), and *Subsurface Transport and Fate Processes* (Lewis, 1993). He has written six books related to environmental impact assessment; one example is *Environmental Impact Assessment* (McGraw-Hill, 1996, second edition). He is also the author or co-author of numerous book chapters, refereed papers, and research reports related to groundwater quality management.

Dr. Canter served on the U.S. Army Corps of Engineers Environmental Advisory Board from 1983 to 1989. Since 1993, he has been a member of the

Consultative Expert Group on Environmental Impact Assessment of the United Nations Environment Program in Nairobi, Kenya. Finally, he currently is Chairman of the Committee on Assessing the Future Value of Groundwater; this Committee is part of the Water Science and Technology Board of the National Research Council.

CONTENTS

DEDICATED TO

Donna

Doug, Carrie, and Haley

Steve

and Greg

Tests being expressed

1 FUNDAMENTAL ASPECTS OF NITRATES IN GROUNDWATER

INTRODUCTION

Groundwater pollution is a growing concern everywhere in the world. The basic problems are twofold: (1) extensive usage of groundwater leading to overdrafts and declining groundwater levels, and (2) pollution of fresh groundwater leading to undesirable effects on users or curtailment of groundwater usage. Groundwater is the source of drinking water to about 50% of the overall population in the U.S., and to over 90% of the rural population (Office of Technology Assessment, 1990). Groundwater is also essential to agricultural operations in many regions of the country.

Nitrates are one of the most problematic and widespread of the vast number of potential groundwater contaminants (Keeney, 1986). Nitrate test results can be expressed as either nitrate–nitrogen (NO_3–N) or as nitrate (NO_3). The drinking water quality standard in the U.S. for NO_3–N is 10 mg/l, whereas, for NO_3 it is 45 mg/l (Chandler, 1989).

To systematically address fundamental aspects of nitrates in groundwater, it is necessary to: (1) understand the nitrogen cycle from a chemical/microbiological perspective; (2) recognize the nitrogen cycle as related to the soil and groundwater environment; (3) delineate natural and man-related sources of nitrate to groundwater; and (4) identify health-related and other effects of excessive concentrations of nitrate in groundwater. This chapter highlights these four topics in addition to the results of a recent nationwide study that addressed, among other topics, nitrates in groundwater.

THE NITROGEN CYCLE

Nitrogen can exist in many forms in the environment. The movement and transformation of these nitrogen compounds through the biosphere can be characterized by the nitrogen cycle; this cycle is depicted in Figure 1.1 (U.S. Environmental Protection Agency, 1994). The atmosphere serves as a reservoir of nitrogen in the form of nitrogen gas (the atmosphere is 79% nitrogen).

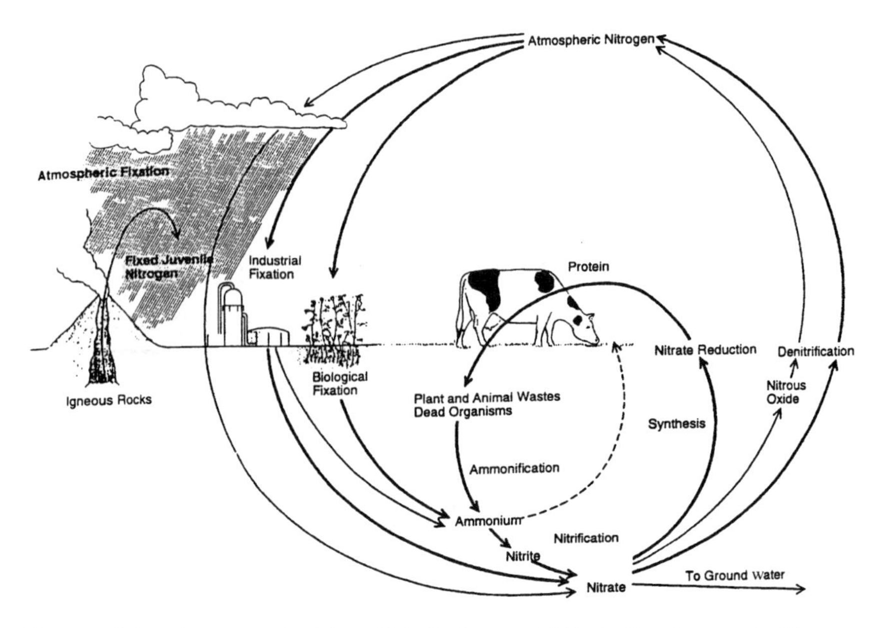

Figure 1.1 The nitrogen cycle. (From U.S. Environmental Protection Agency, in *Nitrogen Control*, Technomic Publishing, Lancaster, PA, 1994, pp. 1–22.)

Atmospheric nitrogen must be combined with hydrogen or oxygen before it can be assimilated by higher plants; the plants, in turn, are consumed by animals. Human intervention through industrial nitrogen fixation processes and the large-scale cultivation of nitrogen-fixing legumes has played a significant role in altering the historical nitrogen cycle. The amount of nitrogen fixed annually by these two mechanisms now exceeds by as much as 10% the amount of nitrogen fixed by terrestrial ecosystems before the advent of agriculture (U.S. Environmental Protection Agency, 1994).

Nitrogen can form a variety of compounds due to its different oxidation states. In the biosphere, most changes from one oxidation state to another are biologically induced. The following nitrogen forms are of interest in relation to the soil/water environment (U.S. Environmental Protection Agency, 1994).

Nitrogen compound	Formula	Oxidation state
Ammonia	NH_3	–3
Ammonium ion	NH^+_4	–3
Nitrogen gas	N_2	0
Nitrite ion	NO^-_2	+3
Nitrate ion	NO^-_3	+5

The un-ionized, molecular ammonia exists in equilibrium with the ammonium ion, the distribution of which depends upon the pH and temperature of the biospheric system; in fact, very little ammonia exists at pH levels less than neutral. Transformation of nitrogen compounds can occur through several mechanisms, including fixation, ammonification, synthesis, nitrification, and denitrification.

Each can be promulgated by particular microorganisms with either a net gain or loss of energy; energy considerations are basic to determining which reactions occur (U.S. Environmental Protection Agency, 1994).

Fixation of nitrogen refers to the incorporation of inert, gaseous nitrogen into a chemical compound such that it can be used by plants and animals. Fixation of nitrogen from N_2 gas to organic nitrogen is predominantly accomplished by specialized microorganisms and the associations between such microorganisms and plants (U.S. Environmental Protection Agency, 1994). Atmospheric fixation by lightning and industrial fixation processes (fertilizer and other chemicals) plays a smaller, but significant, role as a fixation method. The following processes and products are relevant.

Fixation process		**Product**
	→ Biological	→ Organic nitrogen compounds
N_2 gas	→ Lightning	→ Nitrate
	→ Industrial	→ Ammonium, nitrate

Ammonification refers to the change from organic nitrogen to the ammonium form. An important hydrolysis reaction involves urea, a nitrogen compound found in urine:

$$\underset{\text{Urea}}{H_2NCONH_2} + 2H_2O \xrightarrow[\text{urease}]{\text{enzyme}} \underset{\text{Ammonium carbonate}}{(NH_4)_2CO_3}$$

In general, ammonification occurs during decomposition of animal and plant tissue and animal fecal matter and can be expressed as follows:

$$\text{Organic nitrogen} + \text{Microorganisms} \rightarrow NH_3/NH_4^+$$

(Protein, amino acids, etc.)

Synthesis, or assimilation, refers to biochemical mechanisms that use ammonium or nitrate compounds to form plant protein and other nitrogen-containing compounds as follows:

$$NO_3^- + CO_2 + \text{Green plants} + \text{Sunlight} \rightarrow \text{Protein}$$

$$NH_3/NH_4^+ + CO_2 + \text{Green plants} + \text{Sunlight} \rightarrow \text{Protein}$$

Animals require protein from plants and other animals. With certain exceptions, they are not capable of transforming inorganic nitrogen into an organic nitrogen form.

Nitrification refers to the biological oxidation of ammonium ions (U.S. Environmental Protection Agency, 1994). This is accomplished in two steps: first to the nitrite form, then to the nitrate form. Two specific chemoautotrophic bacterial genera are involved, using inorganic carbon as their source of cellular carbon as follows:

$$NH_4^+ + O_2 \xrightarrow[\text{bacteria}]{\text{Nitrosomonas}} NO_2^- + O_2 \xrightarrow[\text{bacteria}]{\text{Nitrobacter}} NO_3^-$$

The transformation reactions are generally coupled and proceed rapidly to the nitrate form; nitrite levels at any given time are relatively low. The nitrate formed may be used in synthesis to promote plant growth, or it may be subsequently reduced by denitrification, as suggested by Figure 1.1 (U.S. Environmental Protection Agency, 1994).

Denitrification refers to the biological reduction of nitrate to nitrogen gas. The reduction can proceed through several steps in a biochemical pathway. A fairly broad range of heterotrophic bacteria are involved in the process, requiring an organic carbon source for energy as follows (U.S. Environmental Protection Agency, 1994):

$$NO_3^- + \text{Organic carbon} \rightarrow NO_2^- + \text{Organic carbon} \rightarrow N_2 + CO_2 + H_2O$$

It should be noted that if both oxygen and nitrate are present the bacteria typically will preferentially use oxygen in the oxidation of the organic matter because it yields more energy. Thus, for denitrification to proceed, anoxic conditions must usually exist, although this is not strictly the case for all bacteria (U.S. Environmental Protection Agency, 1994).

The ammonification, synthesis, nitrification, and denitrification processes are the primary mechanisms employed in the treatment of contaminated groundwater for nitrogen control and/or removal. Additional information on transformation processes is found in Chapter 2; Chapter 8 highlights treatment technologies for contaminated groundwater.

The transport mechanisms primarily responsible for the movement of nitrogen through the environment include precipitation, dustfall, sedimentation in water systems, wind, groundwater movement, stream flow, overland runoff, and volatilization (U.S. Environmental Protection Agency, 1994). Although these are not mechanisms by which transformations take place, they can cause a change in the environment whereby conditions will force changes and transformations will occur. Environmental conditions that affect the above processes include temperature, pH, microbial flora, oxidation/reduction potential, and the availability of substrate, nutrients, and oxygen. Although transport and transformation mechanisms are typically considered as individual processes, it is important to recognize that these comprise a dynamic continuum and there may be no distinct boundary governing the transformation of specific forms of nitrogen (U.S. Environmental Protection Agency, 1994).

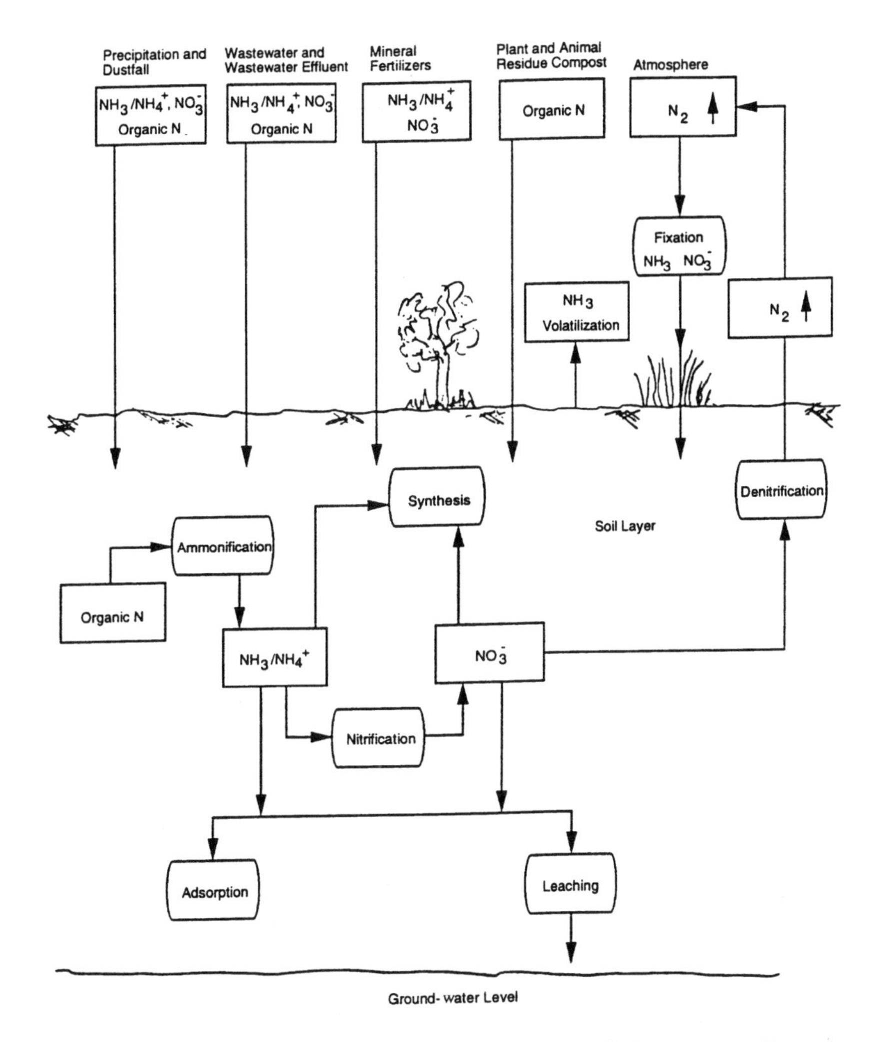

Figure 1.2 The nitrogen cycle in soil and groundwater. (From U.S. Environmental Protection Agency, in *Nitrogen Control*, Technomic Publishing, Lancaster, PA, 1994, pp. 1–22.)

THE NITROGEN CYCLE IN SOIL AND GROUNDWATER

Figure 1.2 delineates the major aspects of the nitrogen cycle associated with the soil-groundwater environment (U.S. Environmental Protection Agency, 1994). Nitrogen can enter the soil from the application of wastewater or sewage treatment plant effluent, commercial fertilizers, plant and animal matter, precipitation, and dustfall. In addition, nitrogen-fixing bacteria in the soil convert nitrogen gas into forms available to plant life. Humans have increased the amount of nitrogen fixed biologically by cultivation of leguminous crops (e.g., peas and beans) (U.S. Environmental Protection Agency, 1994).

Usually, more than 90% of the nitrogen present in soil is organic, either in living plants and animals or in humus originating from decomposition of plant and animal residues. The nitrate content is generally low because it is (1) taken up in synthesis; (2) leached by water percolating through the soil; or (3) subjected to denitrification activity below the aerobic top layer of soil. However, synthesis and denitrification rarely remove all nitrates added to the soil from fertilizers and nitrified wastewater effluents. Accordingly, nitrates leached from soil are a major groundwater quality problem in many areas in the U.S. and elsewhere around the world (U.S. Environmental Protection Agency, 1994).

SOURCES OF NITRATE IN GROUNDWATER

Sources of nitrate in groundwater can be considered in four categories: (1) natural sources, (2) waste materials, (3) row crop agriculture, and (4) irrigated agriculture (Keeney, 1986, 1989). Table 1.1 lists examples for these four categories. The major sources include intensive animal operations, with nitrate from over-application of animal wastes, and irrigated and row crop agriculture, with nitrate from fertilizer-induced mineralization of soil organic nitrogen and from over-application (Keeney, 1989; Canter, 1987). Septic tank systems and other sources such as landfills can be of concern in localized areas. Additional information on nitrate sources is given in Chapter 3. Examples of prevention measures for nitrate contamination of groundwater are found in Chapter 7.

Agrichemicals include both commercial fertilizers and pesticides. In 1986, approximately 75% of all farms in the U.S. were subjected to fertilizer applications (Office of Technology Assessment, 1990). Farming operations that may not use commercial fertilizers include extensive livestock operations, organic farms, and small hobby farms. Nitrate–nitrogen (NO_3–N) concentrations in vadose-zone water below agricultural fields are in the range of 5 to 100 mg/l, with frequent detections in the range of 20 to 40 mg/l (Bouwer, 1990).

Table 1.2 summarizes 1986 data on the national averages of nitrogen fertilizer applications, in selected countries, to arable land and permanent crop land (Conway and Pretty, 1991). Countries where more than 200 kg N/ha was applied are listed in Table 1.2.

Septic tank systems also represent a significant fraction of the nitrogen load to groundwater in the U.S. (U.S. Environmental Protection Agency, 1994). Approximately 25% of the population is served by individual home sewage disposal systems. Effluent from a typical septic tank system has a total nitrogen content of 25 to 60 mg/l. Of this, 20 to 55 mg/l exists as ammonia and less than 1 mg/l exists as nitrate. One specific study characterized a typical system effluent as containing approximately 7 mg/l organic nitrogen, 25 mg/l ammonium-nitrogen, and 0.3 mg/l nitrate-nitrogen (U.S. Environmental Protection Agency, 1994). Septic tank leach fields exhibit rapid nitrification of ammonium-nitrogen under

Table 1.1 Examples of Sources of Nitrate in Groundwater

Sources
Natural sources
Geologic nitrogen which can be mobilized and leached to groundwater via irrigation practices
Unmanaged (natural) climax forests that are normally nitrogen conserving; however, nitrogen losses to groundwater can occur from man-initiated clearcutting and other forest disturbances
Waste materials
Animal manures which may be concentrated in large commercial poultry, dairy, hog, and beef operations
Land application of municipal or industrial sludge or liquid effluent on croplands, forests, parks, golf courses, etc.
Disposal of household wastes or small business wastes into septic tank systems (septic tank plus soil absorption field)
Leachates from sanitary or industrial landfills or upland dredged material disposal sites
Row crop agriculture[a]
Nitrogen losses to the subsurface environment can occur as a result of excessive fertilizer application, inefficient uptake of nitrogen by crops, and mineralization of soil nitrogen
Nitrogen losses to the subsurface environment can occur as a function of fertilizer application rates, seasonal rainfall and temperature patterns, and tillage practices
Irrigated agriculture
Enhanced leaching of nitrogen from excessive fertilizer application rates and inefficient irrigation rates
Associated leaching of nitrogen from soils periodically subjected to leaching to remove salts so that the soils do not become saline and unproductive

[a] Refers to annual crops, tillage of the soil including plowing or some other form of seedbed preparation, and a long period when the soil may be bare, or weed control by herbicides or cultivation; also includes conservation tillage (Keeney, 1989).

aerobic conditions within the leach fields. Ammonium-nitrogen is easily exchanged in many soils below a leach field, whereas nitrate remains soluble and is easily transported to groundwater. If exchange sites become saturated, as in sandy soils, ammonium breaks through to groundwater before it nitrifies. When septic tank leach fields dry out in the summer, or are abandoned, previously adsorbed ammonium can be converted to nitrate, and eventually lost to leaching. Natural tertiary publicly owned treatment works (POTW) treatment systems utilizing soil infiltration or overland flow can typically produce a nitrogen loading to groundwater in a manner similar to septic tank leach fields. A typical final effluent of this type has a total nitrogen of 3 to 10 mg/l (U.S. Environmental Protection Agency, 1994).

Pursuant to Section 305(B) of the Clean Water Act of 1987 in the U.S., each state submits biennially a water quality report to the U.S. Environmental Protection Agency. In a review of 42 such reports in relation to public groundwater supplies, 19 states provided monitoring data on a total of 51 organic and 21 inorganic chemicals. Frequently detected contaminants included nitrates, trichloroethylene (TCE), volatile organic compounds (VOCs), arsenic, and fluorides. The most frequently identified contaminant sources were underground storage tanks (35 states), septic tank systems (31), surface impoundments (29), municipal landfills (27), agricultural activities (26), abandoned hazardous waste sites (24), regulated hazardous waste sites (19), land application/treatment (18), injection

Table 1.2 National Average Nitrogen Fertilizer Applications to Arable Land and Permanent Crops in Selected Countries, 1986

Countries over 500 kg N/ha
The Netherlands (557)
Singapore (550)
Countries 400–500 kg N/ha
Bahrain (470)
Countries 300–400 kg N/ha
Martinique (315)
Countries 200–300 kg N/ha
Egypt (260)
Korean DPR (253)
Belgium/Luxemburg (247)
U.K. (238)
Federal Republic of Germany (212)
Countries 100–200 kg N/ha
Korean Rep (195)
Saudi Arabia (154)
Japan (145)
China (138)
France (135)
Surinam (135)
Countries 50–100 kg N/ha
Cuba (99)
Mauritius (92)
U.S. (75)
Indonesia (64)
Malaysia (57)
Mexico (54)
Sri Lanka (53)
Fiji (50)
Countries 0–50 kg N/ha
Bangladesh (46)
Colombia (40)
India (39)
Philippines (31)
Zimbabwe (30)
Kenya (27)
Chile (20)
Thailand (13)
Nigeria (6)

Note: All countries applying more than 200 kg N/ha are included in this list.

From Conway, G. R. and Pretty, J. N., *Unwelcome Harvest — Agriculture and Pollution*, Earthscan Publications, London, 1991, p. 159. With permission.

wells (18), and on-site industrial and other landfills (16). This composite analysis indicates widespread groundwater contamination problems of a diverse nature, with nitrates being of concern throughout the U.S. (Canter and Maness, 1995).

NATIONAL SURVEY OF NITRATES (AND PESTICIDES) IN DRINKING WATER WELLS

The U.S. Environmental Protection Agency (EPA) recently completed a 5-year National Survey of Pesticides and Nitrates in drinking water wells. The drinking water wells include both community water system (CWS) wells and domestic wells. The EPA designed the survey with two principal objectives corresponding to its two phases. In Phase I, the EPA developed national estimates of the frequency and concentration of pesticides and nitrates in drinking water wells in the U.S. (U.S. Environmental Protection Agency, 1990). In Phase II, the EPA conducted statistical analyses of the collected data as well as information from other databases to enhance understanding of influencing factors on the presence of pesticides and nitrates in drinking water wells (U.S. Environmental Protection Agency, 1992a).

The Phase I survey was designed to yield results that are statistically representative of the nation's CWS wells. The EPA used statistical methods to select a nationally representative subset of CWS wells. The results are not representative of any state or local area. In the first stage, about 51,000 community water systems (CWSs) were identified throughout the country from the EPA's Federal Reporting Data System (FRDS). Then, the EPA characterized all the counties in the U.S. according to pesticide use and relative groundwater vulnerability — the two critical factors affecting the presence of pesticides in drinking water. The counties were categorized into high, moderate, low, and uncommon pesticide use by concentrating on agricultural pesticide use. Groundwater vulnerability scores were determined using the DRASTIC scoring system for each of the counties. The DRASTIC score was used to categorize the counties into areas of relatively high, moderate, and low groundwater vulnerability. It should be noted that fertilizer application rates are related to pesticide usage; thus, the characterization approach also has relevance to nitrates from agricultural sources. The emphasis herein will be on the nitrates results. Additional information on DRASTIC and other aquifer vulnerability mapping or evaluation techniques is given in Chapters 4 and 5.

A sample of 7083 CWSs was selected randomly from the 51,000 CWSs; the sample was stratified by the pesticide usage and groundwater vulnerability of the county in which each CWS was located (see Table 1.3) (U.S. Environmental Protection Agency, 1990). To improve the reliability of estimates made for the subset (domain) of CWS wells in the high vulnerability strata, the survey was designed to collect samples from higher percentages of systems from these strata. In the second stage, after a screening process, the EPA estimated that in the U.S., there are approximately 38,300 CWSs with groundwater wells, which together operate 94,600 wells.

Telephone interviews using a "CWS Screening Questionnaire" were held with the first-stage sample of 7083 CWSs. The results of the screening produced a frame of 5660 eligible systems from which the wells for water sample collection were selected. A sample of 599 eligible systems was then selected, with probabilities proportional to their number of wells for collecting water samples (see

Table 1.3 CWS First-stage Stratification Results

Stratum	Pesticide usage	DRASTIC score	Universe size[a]	Sample rate	Sample size[a]
1	High	High	1,310	0.22	293
2	High	Medium	2,669	0.10	276
3	High	Low	1,432	0.10	148
4	Medium	High	2,183	0.25	555
5	Medium	Medium	3,352	0.10	355
6	Medium	Low	2,559	0.10	265
7	Low	High	3,578	0.24	866
8	Low	Medium	7,199	0.10	771
9	Low	Low	5,545	0.12	647
10	Uncommon	High	3,472	0.22	777
11	Uncommon	Medium	9,511	0.11	1,069
12	Uncommon	Low	7,961	0.13	1,061
			50,771		7,083

[a] Number of systems.

From U.S. Environmental Protection Agency, "National Survey of Pesticides in Drinking Water Wells — Phase I Report," EPA 570/9-90-015, November, 1990, Office of Water, Washington, D.C.

Table 1.4) (U.S. Environmental Protection Agency, 1990). This number was selected with allowances for inability to sample prior to treatment; the minimum desired sample size after dropout was 564. The National Pesticide Survey (NPS) then obtained water samples from 566 selected CWS wells. After technical review of individual well plumbing systems and sample collection procedures, samples from 540 CWS wells were included in the analysis of results. An effort to minimize the temporal effect of the use and occurrence of pesticides in the wells was also undertaken by randomly allocating sampling dates to the wells across the data collection period. The survey was designed so that the number of detections of analytes in the 540 CWS wells provided estimates representative of the 94,600 CWS wells in the U.S. (U.S. Environmental Protection Agency, 1990). In addition to the CWS wells, 827 rural domestic wells (RDWs) were also sampled (Cohen, 1992). It is estimated that there are approximately 10.5 million RDWs in the U.S. (Cohen, 1992).

The well water samples were tested for the presence of 101 pesticides, 25 pesticide degradates, and nitrate. Nitrate and two pesticides, DCPA (dimethyl tetrachloroterephthalate) acid metabolites and atrazine, were detected most frequently. A total of 52.1% (49,300) of all the CWS wells (94,600) was estimated to contain detectable nitrate, of which 45% was due only to nitrate, and 7.1% included both nitrate and pesticides. Based on the results from the RDWs, it was estimated that 57% had detectable nitrate (Cohen, 1992). A total of 1.2% (1130) of the CWS wells was projected to have nitrate–nitrogen concentrations above the MCL/HAL (Maximum Contaminant Level)/(Health Advisory Level), which is equal to 10 mg/l. The median concentration from the Phase I survey was 1.6 mg/l; the maximum concentration was 13 mg/l; and the minimum reporting limit was 0.15 mg/l. Nitrate–nitrogen was projected to exceed the 10 mg/l MCL in 2.4% of the RDWs (Cohen, 1992).

Table 1.4 CWS Second-Stage Stratification Results

Stratum	Pesticide usage	DRASTIC score	Design sample size[a]
1	High	High	27
2	High	Medium	26
3	High	Low	14
4	Medium	High	45
5	Medium	Medium	32
6	Medium	Low	25
7	Low	High	73
8	Low	Medium	68
9	Low	Low	53
10	Uncommon	High	71
11	Uncommon	Medium	90
12	Uncommon	Low	75
			599

[a] Number of systems selected including allowances for inability to sample prior to treatment. Minimum desired sample size after dropout was 564.

From U.S. Environmental Protection Acency, "National Survey of Pesticides in Drinking Water — Phase I Report," EPA 570/9-90-D15, November, 1990, Office of Water, Washington, D.C.

Phase II of the NPS dealt with statistical analyses to test the hypotheses about the relation between individual variables and pesticide and nitrate detections and the degree of association between them. The EPA used a criterion of significance that ensured that there was less than a 5% probability that any single test result occurred by chance alone. The Phase II analyses included tests of association that used chi square analysis, linear and logistic univariate regression, and multiple regression to better understand the association of pesticide and nitrate detections in drinking water wells with different variables. The variables included groundwater sensitivity, pesticide and nitrate use, transport mechanisms, pesticide chemistry, and well condition (U.S. Environmental Protection Agency, 1992a). Variables influencing subsurface movement of contaminants are addressed in Chapters 4 and 5, and specific models related to subsurface transport and fate are delineated in Chapter 6.

The groundwater sensitivity factor was defined as the intrinsic susceptibility of an aquifer to contamination. Sensitivity addresses the hydrogeologic characteristics of the aquifer and the overlying soil and geologic materials (U.S. Environmental Protection Agency, 1992a). The pesticide and fertilizer use factor represents the scope and amount of pesticides and nitrogen fertilizers applied near sampled drinking water wells, and activities, such as cattle raising, that could contribute to the presence of nitrate in places where they could be transported into drinking water wells (U.S. Environmental Protection Agency, 1992a). Transport mechanisms were related to the factors that contribute, either directly or indirectly, to the movements of pesticide and nitrate by water, including precipitation, irrigation, surface water, drainage ditches near wells, other bodies of water such as rivers and lakes, and nearby operating wells. However, transport does not include recharge or groundwater flow factors (U.S. Environmental Protection Agency, 1992a). The chemical characteristics include characteristics of pesticides, such as organic partition coefficients (Koc) and half-life, and characteristics of groundwater, such as water temperature, pH, and conductivity, that could affect the behavior of pesticides or nitrates (U.S. Environmental Protection Agency, 1992a). Finally, the physical characteristics of wells included age, depth, state of repair, and protective devices, such as concrete pads, that could affect the presence of chemicals in the well water (U.S. Environmental Protection Agency, 1992a).

The data used for these variables were procured from both data generated by the NPS and data obtained from other sources such as the U.S. Geological Survey, National Oceanic and Atmospheric Administration (NOAA), census data, etc. Statistical tests were conducted to determine the association of each variable with pesticide or nitrate detection. Key results for tests of association came primarily from the analysis of nitrate detections in CWS wells. The relatively large number of associations involving nitrate detections is partly due to the fact that a significantly larger number of nitrate detections than pesticide detections was obtained. The Phase II analysis also carried out multivariate regression analyses to determine if particular combinations of variables would be strongly associated with the presence of nitrate and pesticides in groundwater.

Groundwater sensitivity at the county and subcounty level was determined using the EPA's agricultural DRASTIC system scores. The statistical analyses showed very little association of groundwater sensitivity with that of the presence of nitrate and pesticides in groundwater. Higher DRASTIC scores were not consistently associated with more frequent detections (the expected direction of association) or with less frequent detections. Different DRASTIC factors were more often associated with detections in CWS wells than with detections in RDWs. County level depth to water table was associated with pesticide detections, and county level hydraulic conductivity was associated with nitrate detections in CWS wells. Subcounty DRASTIC results showed similar inconclusive results. Again, stratification based on combined agricultural DRASTIC scores and county cropping patterns (the "cropped and vulnerable" strata) did not demonstrate a relationship between stratification variables and pesticide or nitrate detections (U.S. Environmental Protection Agency, 1992a).

Several factors related to pesticide use and agronomic activity were correlated with detections of nitrate and pesticides as well as concentrations of nitrate in groundwater. Significant correlations were found between sales of nitrogen fertilizer and nitrate concentrations in both CWS wells and RDWs. However, an analysis of the relationship between nitrogen fertilizer sales and detections of nitrate did not indicate an association between these variables. Crop value, acreage of cropland fertilized, and market value of livestock were all shown to be associated with nitrate concentrations above 0.15 mg/l in both CWS wells and RDWs. A correlation between the market value of livestock and nitrate detections was also observed. Multivariate analyses further supported that nitrate concentrations are affected by fertilizer use and agronomic activity. Two surrogate measures of agronomic activity that strongly related to pesticide and nitrate detections were the market value of crops and acres of fertilized rangeland (U.S. Environmental Protection Agency, 1992a).

Analyses of variables relating to transport of chemicals to groundwater, including several precipitation measures, indicated that increased precipitation was inversely related to detections and concentrations of nitrate. This result suggests that high levels of rainfall may cause the fertilizer to run off before entering groundwater, or to dilute the concentrations of nitrate to the point that they fall below the detection limits used in the survey. In contrast, flood irrigation was associated with a greater likelihood of detection in both CWS wells and RDWs. Nitrate detections were more likely in CWS wells drawing water from an unconfined aquifer. Some evidence was also obtained that nitrate detections in CWS wells were associated with moist conditions (U.S. Environmental Protection Agency, 1992a).

Some of the chemical characteristics factors were related with detections and concentrations of nitrates. The likelihood of detecting nitrate is greater in groundwater with low temperature or low pH. Also, the likelihood of detecting nitrate is greater in wells delivering water with low electrical conductivity. However, higher conductivity of well water was found to correspond to higher nitrate concentrations. In terms of well condition, shallower wells were found associated with nitrate detections in both CWS wells and RDWs, and with pesticide detections in CWS wells (U.S. Environmental Protection Agency, 1992a).

In addition to analyzing single factors, the EPA also studied whether combinations of variables are good predictors of the occurrence of nitrate. Table 1.5 shows that several factors, when analyzed together, are statistically related (U.S. Environmental Protection Agency, 1992b).

Table 1.5 Factors Associated with Contamination of CWS Wells

Nitrate Detections	Nitrate Concentrations
Fertilized pasture and rangeland	Monthly precipitation
Monthly precipitation	Well-water conductivity
Well-water pH	Total nitrogen fertilizer sales
Properly farmed	Well depth
	Palmer drought index score
	Market value of crops

From U.S. Environmental Protection Agency, "National Pesticide Survey — Update and Summary of Phase II Results," EPA 570/9-91-021, Winter, 1992, Office of Water, Washington, D.C.

Owing to the many limitations in the NPS and analysis process, it was suggested that the reported results should not be interpreted as conclusions drawn from controlled experiments or as evidence of causation. The results are statistical measures of association or correlation that were designed to test hypotheses developed from several scientific and policy questions. It was noted that the results should be used with caution as indicators of topics that may be suitable for further careful investigation. The factors that did not show statistical relationships should not be assumed to be unimportant.

The Phase II survey report included estimates of national population exposure and resultant health risks due to nitrate for RDWs and for CWSs (U.S.

Environmental Protection Agency, 1992a). Estimates were provided of the populations corresponding to quantiles of general interest (e.g., 95th and 99th percentiles), and of the number of individuals exposed above health-based levels. Summarized below is information related to CWSs. The results presented are based on the following nationwide data:

1. Approximately 136,000,000 people drink groundwater from CWS wells nationally.
2. There are approximately 94,600 wells in 35,800 CWSs that provide groundwater for drinking. On average, each system consists of approximately 2.6 wells.

Tables 1.6 and 1.7 present estimates of the number of people exposed to nitrate in CWS wells (U.S. Environmental Protection Agency, 1992a). The following comments are pertinent.

1. Aproximately 3 million people drink water from CWS wells that contain nitrate at a concentration of at least 10 mg/l, the MCL for nitrate. Of the people exposed to nitrate above this level, approximately 43,500 are expected to be infants at possible risk of developing methemoglobinemia, particularly if the water is also contaminated with bacteria.
2. Approximately 85 million people drink water from CWS wells that contain nitrate.
3. The median nitrate concentration to which people in the United States are exposed is approximately 0.63 mg/l.

Table 1.6 Estimates of Population Exposed to Nitrate in Community Water System Wells by Distribution Percentile

Percentile	People exposed	Concentration (mg/l)	95% Confidence interval Lower bound (mg/l)	95% Confidence interval Upper bound (mg/l)
Median	68,000,000	0.63	0.45	0.95
95	6,800,000	6.52	5.34	7.60
99	1,360,000	14.2	10.6	17.7

From U.S. Environmental Protection Agency, "Another Look: National Survey of Pesticides in Drinking Water Wells — Phase II Report," EPA 579/09-91-020, January, 1992, Office of Water, Washington, D.C.

Table 1.7 Estimates of Population Exposed to Nitrate by Concentration in Community Water System Wells

Concentration (mg/l)	Population exposed	95% Confidence interval Lower bound	95% Confidence interval Upper bound
All concentrations >0	85,300,000	78,100,000	98,900,000
≥10	2,980,000	1,600,000	4,260,000

From U.S. Environmental Protection Agency, "Another Look: National Survey of Pesticides in Drinking Water Wells — Phase II Report," EPA 579/09-91-020, January, 1992, Office of Water, Washington, D.C.

HEALTH CONCERNS RELATED TO NITRATES

The concentration of nitrates in groundwater is of primary concern due to potential human health impacts from groundwater usage. Depending on the use of the groundwater, animals, crops, or industrial processes could also be affected. The toxicity of nitrate to humans is due to the body's reduction of nitrate to nitrite. This reaction takes place in the saliva of humans of all ages and in the gastrointestinal tract of infants during their first 3 months of life. The toxicity of nitrite has been demonstrated by vasodilatory/cardiovascular effects at high dose levels and methemoglobinemia at lower dose levels (*Federal Register,* 1985).

Methemoglobinemia refers to an effect in which hemoglobin is oxidized to methemoglobin. When amounts of methemoglobin in blood increase, oxygen levels in blood dwindle. Cyanosis, the illness of oxygen starvation, results, and that particular cyanosis is referred to as methemoglobinemia (Winneberger, 1982). However, the effects of methemoglobinemia are rapidly reversible; and there are, therefore, no cumulative effects.

Infants up to 3 months of age are the most susceptible subpopulation with regard to nitrate. This is due to the fact that about 10% of ingested nitrate is transformed to nitrite in the infant (*Federal Register,* 1985). When nitrites combine with hemoglobin to form methemoglobin, the result is a diminished oxygen transport/transfer capability in the blood. The infant then suffers from cellular anoxia and clinical cyanosis (turns blue, hence the terms "blue baby" or "blue baby syndrome"). This phenomena can occur in infants when approximately 10% of the total hemoglobin has been converted to methemoglobin (World Health Organization, 1985). Additional reasons for concern include the low activity of the enzyme that reduces methemoglobin, higher susceptibility of infant methemoglobin to oxidation, and higher pH in the stomach and intestines that promotes bacterial reduction of nitrate to nitrite (Keeney, 1986).

In the U.S., the current maximum contaminant limit (MCL) for nitrate-nitrogen under the Primary Drinking Water Regulations is 10 mg/l (*Federal Register,* 1985). This level was based upon human case studies in which fatal poisonings in infants occurred following ingestion of groundwater containing nitrate concentrations greater than 10 mg/l nitrate–nitrogen. The World Health Organization (WHO) guidelines are 10 mg/l for nitrate–nitrogen and 1 mg/l for nitrite-nitrogen. The basis is that undesirable increases in methemoglobin levels in blood occur at levels from 10 to 20 mg/l nitrate; the ingestion of nitrite leads to a more rapid onset of clinical effects, thus the guideline value is correspondingly lower (1 mg/l). Additional information on water quality standards is summarized in Table 1.8 (World Health Organization, 1985).

An additional public health issue of potential concern is that several studies have shown that simultaneous ingestion of nitrite (or nitrate with amines) results in cancers of many organ systems. The N-nitroso compounds are presumed to be the ultimate carcinogenic substances (Tannenbaum and Green, 1985). Several epidemiological studies have indicated significant positive correlations between exposure to nitrate and cancer risk; for example, nitrate in drinking water has been correlated with gastric cancer risk in Colombia and England; and exposure

Table 1.8 Limits for Nitrate in Drinking Water

Organization	Year	Limit specification	mg/l of NO_3–N
WHO European standards	1970	Recommended	11.3
		Acceptable	11.3–22.6[a]
		Not recommended	>22.6
WHO International standards	1971		10.2
WHO Working Group	1977	General population	
		Acceptable	11.3
		Borderline	11.3–22.6
		Unacceptable	>22.6
		Infants <6 months	
		Unacceptable	11.3
U.S. EPA	1977	Maximum contaminant level	10.2
Health and Welfare Canada	1978	Maximum acceptable concentration	10.2
EEC (Directive on quality of surface waters intended for abstraction of drinking water)	1977	Imperative limit	11.3
		Guide level	5.6
EEC (Directive on quality of water for human consumption)	1980	Maximum admissible concentration	11.3
		Guide level	5.6
WHO guidelines for drinking water quality	1984	Guideline value	10

[a] Within this range, although a problem may not be apparent, physicians should be notified of the possible occurrence of infantile methaemoglobinaemia.

Note: WHO, World Health Organization; EEC, European Economic Community; EPA, Environmental Protection Agency.

From World Health Organization, "Health Hazards from Nitrates in Drinking Water — Report on a WHO Meeting in Copenhagen, March 5–9, 1984," Regional Office for Europe, Copenhagen, 1985, pp. 3, 49–66. With permission.

to nitrate-containing fertilizer appeared to be linked to gastric cancer mortality in Chile. However, it should be noted that high risk for gastric cancer correlate not only with nitrate, but also with several other dietary or environmental factors, and whether any of these associations actually involve causation is far from clear (Tannenbaum and Green, 1985; Kleinjans, et al., 1991).

The question of whether nitrite itself is a carcinogen is currently unanswerable; however, research is ongoing on this issue (*Federal Register,* 1985). The role of nitrite as a precursor to carcinogenic nitrosamines and other N-nitroso compounds is firmly established. Nitrite reacts with amines or amides under a variety of conditions to yield the N-nitroso derivatives, the preponderance of which are carcinogenic to animals. The expectation that these N-nitroso compounds are also human carcinogens suggests a mechanism whereby exposure to nitrite might result in carcinogenesis (Tannenbaum and Green, 1985).

Detailed information on the toxicological effects of nitrate and nitrite is provided by Life Systems, Inc. (1987). Topics addressed include toxicokinetics and human exposure, health effects of nitrate and nitrite in humans and animals, and mechanisms and quantification of toxicological effects. Additional health-related information is given by Dahab and Bogardi (1990) and the World Health Organization (1985).

Excessive nitrates in groundwater have also caused problems with ruminants (cud-chewing animals with divided stomachs). Sheep and cattle can be seriously affected by nitrates from birth through adulthood. Monogastric (single stomach) infants such as horses, pigs, and chickens are also candidates for health problems. As chickens and pigs mature, they are much less affected by the presence of nitrates. However, horses can be affected through adulthood (Chandler, 1989).

Symptoms of nitrate–nitrite poisoning in livestock include cyanosis in and about the non-pigmented areas (mouth and eyes), shortness of breath, rapid heartbeat, staggered gait, frequent urination, and collapse. In severe cases, convulsions, coma, and death may result within hours. A loss of milk production in cows and aborted calves are also signs of possible nitrate poisoning (Chandler, 1989).

SUMMARY

Nitrates in groundwater represent a widely distributed pollution concern; nitrates are perhaps the most ubiquitous of all groundwater contaminants. Natural and man-induced sources of nitrates in groundwater are a result of water usage for irrigation; excessive applications of commercial fertilizers or manure; and waste disposal practices associated with land application of sludges or wastewater effluents, municipal or industrial landfills, and septic tank systems. The key concern regarding usage of groundwater with excessive concentrations of nitrates is related to human health effects, particularly with regard to infants. The major effects are associated with losses in oxygen transport/transfer capabilities in the blood.

SELECTED REFERENCES

Bouwer, H., "Agricultural Chemicals and Groundwater Quality," *Journal of Soil and Water Conservation*, March-April, 1990, pp. 184-189.

Canter, L.W., "Nitrates in Ground Water from Agricultural Practices — Causes, Prevention, and Clean-up," July, 1987, report to United Nations Development Program, University of Oklahoma, Norman, Oklahoma.

Canter, L.W. and Maness, K.M., "Ground Water Contaminants and Their Sources," *International Journal of Environmental Studies*, Vol. 47, 1995, pp. 1-17.

Chandler, J., "Nitrates in Water," *Water Well Journal*, Vol. 43, No. 5, May, 1989, pp. 45-47.

Cohen, S., "Results of the National Drinking Water Survey: Pesticides, Nitrates, and Well Characteristics," *Water Well Journal*, Vol. 46, No. 8, August, 1992, pp. 35-38.

Conway, G.R. and Pretty, J.N., *Unwelcome Harvest — Agriculture and Pollution*, Earthscan Publications, Ltd., London, England, 1991, p. 159.

Dahab, M.F. and Bogardi, I., "Risk Management for Nitrate-Contaminated Groundwater Supplies," November, 1990, University of Nebraska, Lincoln, Nebraska, pp. 9-16.

Federal Register, "National Primary Drinking Water Standards," Vol. 50, No. 219, November 13, 1985, pp. 46880-47022.

Keeney, D., "Sources of Nitrate to Ground Water," *CRC Critical Reviews in Environmental Control*, Vol. 16, No. 3, 1986, pp. 257-304.

Keeney, D.R., "Sources of Nitrate to Ground Water," in *Nitrogen Management and Ground Water Protection*, Follett, R.F., Ed., Elsevier Science Publishers B.V., Amsterdam, The Netherlands, 1989, chap. 2, pp. 23-34.

Kleinjans, J.C.S., Albering, H.J., Marx, A., van Maanen, J.M.S., van Agen, B., ten Hoor, F., Swaen, G.M.H., and Mertens, P.L.J., "Nitrate Contamination of Drinking Water: Evaluation of Genotoxic Risk in Human Populations," *Environmental Health Perspectives*, Vol. 94, August, 1991, pp. 189-193.

Life Systems, Inc., "Drinking Water Criteria Document for Nitrate/Nitrite," TR-832-77, May, 1987, Cleveland, Ohio.

Office of Technology Assessment, "Beneath the Bottom Line: Agricultural Approaches to Reduce Agrichemical Contamination of Groundwater," OTA-F-418, November, 1990, U.S. Congress, U.S. Government Printing Office, Washington, D.C., pp. 3-20.

Tannenbaum, S.R. and Green, L.C., "Selected Abstracts on the Role of Dietary Nitrate and Nitrite in Human Carcinogenesis," 1985, International Cancer Research Data Bank Program, National Cancer Institute, Washington, D.C.

U.S. Environmental Protection Agency, "National Survey of Pesticides in Drinking Water Wells — Phase I Report," EPA 570/9-90-015, November, 1990, Office of Water, Washington, D.C.

U.S. Environmental Protection Agency, "Another Look: National Survey of Pesticides in Drinking Water Wells — Phase II Report," EPA 579/09-91-020, January, 1992a, Office of Water, Washington, D.C.

U.S. Environmental Protection Agency, "National Pesticide Survey — Update and Summary of Phase II Results," EPA 570/9-91-021, Winter, 1992b, Office of Water, Washington, D.C.

U.S. Environmental Protection Agency, *Nitrogen Control*, Technomic Publishing Company, Inc., Lancaster, Pennsylvania, 1994, pp. 1-22.

Winneberger, J.H., *Nitrogen, Public Health, and the Environment*, 1982, Ann Arbor Science Publishers, Inc., Ann Arbor, Michigan, p. 5.

World Health Organization, "Health Hazards from Nitrates in Drinking Water — Report on a WHO Meeting in Copenhagen, March 5-9, 1984," Regional Office for Europe, Copenhagen, Denmark, 1985, pp. 3, 49-66.

2 INFLUENCE OF SUBSURFACE PROCESSES

INTRODUCTION

Nitrate contamination of groundwater is typically a result of nitrogen movement into the unsaturated (vadose) zone, the occurrence of transformation processes in either the unsaturated or saturated zones, and nitrate movement with groundwater. This chapter will summarize nitrogen transformation, and transport and fate processes in the subsurface environment; factors influencing nitrogen leaching from fertilizers; and the use of laboratory studies for examination of both the transformation processes and leaching.

NITROGEN TRANSFORMATIONS

Table 2.1 contains summary comments on 14 references dealing with nitrogen transformations in the subsurface environment. Only selected information from several of the 14 references will be included herein. The transport and fate of nitrogen in the subsurface environment depends upon the form of entering nitrogen and various biochemical and physicochemical processes involved in transforming one form of nitrogen to another form.

Nitrogen can enter the subsurface environment in either organic or inorganic forms, depending upon the source. Organic nitrogen consists of compounds from amino acids, amines, proteins, and humic compounds with low nitrogen content (Reddy and Patrick, 1981). Inorganic nitrogen consists of ammonium, nitrite, and nitrate forms. Nitrogen from untreated or partially treated wastewater discharges or human waste fertilizers may be in either organic or ammonium form, while nitrogen from chemical fertilizers will typically be in ammonium or nitrate form. Transformation processes in the subsurface include: (1) ammonification, (2) ammonia volatilization, (3) nitrification, and (4) denitrification. Transport processes that may be involved in subsurface movement of various forms include: (1) diffusion of ammonium forms, (2) diffusion of nitrate forms, and (3) movement of either form with the water phase. Abiotic processes such as adsorption and cation exchange may cause retention of some ammonium forms in the

Table 2.1 References Related to Nitrogen Transformations in the Subsurface Environment

Author(s) (year)	Comments
Alfoldi (1983)	Discussion of transformation and transport processes in three main hydrogeological soil types
Behnke (1975)	Description of the biogeochemistry of nitrogen compounds in groundwater
Bradley, Aelion, and Vroblesky (1992)	Study of factors affecting denitrification in sediment contaminated with JP-4 jet fuel
Chopp, Clapp, and Schmidt (1982)	Effects of treated municipal wastewater effluents on ammonia-oxidizing bacteria in soil
Devitt, et al. (1976)	Effects of soil types and irrigation practices on nitrates in the subsurface environment
Howard (1985)	Influence of denitrification on nitrate concentrations in a major limestone aquifer
Keeney (1983)	Review of transformations and transport of nitrogen in the subsurface environment
Keeney (1986)	Review of the chemical properties of nitrate and nitrite, and the soil nitrogen cycle (including mineralization-immobilization, nitrification, denitrification, and plant uptake and recycling)
Lowrance and Pionke (1989)	Review of physical, biological, and chemical processes leading to changes in nitrate concentration in groundwater; and discussion of case studies illustrating the effects of these processes on different aquifers
Lowrance and Smittle (1988)	Measurement of denitrification losses of nitrogen, crop uptake of nitrogen, and resultant availability of nitrogen for leaching in intensive vegetable crop production systems in a sandy soil
Reddy and Patrick (1981)	Review of the agronomic and ecologic significance of nitrogen transformations and loss in flooded soils and sediments
Rolston and Broadbent (1977)	Field studies of denitrification in the subsurface environment
Rolston et al. (1980)	Field studies of the effects of irrigation frequency on denitrification in the subsurface environment
Schubauer-Berigan et al. (1990)	Examination of microbial degradation pathways and degradation rates, under both aerobic and anaerobic conditions, for nitrates in Long Island groundwaters

subsurface; in addition, biotic processes such as incorporation into microbial or plant biomass can occur. Figure 2.1 summarizes the nitrogen cycle in the subsurface environment (Reddy and Patrick, 1981). As suggested in Figure 2.1, the occurrence of certain transformation processes is a function of the presence of oxidizing or reducing conditions.

Ammonification is the first step in the mineralization of organic nitrogen. Ammonification is defined as the biological conversion of organic nitrogen to ammonium-nitrogen (Reddy and Patrick, 1981). Under anaerobic soil conditions, ammonium-nitrogen accumulation occurs because of the suppression of nitrification.

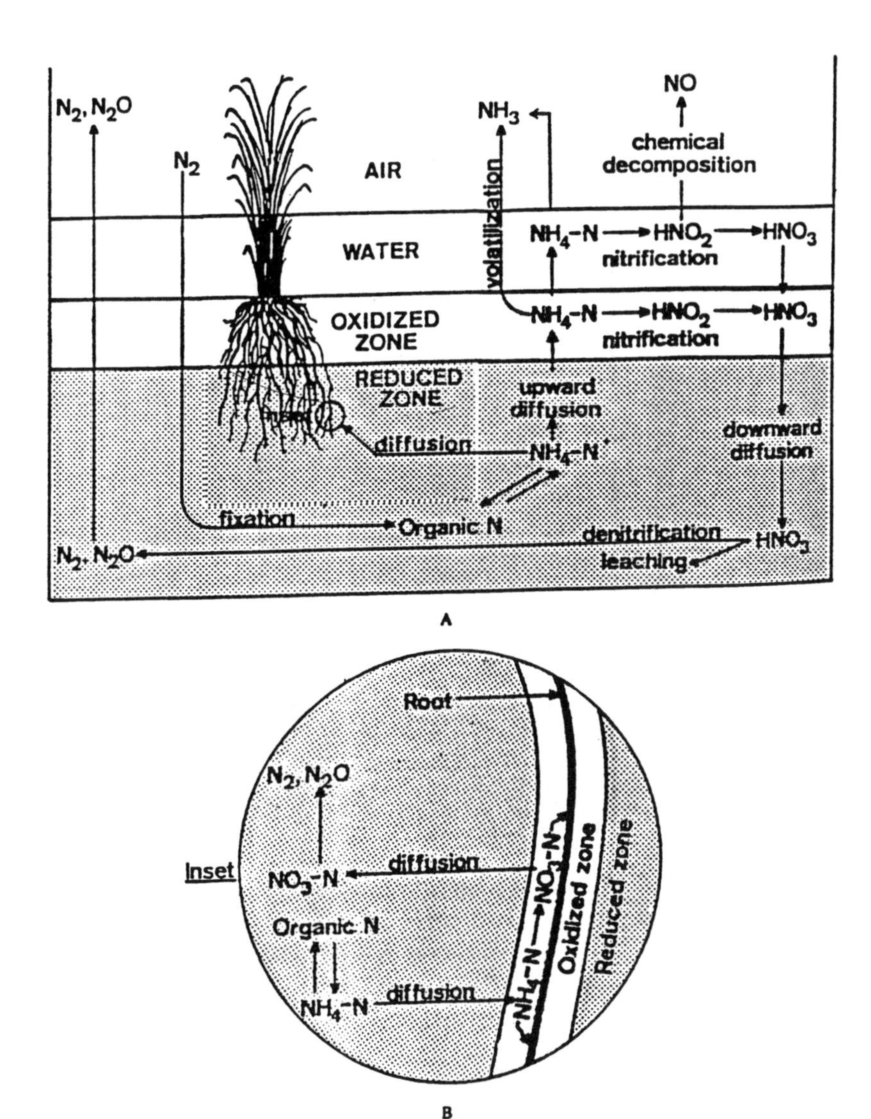

Figure 2.1 The nitrogen cycle in the subsurface environment. (From Reddy, K. R. and Patrick, W. H., in *CRC Critical Reviews in Environmental Control*, Vol. 13, No. 4, 1981, pp. 273–303.)

Ammonia volatilization is a physicochemical process where ammonium-nitrogen is known to be in equilibrium between the gaseous and hydroxyl forms as follows (Reddy and Patrick, 1981):

$$NH_3\ (aq) + H_2O \rightarrow NH_4^+ + OH^- \qquad (1)$$

Reaction (1) is pH dependent, with an alkaline pH favoring the presence of aqueous forms of NH_3 in solution; while at acidic or neutral pH, the ammonium–nitrogen is predominantly in the ionic form. Figure 2.2 shows the effect of pH and temperature on the distribution of ammonia and ammonium ions in water (Behnke, 1975).

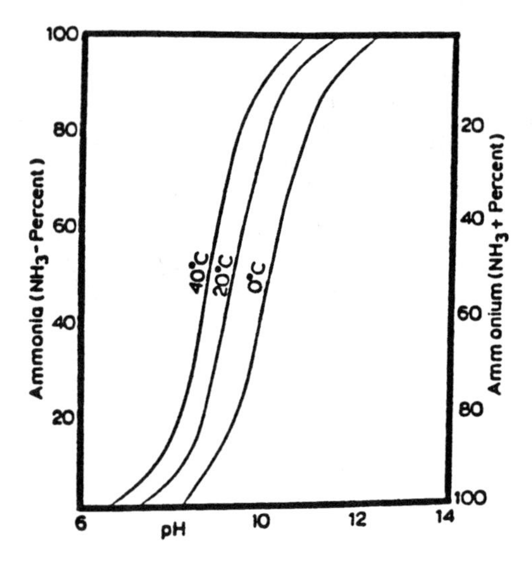

Figure 2.2 Effect of pH and temperature on the distribution of ammonia and ammonium ion in water. (From Behnke, J., *Journal of Hydrology,* Vol. 27, 1975, pp. 155–167. With permission of Elsevier Science, Amsterdam.)

Nitrification is defined as the biological oxidation of ammonium–nitrogen to nitrate–nitrogen (Reddy and Patrick, 1981). Nitrification is known to take place in two stages as a result of the activity of chemoautotrophic bacteria of the genera Nitrosomonas ($NH_4^+ \rightarrow NO_2$) and Nitrobacter ($NO_2 \rightarrow NO_3$). Both organisms are Gram negative, aerobic, chemoautotrophic rods. These nitrifiers derive energy from the oxidation of ammonium–nitrogen and/or nitrite–nitrogen. These organisms require O_2 during ammonium–nitrogen oxidation to nitrite–nitrogen and nitrite–nitrogen oxidation to nitrate–nitrogen. Ammonium oxidation of nitrite can be written as (Reddy and Patrick, 1981):

$$NH_4^+ + 1^1/_2O_2 \rightarrow NO_2^- + 2\ H^+ + H_2O \quad (2)$$

Nitrite oxidation to nitrate can be written as (Reddy and Patrick, 1981):

$$NO_2^- + {}^1/_2O_2 \rightarrow NO_3^- \quad (3)$$

Combining reactions (2) and (3) yields:

$$NH_4^+ + 2\ O_2 \rightarrow NO_3^- + 2\ H^+ + H_2O \quad (4)$$

Therefore, it takes 2 moles oxygen to oxidize each mole of ammonium-nitrogen to nitrate. Stoichiometrically, the organisms require 4.57 g O_2 per gram of ammonium-nitrogen oxidized (Reddy and Patrick, 1981).

Several kinetic rate equations have been proposed for nitrification, including zero order, first order, and the Monod equation of population dynamics. Table 2.2 contains some first-order kinetic rate coefficients for nitrification in soils (Reddy and Patrick, 1981). Environmental factors influencing the nitrification rate include: (1) temperature, (2) pH, (3) alkalinity of water, (4) inorganic carbon source, (5) microbial population, and (6) ammonium-nitrogen concentration.

Table 2.2 First-Order Kinetic Rate Coefficients for Nitrification in Soils

Type of system and process	Temperature (°C)	First-order rate constant (k day^{-1})	Remarks
Salinas clay	24		Laboratory incubation
$NH_4 \rightarrow NO_2$		0.22	
$NO_2 \rightarrow NO_3$		9.0	
Milvill loam	22		Laboratory incubation
$NH_4 \rightarrow NO_2$		0.14	
$NO_2 \rightarrow NO_3$		9.00	
Tippera clay loam $NH_4 \rightarrow NO_3$	20	0.003	Laboratory incubation at varying temperatures
Columbia silt loam $NH_4 \rightarrow NO_3$		0.24–0.72	Steady-state flow through soil column maintained at –85 cm suction
Hanford sandy loam $NH_4 \rightarrow NO_3$		0.76–1.11	Steady-state flow through soil columns, using ^{15}N
Webster silt loam $NH_4 \rightarrow NO_3$	35	0.052	Laboratory incubation study with 35-day incubation
Marshall silt loam $NH_4 \rightarrow NO_3$	35	0.048	
Colo silty clay loam $NH_4 \rightarrow NO_3$	35	0.051	
Monona silty clay loam $NH_4 \rightarrow NO_3$	35	0.047	
Grundy silt loam $NH_4 \rightarrow NO_3$	35	0.072	
Lentic System Clinton River $NH_4 \rightarrow NO_3$		3.10	Field study conducted on the Clinton River between Pontiac, MI, and Rochester, MI. Nitrification rates calculated based on changes in NH_4–N and NO_3–N concentration

From Reddy, K. R. and Patrick, W. H., in *CRC Critical Reviews in Environmental Control*, Vol. 13, No. 4, 1981, pp. 273–303.

Denitrification is defined as the biological reduction of nitrate to gaseous end-products such as molecular N_2 or N_2O. A commonly accepted pathway for biological denitrification is (Keeney, 1986):

$$NO_3^- \rightarrow NO_2^- \rightarrow NO \rightarrow N_2O \rightarrow N_2 \quad (5)$$

Under anaerobic or oxygen-free conditions and in the presence of available organic substrate, the denitrifying organisms can use nitrate as an electron acceptor during respiration. Genera involved in the denitrification process include *Pseudomonas, Acinetobacter, Bacillus, Micrococcus, Gluconobacter, Alcaligenes, Halobacteriums, Thiobacillus, Xanthomonas, Moraxella, Paracoccus, Spirillum,* and *Rhodopseudomonas*. When oxygen is available, these organisms oxidize a carbohydrate substrate to CO_2 and H_2O as follows (Reddy and Patrick, 1981):

$$C_6H_{12}O_6 + 6\ O_2 \rightarrow 6\ CO_2 + 6\ H_2O \qquad (6)$$

Under oxygen-free conditions, some microorganisms oxidize a carbohydrate substrate to CO_2 and H_2O using nitrate instead of oxygen as an electron acceptor and converting the nitrate to N_2 gas, as shown by the following overall reaction (Reddy and Patrick, 1981):

$$5(CH_2O) + 4\ NO_2 + 4\ H^+ \rightarrow 5\ CO_2 + 2\ N_2 + 7\ H_2O \qquad (7)$$

Several kinetic rate equations have also been proposed for denitrification, including zero-order and first-order relationships. Table 2.3 contains some first-order rate constants for denitrification in soils and sediments (Reddy and Patrick, 1981). Several factors are known to influence the rate of denitrification, either directly or indirectly. Among these are the absence of oxygen, presence of readily available carbon, temperature, soil moisture, pH, presence of denitrifiers, and soil texture.

As an example of the influence of denitrification in groundwater, Howard (1985) described the importance of this transformation process in a major limestone aquifer. The Chalk aquifer of eastern-central England is seriously affected by nitrates leached from agricultural land and, like many similar aquifers, its future as a potable resource depends on the nature and rates of denitrification. The regional distribution of nitrate supports a viewpoint based on thermodynamic criteria that denitrification is actively occurring and that the problem will be short-lived. More detailed considerations based on major-ion and environmental isotope data, however, indicate that denitrification is not significant and that the apparent lowering of the nitrate concentration (from >10 to <2 mg/l NO_3–N) in the apparent direction of flow is primarily due to mixing between waters of different origins. The presence of reduced nitrogen species is possible evidence of denitrification in some older waters. Howard (1985) concluded that denitrification cannot be relied upon to reduce elevated concentrations of nitrate in modern recharge waters.

Rolston et al. (1980) described a study focused on the influence of denitrification on the nitrate in irrigation return waters. Absolute amounts and rates of denitrification from a Yolo loam field profile at Davis, California, were studied in relation to the influence of irrigation frequency and soil incorporation of crop residue. Application frequencies of three irrigations per week, one irrigation per week, and one irrigation every 2 weeks were established on areas cropped with

Table 2.3 First-Order Rate Constants for Denitrification in Soils and Sediments

Soil/sediment	Temperature (°C)	First-order rate constant (k day^{-1})	Remarks
Peat soil	5–36	0.1–0.5	Laboratory incubation study conducted at various temperatures; rates based on NO_3–N loss
Florida organic soils (8 soils)	30	0.1–1.5	Organic soils obtained from several locations in south and central Florida, soils incubated under anoxic conditions, and stirred continuously; rates based on NO_3–N loss
Huron soil (silty clay loam)	30	2.15	Soil obtained from cropped field
	15	0.72	
	10	0.54	
30 Soil types	35	0.03–0.90	Soils collected from several locations in the U.S.; rates are influenced by diffusion of NO_3–N from floodwater; rates based on NO_3–N loss
8 Soil types	30	0.03–0.76	Soils collected from several locations in the U.S.; soils incubated under anoxic conditions, and stirred continuously; rates based on NO_3–N loss
Crowley silt loam	30		Soil incubated to an O_2-free atmosphere, with NO_3–N present in both floodwater and underlying soil; rates based on NO_3–N disappearance
No floodwater		0.37	
3-cm floodwater		0.18	
6-cm floodwater		0.13	
Handford sandy loam	20–22	0.07	Rate constants measured under steady-state conditions
Columbia silt loam		0.17	
Moreno clay loam		0.05	
		0.14	
Yolo loam		0.004–0.03	Laboratory soil columns using ^{14}N
Norfolk sandy loam	30		Laboratory incubation study with no excess floodwater; soils treated with organic wastes that underwent various degrees of decomposition
No waste added		0.07	
Beef waste added		1.06–2.57	
Poultry waste added		0.10–1.34	
Swine waste added		0.18–0.88	
Danish lake sediments (Copenhagen and Denmark)		0.04–0.76	Sediments obtained from various times during seasons; ^{15}N was used to determine denitrification; rates measured based on N_2 production

From Reddy, K. R. and Patrick, W. H., in *CRC Critical Reviews in Environmental Control*, Vol. 13, No. 4, 1981, pp. 273–303.

perennial rye grass. Fertilizer was applied at the rate of 300 kg N per hectare as KNO_2 enriched with 56 to 58% ^{15}N to 1-m^2 plots. The flux of volatile gases at the soil surface was measured from the accumulation of N_2O and $^{15}N_2$ beneath air-tight covers placed over the soil surface for 1 to 4 hours at several times immediately after irrigation and at less frequent intervals as denitrification fluxes decreased. Small rates of total denitrification were measured in this well-drained alluvial soil under normal cyclic applications of irrigation water. For plots without carbon addition, the largest denitrification of only 1.5% of the applied fertilizer was measured in the most frequently irrigated plot. For the least frequently irrigated plot of one irrigation every 2 weeks, only 0.7% of the applied fertilizer was denitrified. Denitrification rates decreased to near-zero values within 1 or 2 days after irrigation.

The application frequency of three irrigations per week gave higher NO_3^- concentrations as measured by both soil solution and soil samples within the root zone of the crop than those of the other two irrigation frequencies (Rolston et al., 1980). Thus, frequent, small irrigations tended to result in less leaching losses than frequent, large irrigations. Denitrification of nitrate fertilizer was simulated using a mathematical model that included transport and plant uptake of water and nitrogen in soil. The rate of denitrification was demonstrated to be a function of nitrate concentration, available carbon concentration, degree of soil-water saturation, and temperature.

In 1975, the loss of about 83,000 gallons JP-4 jet fuel resulted in contamination of a shallow aquifer near Charleston, South Carolina (Bradley, Aelion, and Vroblesky, 1992). A site-specific study was then conducted to examine influencing factors related to an *in situ* bioremediation program involving nitrate additions. The fate of the amended nitrate; the effect of pH, nitrate, and phosphate on denitrification; and the variability of denitrification in sediments collected at the site were examined. Denitrification (N_2O–N production) accounted for 98% of the depletion of nitrate-nitrogen under anaerobic conditions. Both carbon mineralization and denitrification rates increased asymptotically with increasing nitrate to a maximum at approximately 1 m*M* nitrate. Addition of up to 1 m*M* phosphate did not significantly increase the N_2O and CO_2 production. Denitrification rates were at least 38% lower at pH 4 than those observed at pH 7. Comparisons of samples with differing degrees of hydrocarbon contamination indicated that at least a tenfold variation in sediment denitrification occurred at the contaminated site (Bradley, Aelion, and Vroblesky, 1992).

NITROGEN TRANSPORT AND FATE

Diffusion of ammonium-nitrogen and nitrate in the subsurface environment plays a role in subsurface transport along with abiotic processes such as adsorption and ion exchange. The quantity of ammonium-nitrogen transferred by diffusion phenomena per unit area per unit time is proportional to the diffusion coefficient and the concentration gradient. At the soil/water interface, concentration gradients can be relatively large and, hence, diffusion of ammonium-nitrogen from the anaerobic layer can be rapid. Molecular diffusion is assumed to follow Fick's law and is expressed in terms of the whole soil or sediment diffusion coefficient, D, which includes the effects of tortuosity but not those of adsorption. Data in Table 2.4 show the D values for ammonium-nitrogen moving in several anaerobic

Table 2.4 Diffusion Coefficients for NH_4–N Movement in Soils and Sediments

Soil/sediment		Temperature (°C)	Diffusion coefficients (cm^2/day)	Remarks
Flooded soil				Experiment conducted with an initial ammonium-N concentration of 200 μg/g soil; incubation period of 2 days
Crowley silt loam		30	0.216	
Midland silt loam		30	0.172	
Mhoon silty clay loam		30	0.059	
Marine sediments (Long Island Sound)		20	0.850	Sediments with additional ammonium-N added; diffusion coefficients adjusted for adsorption; incubation period of 5.9 days
Volumetric water content (cm^3/cm^3)				Experiment conducted with soils treated with 14.3 meq/l, of $(NH_4)_2SO_4$ solution, and adjusted to various water contents; soils were sterilized using γ-radiation before use; incubation period of 5–17 days
Urrbrae loam	0.10	21	0.003	
	0.15	21	0.009	
	0.20	21	0.013	
	0.25	21	0.019	
Wanbi sand	0.10	21	0.019	
	0.15	21	0.022	
	0.20	21	0.030	
Volumetric water content (cm^3/cm^3)				Experiment conducted with an initial ammonium-N concentration of 200 μg/g soil; nitrification inhibitor was used to prevent ammonium-N oxidation; incubation period of 7 to 15 days
Crowley silt loam	0.06	30	0.004	
	0.11	30	0.010	
	0.27	30	0.050	
Flood Maahas clay				Laboratory study conducted using undisturbed wetland soil cores; incubation period of 7 to 28 days
Prilled urea		25	0.019	
Super granule urea			0.022	
Sulfur-coated urea			0.016	

From Reddy, K. R. and Patrick, W. H., in *CRC Critical Reviews in Environmental Control*, Vol. 13, No. 4, 1981, pp. 273–303.

soil or sediment systems (Reddy and Patrick, 1981). Nitrate diffusion coefficient data into underlying anaerobic soil is summarized in Table 2.5 (Reddy and Patrick, 1981). The D values for nitrate are typically greater than those for ammonium-nitrogen since nitrate is an anion and is not adsorbed onto the subsurface exchange complex.

Ammonium-nitrogen may not be transported through the unsaturated zone and into groundwater due to adsorption, cation exchange, incorporation into microbial biomass, or release to the atmosphere in the gaseous form. Adsorption is probably the major mechanism of removal in the subsurface environment. Under anaerobic conditions in the subsurface environment, positively charged ammonium ions (NH_4^+) are readily adsorbed onto negatively charged soil particles. Since anaerobic conditions in soils are usually associated with saturated soils, some movement of ammonia with groundwater can occur. However, this movement will be slow since adsorption will continue to occur on soil particles in the aquifer.

To serve as an example of the influence of adsorption, a study conducted by Sparks (1987) will be summarized. Nitrate retention as it affects groundwater

Table 2.5 Diffusion Coefficients for NO_3–N Movement in Soils

Type of system		Temperature (°C)	Diffusion coefficients (cm^2/day)	Remarks
Flooded soil				Experiment conducted with an initial nitrate-N concentration of 300 μg/ml, incubation period of 24 hr, saturated soil moisture content
Crowley silt loam		30	1.33	
Midland silt loam		30	1.94	
Mhoon silty clay loam		30	0.96	
Flooded organic soil		8	0.25	Experiment conducted with an initial nitrate-N concentration of 100 μg/ml, incubation period of 78 hr
		28	0.44	
Volumetric water content (cm^3/cm^3)				Experiment conducted with the soils amended with 28.6 meq/l of KNO_3 solution, and adjusted to various water contents; soils were sterilized using γ-radiation before use; incubation period of 1–5 days
Urrbrae loam	0.10	21	0.04	
	0.20	21	0.17	
	0.25	21	0.32	
Wanbi sand	0.05	21	0.13	
	0.10	21	0.23	
	0.15	21	0.34	
	0.20	21	0.42	
Saturated water content (cm^3/cm^3)				Experiment conducted with silt, sand, or glass beads saturated with 0.1 *N* $NaNO_3$
Silt	0.43	30	1.15	
Sand	0.42	30	1.20	
Glassbeads	0.38	30	1.21	
Flooded soils				
Everglades muck		25	0.51	Nitrate diffusion from flood water into the underlying soil column was measured; diffusion coefficients for nitrate-N were measured, using the changes in chloride concentration of flood water
Floridana fine sand			0.33	
Astor sand			0.68	
Surrency sand			1.80	
Samsula muck			1.70	
Pickney fine sand			0.81	
Riviera fine sand			1.30	
Delray fine sand			1.20	
Chastain silt loam			0.35	
Brighton peat			0.42	
Iberia silty clay			0.45	
Chobee fine sandy loam			0.38	
Eureka fine sandy loam			0.72	
Valkaria fine sand			0.26	

From Reddy, K. R. and Patrick, W. H., in *CRC Critical Reviews in Environmental Control*, Vol. 13, No. 4, 1981, pp. 273–303.

pollution was investigated on nine major mid-Atlantic soil types. The soils had a wide range in organic matter, clay, and oxide content. Charge properties including anion exchange capacity (AEC) and point of zero salt effect (PZSE) were determined. The PZSE values were low, indicating little anion adsorption capacity; however, the AEC values were often significant and increased with profile depth as oxide and clay contents increased. The kinetics of nitrate retention and release and the effect of competitive anions on nitrate retention were investigated at pH 4.0, 5.5, and 7.0. Nitrate retention was highest at pH 4.0 and was strongly correlated with clay and oxide contents. The adsorption kinetics were rapid and completely reversible, indicating that the nitrate adsorption mechanism was electrostatic. Whenever sulfate was present, nitrate retention significantly decreased and was even depressed when chloride was added (Sparks, 1987).

Cation exchange may be involved, along with adsorption in the retention of ammonium ions in soils. However, just as the adsorption capacity of a soil can be exceeded, the cation exchange capacity can also be exhausted. Under these conditions, the cation exchange sites in the soil beneath an ammonium input source would become equilibrated with the cations in the leachate. The leachate would then move to the groundwater with its cation composition essentially unchanged. Ammonia-nitrogen can be incorporated into microbial or plant biomass in the subsurface environment; however, this is probably not a major removal mechanism relative to nitrogen in the subsurface. Finally, as noted earlier, ammonia gas can be released to the atmosphere as a function of the soil/liquid pH conditions. When the pH is neutral or below, most of the nitrogen is in the ammonium ion form. As the pH becomes basic, the ammonium ion is transformed into ammonia and can be released from the soil as a gas.

Nitrate is typically more mobile in the subsurface environment than ammonium-nitrogen. When nitrogen in the form of nitrate reaches groundwater, it becomes very mobile because of its solubility and anionic form. Nitrates can move with groundwater with minimal transformation. They can migrate long distances from input areas if there are highly permeable subsurface materials that contain dissolved oxygen. The only condition that can affect this process is a decline in the redox potential of the groundwater. In this case, and as noted earlier, the denitrification process can occur.

FACTORS INFLUENCING NITROGEN LEACHING FROM FERTILIZERS

Nitrogen leaching from applied fertilizers is the precursor event to the transformation processes described in the previous section. Table 2.6 contains summary comments on 17 references dealing with factors influencing the leaching of nitrogen into the subsurface environment. To serve as examples, the amount of leaching and hence the amount of nitrates in groundwater have been found to be a function of the timing of fertilizer application, vegetative cover, and soil porosity (Borneff and Adabe, 1973); fertilizer application method and amount added (Gillings, 1973); fertilizer amount added and drainage volume from various water treatments (Letey et al., 1978); and fertilizer amount added and irrigation rate (Ludwick, Reuss, and Langin, 1976).

Borneff and Adabe (1973) described a 2-year sampling program, at 14-day intervals, for three wells in the Mainz, West Germany, area. The highest amount of nitrate in soil was found in the vicinity of two of the wells situated in a wine-growing area and corresponded with the fertilization period (June and July). A minimum amount was recorded in the winter. Groundwater abstracted from wells exhibited the highest concentrations of nitrates 6 months after fertilization. Nitrate fluctuation in the third well situated in a fruit- and vegetable-growing area was

Table 2.6 References Related to Leaching of Nitrogen into the Subsurface Environment

Author(s) (year)	Comments
Anonymous (1984)	Summaries of papers presented at a technical symposium; some deal with nitrogen leaching
Bergstrom (1987)	Study of effects of plowing and weather conditions on nitrate leaching from fertilizer applied to annual and perennial crops
Bergstrom and Brink (1986)	Influence of fertilizer application rate, season, and crop type on nitrogen leaching losses and distribution of inorganic nitrogen in soil in central Sweden
Berndt (1990)	Study of the contribution of land application of sewage plant effluent to nitrates in groundwater
Borneff and Adabe (1973)	Factors influencing nitrogen leaching in a fertilized agricultural area
Bouwer (1983)	Nitrate leaching through the vadose zone as a function of irrigation water application
Bumb et al. (1987)	Determination of first-order rate constant for the leaching of ammonia and nitrate from the storage area at a fertilizer plant
Desprez, Landreau, and Vogt (1982)	Unconfined aquifer zones vulnerable to nitrogen leaching
Gillings (1973)	Field experiment on nitrogen leaching as a function of normal agricultural practices
Letey et al. (1978)	Field experiments on nitrate leaching at commercial farming sites in California
Ludwick, Reuss, and Langin (1976)	Soil nitrate concentrations as a function of fertilizer applications
Martin et al. (1982)	Computer model for studying the interaction of nitrogen and water management on corn production
Morton, Gold, and Sullivan (1988)	Study of the influence of overwatering and fertilization rate on nitrogen losses from lawns in urban areas
Pal and Broadbent (1981)	Use of soil columns to study leaching of nitrogen from applied sewage
Spalding and Kitchen (1988)	Determination of influence of fertilization rate on vadose-zone nitrate content beneath irrigated cropland
Stone (1982)	Field study of water and nitrogen balances at a site in Kansas
Tolman (1985)	Survey of seasonal fluctuations of nitrate in groundwater at an Idaho site subjected to irrigation and precipitation influences

less evident; but concentrations were somewhat higher, even though the fertilization was not as intense as that in the wine-growing area. The extent of vegetative cover and the porosity of the soil were also found to be decisive factors for the fertilization-dependent nitrate concentration in groundwater. The leaching of naturally formed nitrate in the soil was also of considerable importance.

Gillings (1973) described an experiment in which nitrate and ammonium fertilizers were applied to field plots containing sorghum and beets. Suction lysimeters were used to sample soil at the 1-, 4-, and 9-ft depths, and surface runoff was also sampled. Nitrate, nitrite, and ammonium were determined in these samples. More nitrogen was lost to both surface runoff and deep leaching from broadcast fertilizers than from drilled fertilizers. Little difference in the levels of nitrogen (nitrate plus ammonium) in soil water was found between nitrate and ammonium fertilizers. The largest deep leaching of nitrates occurred in the fall and in the spring. The amount leached was proportional to the amount of nitrogen applied. When 175 kg N/ha was applied, the average (of three replications) amount of nitrate found in soil water at the 9-ft depth was 6 to 9 mg/l as nitrate-nitrogen. At the 4-ft depth, nitrate-nitrogen varied from 1 to 9 mg/l when treatment ranged from no nitrogen to 175 kg N/ha.

The amounts of leached nitrates for a given time period were determined by Letey et al. (1978) at various commercial farming sites in California, and in a carefully controlled experimental plot receiving various water and fertilizer application treatments. Some of the agricultural sites had tile drainage systems and others had "free drainage" to the groundwater. Linear regression analyses were conducted on the data. Similar results were observed for the tile and free drainage systems. The highest correlation coefficient was of the drainage volume and fertilizer nitrogen application. The next highest correlation coefficient was for amount leached versus drainage volume, followed by amount leached versus fertilizer nitrogen application. A significant linear relationship between amount of leached nitrate and drainage volume was also obtained at the experimental plot.

The purpose of the study by Ludwick, Reuss, and Langin (1976) was to evaluate soil nitrate accumulations following 4 years of continuous corn grown with different nitrogen and irrigation regimes, and to compare these results to present nitrate concentrations found in irrigated farm fields of central and eastern Colorado. Soil nitrate content in the 300-cm sampled profile was significantly influenced by both fertilizer nitrogen and irrigation treatments; the greater accumulations were associated with the two higher fertilizer nitrogen application rates and the two lower irrigation rates. Nitrate content increased linearly in relation to fertilizer nitrogen between 67 and 269 kg N/ha, and could be described by two simple regression equations separating the irrigation treatments into low rates and high rates.

USAGE OF LABORATORY STUDIES

Laboratory studies can aid in the development of information on transformation and transport and fate processes, and on factors influencing nitrogen leaching from fertilizers. Table 2.7 contains summary comments on ten references describing potentially relevant laboratory studies. General nitrogen transformations were the focus of two studies (Lindley, Dale, and Mannering, 1974; Williford et al., 1969), and denitrification was emphasized in four studies (Biswas and Warnock, 1985; Davenport, Lembke, and Jones, 1975; Schwan, Kramer, and Gericke, 1984; Willardson and Meek, 1969).

Lindley, Dale, and Mannering (1974) described a laboratory study of nitrogen transformation in wastewater percolating through soil columns. The study

Table 2.7 References Related to Laboratory Studies of Nitrogen Transformations in the Subsurface Environment

Author(s) (year)	Comments
Bachelor, et al. (1991)	User guide for a mathematical model (LT3VSI) developed to describe transport and fate processes, including denitrification, in nonhomogeneous laboratory scale-model aquifers
Biswas and Warnock (1985)	Use of laboratory soil columns as denitrifying reactors subjected to varying carbon-to-nitrogen ratios
Davenport, Lembke, and Jones (1975)	Study of denitrification in laboratory sandy columns with methanol added as a carbon source
El Etreiby and Laudelout (1988)	Use of columns of undisturbed loess soil to study the movement of tritiated water, chloride, and nitrate as related to the transfer of nitrate toward an aquifer
Lindley, Dale, and Mannering (1974)	Laboratory study of nitrogen transformations in wastewater percolating through soil columns
Lindstrom and Boersma (1990)	Development of a two-dimensional mathematical model for simulating denitrification in a nonhomogeneous laboratory scale, single-layer aquifer
Lindstrom et al. (1991)	Study of the influence of hydraulics, nitrogen chemistry, and microbiology in a nonhomogeneous laboratory-scale, single-layer aquifer; and use of model by Lindstrom and Boersma (1990)
Schwan, Kramer, and Gericke (1984)	Development of mathematical model of denitrification based on laboratory reaction vessels
Willardson and Meek (1969)	Use of large soil columns to study anaerobic reduction of nitrate
Williford et al. (1969)	Use of soil lysimeters to study the movement of nitrates in soil systems

addressed varying applied wastes, soil moisture contents, and soil types. Williford et al. (1969) reported on a study of the movement of nitrates in soil systems by using 14 lysimeters made of techite (fiber glass) pipe 15 in. in diameter and 6.7 ft in length. The lysimeters were filled with four major soil types from the west side of the San Joaquin Valley in California. The volumes extracted from the probes and collected in the leachates were recorded. The samples were analyzed for chlorides, ammonia, total nitrogen, and atom percent excess nitrogen. The nitrate levels in leachates collected during initial leaching and before fertilization ranged from 20,000 mg/l in oxalis clay to 2500 mg/l in panoche fine sandy loam. These high nitrate levels indicate how relatively low levels of native nitrates can be concentrated in the groundwater by leaching. The highest percentage of fertilizer nitrogen was leached from soils to which KNO_3 was applied. In panoche fine sandy loam and lenthent sandy clay loam, respectively, 82 and 65% of the total nitrogen collected in the soil extract was fertilizer nitrogen. By comparison, only 14 to 24% of the nitrogen in the extract was fertilizer nitrogen when ammonium sulfate and sulfur-coated urea were the nitrogen sources. Much of the added ammonium-nitrogen was tied up by the clay complex of the soil near the soil surface.

As noted in Chapter 1, septic tank systems may be a source of nitrates in groundwater systems. Biswas and Warnock (1985) conducted a laboratory study related to denitrification of septic tank system effluents in the subsurface environment. The study was conducted using kitchen waste as the source of organic carbon for denitrification in an attached growth system. Soil columns were used as denitrifying reactors, with the feed solutions composed of a nitrified solution and kitchen waste. Five different carbon (expressed as methanol) to nitrogen ratios ranging from 2.2 to 6:1 were applied to 20 soil columns. A high degree of denitrification was obtained from C:N ratios of 4:1 or higher.

Davenport, Lembke, and Jones (1975) found that nitrate was effectively reduced when methanol was added as a substrate material to a slowly moving solution in porous columns. Applied nitrate exhibited 87% removal during 24 days at 24°C, and 62% removal following 27 days at 13°C. The passage of nitrate and chloride through the columns was accompanied by an increase in redox potential and, in some cases, discoloration of the effluent. The removal of a high percentage of nitrate at relatively large pore velocities was encouraging for the prospect of removing excess nitrate from soil water in the vicinity of tile drains. While a technique was not described, it could involve a system for water table control with additions of substrate material introduced by surface application or deep plowing.

Schwan, Kramer, and Gericke (1984) also studied denitrification using ten reaction vessels 1 m^3 in size. The vessels were filled with coarse sand or fine gravel. The pore volume (water saturation) was 227 to 260 liters. Anaerobic conditions were established by the addition of 6 g glucose. The reactors were fed tap water with 50 and 200 mg/l nitrate from KNO_3 in such a way that a volumetric rate of flow of 0.2, 0.4, and 0.8 l/d was created. The volumetric rates of flow approximated the natural recharge of groundwater, the recharge of groundwater under conditions of irrigation, and the recharge conditions of an intensive wastewater land treatment system. The effluent from the reaction vessels were monitored for nitrate concentration on a monthly basis. The concentration was stationary from the 7th to the 55th month after the beginning of the experiment.

Willardson and Meek (1969) described some laboratory studies on large soil columns with a controlled water table and submerged drains. These studies demonstrated that very low nitrate content water can be delivered from agricultural drains. A field experiment on drain submergence that intercepts groundwater high in native nitrate indicated that denitrified water from irrigated agriculture can reduce nitrate concentrations in drainage water by dilution. Where proper conditions can be established (including oxygen shortage, adequate organic carbon as an energy source, and a bacterial population), anaerobic conditions develop and result in denitrification. Based on the results obtained in the columns, a nitrate-nitrogen content of 500 mg/l in the groundwater would require a surface application of 1000 pounds per acre nitrogen as a nitrate fertilizer in a single application.

Finally, Lindstrom and Boersma (1990) described a two-dimensional mathematical model for simulating the transport and fate of organic chemicals in a laboratory-scale, single-layer aquifer. Two large (4 ft wide, 4 ft high, and 16 ft long) physical aquifers were used, with each aquifer containing three horizontal layers of material. Each layer was assumed to be homogenous and isotropic with respect to water flow. The systems were used for the study of transport and fate of chemicals, and for evaluation of growth characteristics of indigenous microbial populations. The physical aquifers can also be used for the study of proposed physical and biological remediation schemes. The developed mathematical model accounts for the major physical processes of storage, dispersion, and advection, and can also account for linear equilibrium sorption, three first-order loss processes (including microbial degradation, irreversible sorption and/or dissolution into the organic phase, metabolism in the sorbed state), and first-order loss in the sorbed state. A broad range of remediation scenarios may be considered by the use of the model, including placement of injection/extraction wells to induce plume spreading or plume shaping and the effects of regions of varying hydraulic conductivity on the shape of the plumes.

SUMMARY

Chemical and microbiological processes can influence the movement of various forms of nitrogen in the subsurface environment. Nitrification and denitrification processes, along with other microbial, abiotic, and hydrodynamic phenomena, influence nitrate concentrations in groundwater; such influences can be exacerbated spatially, vertically, and temporally. An understanding of subsurface processes is fundamental to identifying natural and man-made sources of nitrate contamination, evaluating influencing factors, modeling transport and fate, delineating appropriate protection measures, and selecting nitrate *in situ* remediation and/or pump and treatment cleanup schemes.

SELECTED REFERENCES

Alfoldi, L., "Movement and Interaction of Nitrates and Pesticides in the Vegetation Cover-Soil Groundwater-Rock System," *Environmental Geology*, Vol. 5, No. 1, 1983, pp. 19–25.

Anonymous, "Agriculture Group Symposium: Agriculture and Water Quality, Part 2," *Journal of the Science of Food and Agriculture*, Vol. 35, No. 8, 1984, pp. 855–862.

Bachelor, G.A., Cawlfield, D.E., Lindstrom, F.T., and Boersma, L., "User's Guide for the Mathematical Model LT3VSI for Denitrification in Nonhomogeneous Laboratory Scale Aquifers," EPA/600/2-91/034, March, 1991, Oregon State University, Corvallis, Oregon.

Behnke, J., "Summary of the Biochemistry of Nitrogen Compounds in Ground Water," *Journal of Hydrology*, Vol. 27, No. 1-2, October, 1975, pp. 155–167.

Bergstrom, L., "Nitrate Leaching and Drainage from Annual and Perennial Crops in Tile-Drained Plots and Lysimeters," *Journal of Environmental Quality*, Vol. 16, No. 1, January-March, 1987, pp. 11–18.

Bergstrom, L. and Brink, N., "Effects of Differentiated Applications of Fertilizer N on Leaching Losses and Distribution of Inorganic N in the Soil," *Plant and Soil*, Vol. 93, No. 3, 1986, pp. 333–345.

Berndt, M.P., "Sources and Distribution of Nitrate in Ground Water at a Farmed Field Irrigated with Sewage Treatment Plant Effluent," USGS/WRI-90-4006, 1990, U.S. Geological Survey, Tallahassee, Florida.

Biswas, N. and Warnock, R.G., "Nitrogen Transformations and Fate of Other Parameters in Columnar Denitrification," *Water Research*, Vol. 19, No. 8, 1985, pp. 1065–1071.

Borneff, J. and Adabe, B., "Nitrate in Ground Water and its Relation to Fertilization," *Zentralbl. Bakteriol. Parasitenkd. Infektionskr. Hyg. Erste Abt. Orig. Reihe B. Hyg. Praev. Med.*, Vol. 157, No. 4, 1973, pp. 337–345.

Bouwer, H., "Effect of Irrigated Agriculture on Groundwater," *Advances in Irrigation and Drainage: Surviving External Pressures*, Borrelli, J., Hasfurther, V.R. and Burman, R.D., Eds, American Society of Civil Engineers, New York, 1983, pp. 175–182.

Bradley, P.M., Aelion, C.M., and Vroblesky, D.A., "Influence of Environmental Factors on Denitrification in Sediment Contaminated with JP-4 Jet Fuel," *Ground Water*, Vol. 30, No. 6, November-December, 1992, pp. 843–848.

Bumb, A.C., McKee, C.R., Way, S.C., Drever, J.T., and Halepaska, J.C., "Ammonia and Nitrate Migration from the Vadose Zone to the Ground Water System: Containment, Recovery, and Natural Restoration," *Proceedings of the First National Outdoor Action Conference on Aquifer Restoration*, National Water Well Association, Dublin, Ohio, 1987, pp. 95–123.

Chopp, K.M., Clapp, C.E., and Schmidt, E.L., "Ammonia Oxidizing Bacteria Populations and Activities in Soils Irrigated with Municipal Wastewater Effluent," *Journal of Environmental Quality*, Vol. 11, No. 2, April-June, 1982, pp. 221–226.

Davenport, L.A., Lembke, W.D., and Jones, B.A., "Denitrification in Laboratory Sandy Columns," *Transactions of the American Society of Agricultural Engineers*, Vol. 18, No. 1, January-February, 1975, pp. 195–206.

Desprez, N., Landreau, A., and Vogt, D., "Mineralization of Groundwater by Agricultural Practices: Mapping of Sensitive Areas — Application to the Indre et D'Loire District (France)," *Memoires, International Association of Hydrogeologists*, Vol. 16, No. 1, 1982, pp. 195–206.

Devitt, D. et al., "Nitrate Nitrogen Movement Through Soil as Affected by Soil Profile Characteristics," *Journal of Environmental Quality*, Vol. 5, No. 3, 1976, pp. 283–287.

El Etreiby, F. and Laudelout, H., "Movement of Nitrate Through a Loess Soil," *Journal of Hydrology*, Vol. 97, No. 3/4, February, 1988, pp. 213–224.

Gillings, O.J., "Nitrate Leaching in Soil on Rutgers Agricultural Research Center at Adelphia, New Jersey," M.S. Thesis, 1973, Rutgers, The State University, New Brunswick, New Jersey.

Howard, K.W.F., "Denitrification in a Major Limestone Aquifer," *Journal of Hydrology*, Vol. 76, No. 3-4, February, 1985, pp. 265–280.

Keeney, D.R., "Sources of Nitrate to Ground Water," *CRC Critical Reviews in Environmental Control*, Vol. 16, No. 3, 1986, pp. 257–304.

Keeney, D.R., "Transformations and Transport of Nitrogen," in *Agricultural Management and Water Quality*, Schaller, F.W. and Bailey, G.W., Eds, Iowa State University Press, Ames, Iowa, 1983, pp. 48–64.

Letey, J. et al., "Effect of Water Management on Nitrate Leaching," *Proceedings of National Conference on Management of Nitrogen in Irrigated Agriculture*, 1978, University of California at Riverside, Riverside, California, pp. 231–249.

Lindley, J.A., Dale, A.C., and Mannering, J.V., "Animal Waste and Nitrate Movement Through Soil," Paper 74-2017, 1974, American Society of Agricultural Engineers, St. Joseph, Missouri.

Lindstrom, F.T. and Boersma, L., "Denitrification in Nonhomogeneous Laboratory Scale Aquifers: 1. Preliminary Model for Transport and Fate of a Single Compound," EPA/600/2-90/009, March, 1990, U.S. Environmental Protection Agency, R.S. Kerr Environmental Research Laboratory, Ada, Oklahoma.

Lindstrom, F.T., Boersma, L., Myrold, D., and Barlaz, M., "Denitrification in Nonhomogeneous Laboratory Scale Aquifers: 4. Hydraulics, Nitrogen, Chemistry and Microbiology in a Single Layer," EPA/600/S2-91/014, May, 1991, U.S. Environmental Protection Agency, R.S. Kerr Environmental Research Laboratory, Ada, Oklahoma.

Lowrance, R.R. and Pionke, H.B., "Transformations and Movement of Nitrate in Aquifer Systems," Ch. 13 in *Nitrogen Management and Ground Water Protection*, Follett, R.F., Ed., Elsevier Science Publishers, B.V., Amsterdam, The Netherlands, 1989, pp. 373–392.

Lowrance, R.R. and Smittle, D., "Nitrogen Cycling in a Multiple Crop Vegetable Production System," *Journal of Environmental Quality*, Vol. 17, No. 1, January-March, 1988, pp. 158–162.

Ludwick, A.E., Reuss, J.O., and Langin, E.J., "Soil Nitrates Following Four Years Continuous Corn and As Surveyed in Irrigated Farm Fields of Central and Eastern Colorado," *Journal of Environmental Quality*, Vol. 5, No. 1, January-March, 1976, pp. 82–86.

Martin, D.L. et al., "Evaluation of Nitrogen and Irrigation Management for Corn Production Using Water High in Nitrate," *Soil Science Society of America Journal*, Vol. 46, No. 5, September-October, 1982, pp. 1056–1062.

Morton, T.G., Gold, A.J., and Sullivan, W.M., "Influence of Overwatering and Fertilization on Nitrogen Losses from Home Lawns," *Journal of Environmental Quality*, Vol. 17, No. 1, January-March, 1988, pp. 124–130.

Pal, D. and Broadbent, F.E., "Leaching of Calcium and Magnesium from Soil Columns as Affected by the Form of Nitrogen in Applied Sewage," *Soil Science Society of America Journal*, Vol. 45, No. 1, January-February, 1981, pp. 56–60.

Reddy, K.R. and Patrick, W.H., "Nitrogen Transformations and Loss in Flooded Soils and Sediments," in *CRC Critical Reviews in Environmental Control*, Vol. 13, No. 4, 1981, pp. 273–303.

Rolston, D.E. and Broadbent, F.E., "Field Measurement of Denitrification," EPA-600/2-77-233, November, 1977, U.S. Environmental Protection Agency, R.S. Kerr Environmental Research Laboratory, Ada, Oklahoma.

Rolston, D.E. et al., "Denitrification as Affected by Irrigation Frequency of a Field Soil," EPA-600/2-80-066, April, 1980, U.S. Environmental Protection Agency, R.S. Kerr Environmental Research Laboratory, Ada, Oklahoma.

Schubauer-Berigan, J.P., Capone, D.G., Cochran, J.K., Kazumi, J., and Epler, N., "Bacterial Transformations of Nitrate and Aldicarb (TEMIK) in Anoxic Groundwaters of Long Island," USGS/G-1282, October, 1990, Marine Sciences Research Institute, State University of New York at Stony Brook, Stony Brook, New York.

Schwan, M., Kramer, D., and Gericke, C., "Simulation of Nitrate Degradation in Groundwater," *Acta Hydrochimica et Hydrobiologica*, Vol. 12, No. 2, 1984, pp. 163–171.

Spalding R.F. and Kitchen, L.A., "Nitrate in the Intermediate Vadose Zone Beneath Irrigated Cropland," *Ground Water Monitoring Review*, Vol. 8, No. 2, Spring, 1988, pp. 89–95.

Sparks, D.L., "Nitrate Retention as it Affects Groundwater Pollution in Mid-Atlantic Soils," September, 1987, Department of Plant Science, University of Delaware, Newark, Delaware.

Stone, L.R., "Movement of Water and Nitrate-Nitrogen in a Typical Silt Loam Soil of Western Kansas," OWRT-A-101-KAN (1), September, 1982, Kansas Water Resources Research Institute, Kansas State University, Manhattan, Kansas.

Tolman, J., "Variations in Nitrate Concentrations at Idaho Site Due to Seasonal Influences," *Journal of Environmental Health*, Vol. 48, No. 1, July/August, 1985, pp. 22–25.

Willardson, L.S. and Meek, B.D., "Agricultural Nitrate Reduction at a Water Table," *Collected Papers Regarding Nitrates in Agricultural Wastewaters*, Report No. 13030ELY, December, 1969, Federal Water Quality Administration, Washington, D.C., pp. 41–52.

Williford, J.W. et al., "The Movement of Nitrogenous Fertilizers Through Soil Columns," *Collected Papers Regarding Nitrates in Agricultural Wastewaters*, Report No. 13030ELY, December, 1969, Federal Water Quality Administration, Washington, D.C., pp. 29–39.

3 ILLUSTRATIONS OF NITRATE POLLUTION OF GROUNDWATER

INTRODUCTION

This chapter summarizes selected illustrations of nitrate pollution of groundwater, particularly as a consequence of agricultural usage of fertilizers or land disposal of domestic wastewaters. The purpose of this review is twofold: (1) to demonstrate the widespread problem of groundwater pollution by nitrates, and (2) to identify the influence of various climatological and hydrogeological factors, fertilizer application practices, and subsurface transformation and transport and fate processes on resultant nitrate concentrations in groundwater. It should be noted that Chapter 1 contains a summary of a recent 5-year survey of pesticides and nitrates in drinking water wells in the U.S. The reviewed illustrations in this chapter are divided into those from: (1) the U.S., and (2) Europe and other countries. A summary section concludes this chapter.

ILLUSTRATIONS FROM THE UNITED STATES

Table 3.1 contains summary comments on 29 references dealing with nitrate groundwater pollution studies in the U.S. Fifteen of the 50 states are represented as follows: (1) California (Anton, Barnickol, and Schnaible, 1988; Klein and Bradford, 1979; Lawrence et al., 1977); (2) Connecticut (DeRoo, 1980); (3) Delaware, Maryland, and Virginia (Hamilton and Shedlock, 1992); (4) Iowa (Hallberg, 1986; Hallberg, Libra, and Hoyer, 1985; Walker and Hoehn, 1989); (5) Kansas (Spruill, 1982, 1983); (6) Maryland (Bachman, 1984); (7) Michigan (Cummings, Twenter, and Holtschlag, 1984); (8) Missouri (Tryon, 1976; Sievers and Fulhage, 1992); (9) Nebraska (Exner and Spalding, 1979, 1990; Hergert, 1982; Spalding et al., 1978); (10) New York (Baier and Rykbost, 1976; Buller, Nichols, and Harsch, 1978); (11) North Carolina (Jennings et al., 1991); (12) Pennsylvania (Pionke and Urban, 1985); (13) Texas (Kreitler and Browning, 1983; Reeves and Miller, 1978; Jensen, 1991); and (14) Wisconsin (Saffigna and Keeney, 1977). In addition, information from 12 midcontinental states (Kolpin,

Table 3.1 References Related to United States Case Studies

Author(s) (year)	Comments
Anton, Barnickol, and Schnaible (1988)	Review of sources of nitrates in California groundwater along with summaries of several case histories of the seriousness of such pollution
Bachman (1984)	Factors affecting nitrates in groundwater in eastern Maryland
Baier and Rykbost (1976)	Influence of potato fertilization on nitrates in groundwater in Long Island, New York
Buller, Nichols, and Harsch (1978)	Sources of chloride and nitrate inputs to groundwater in Cortland County, New York
Cummings, Twenter, and Holtschlag (1984)	Nitrogen inputs to the hydrologic system in southwest Michigan
DeRoo (1980)	Nitrate fluctuations in Connecticut groundwater as influenced by fertilizer usage
Exner and Spalding (1979)	Effects of fertilization and irrigation on chlorides, sulfates, and nitrates in groundwater in Holt County, Nebraska
Exner and Spalding (1990)	Summary of the nature and geographical extent of nitrate pollution of groundwater in Nebraska
Hallberg (1986)	Review of nitrates in Iowa groundwater from routine agricultural practices
Hallberg, Libra, and Hoyer (1985)	Discussion of the role of infiltration in nitrate contamination of groundwater in karst-carbonate aquifers in Iowa
Hamilton and Shedlock (1992)	Review of the sources and geographical extent of nitrate pollution of groundwater in the Delmarva Peninsula of Delaware, Maryland, and Virginia
Hergert (1982)	Agricultural contributions of nitrates to groundwater in the Nebraska sandhills area
Jennings, Sneed, Huffman, Humenik, and Smolen (1991)	Presentation of the results of a study of the occurrence of nitrates and pesticides in over 9000 rural domestic wells in North Carolina
Jensen (1991)	Review of the occurrence of nitrates in groundwater in agricultural areas of Texas
Klein and Bradford (1979)	Nitrate distribution in the unsaturated soil zone in San Bernardino County, California
Kolpin, Burkart, and Thurman (1994)	Report on the occurrence and distribution of selected herbicides, atrazine metabolites, and nitrate in near surface aquifers in 12 states in the midcontinental United States
Kreitler and Browning (1983)	Use of nitrogen isotope analysis to determine natural vs. human sources of nitrate pollution in the Edwards aquifer in Texas
Lawrence et al. (1977)	Depth profiles of nitrates in groundwater in the Redlands, California area
Pionke and Urban (1985)	Land usage and geological influences on groundwater nitrates in a small Pennsylvania watershed
Reeves and Miller (1978)	Causative factors for nitrates, chlorides, and dissolved solids in the Ogallala aquifer in west Texas
Saffigna and Keeney (1977)	Contributions of irrigated agriculture to nitrate and chloride concentrations in the groundwater of the central Wisconsin sand plains
Sievers and Fulhage (1992)	Survey results from a study of 25 pesticides and nitrate in 201 rural wells in 8 agricultural areas in Missouri; 22% of the wells exceeded the drinking water standard for nitrate
Spalding et al. (1978)	Commercial fertilizer contributions to groundwater nitrates in Merrick County, Nebraska

Table 3.1 References Related to United States Case Studies (Continued)

Author(s) (year)	Comments
Spruill (1982)	Nitrate variations in groundwater in 3 areas of Kansas
Spruill (1983)	Nitrate variations with groundwater depth in 2 areas of Kansas
Swanson (1992)	Report on nitrate–nitrogen concentrations in groundwater from 8 midwestern states
Taylor and Bigbee (1973)	Fluctuations of groundwater nitrate concentrations with fertilizer applications and irrigation season
Tryon (1976)	Agricultural land use influences on nitrates and coliform bacterial density in groundwater in and around Phelps County, Missouri
Walker and Hoehn (1989)	Synopsis of nitrate analyses on 44,000 groundwater samples in Iowa

Burkart, and Thurman, 1994) and eight midwestern states (Swanson, 1992) is also noted.

Information from all 29 references listed in Table 3.1 will not be presented in detail; those to be excluded include four studies related to the influence of geological formations on nitrates in groundwater (Spruill, 1982 and 1983; Hallberg, Libra, and Hoyer, 1985; Sievers and Fulhage, 1992); three studies related to the influence of irrigation practices (Exner and Spalding, 1979; Hergert, 1982; Taylor and Bigbee, 1973); three studies related to multiple sources of nitrates in groundwater (Buller, Nichols, and Harsch, 1978; Hamilton and Shedlock, 1992; and Jensen, 1991); two studies related to the influence of groundwater pumping (Klein and Bradford, 1979; Lawrence et al., 1977); and one study on surface and groundwater relationships (Cummings, Twenter, and Holtschlag, 1984). Information from the remaining references will be presented in terms of four topical issues: (1) problem assessment surveys; (2) factors and practices influencing nitrates in groundwater; (3) nitrogen budgets; and (4) useful tools for groundwater nitrate surveys.

Problem Assessment Surveys

A study of the nitrate, chloride, and dissolved solids in the groundwater of the Ogallala aquifer in a 27-county area of west Texas was conducted in the 1970s (Reeves and Miller, 1978). A total of 1597 analyses from the 27 counties was used. Figure 3.1 shows that the areas having nitrate concentrations >45 mg/l are concentrated in the southeastern part of the southern high plains, essentially south of a line from Clovis, New Mexico, to Lubbock, Texas (the heavy dashed line trending NW-SE across Figures 3.1, 3.2, and 3.3) (Reeves and Miller, 1978). North of this line, nitrate values are usually <10 mg/l; whereas south of the line, nitrate may exceed 60 mg/l and in some localities is >170 mg/l. Figure 3.2 shows that most areas having chloride concentrations >100 mg/l are south of the line joining Clovis, New Mexico, and Lubbock, Texas. North of the line, chlorides are usually <20 mg/l whereas south of the line, chloride commonly exceeds 500 mg/l and may, in some areas, be >2000 mg/l. Figure 3.3 shows that areas having dissolved solids values >500 mg/l are located principally south of the line joining Clovis, New Mexico, with Lubbock, Texas. North of the line mean dissolved solids values are in the range of 226 to 476 mg/l, whereas south of the line dissolved solids values often exceed 1000 mg/l and, in some local areas, may be >8000 mg/l.

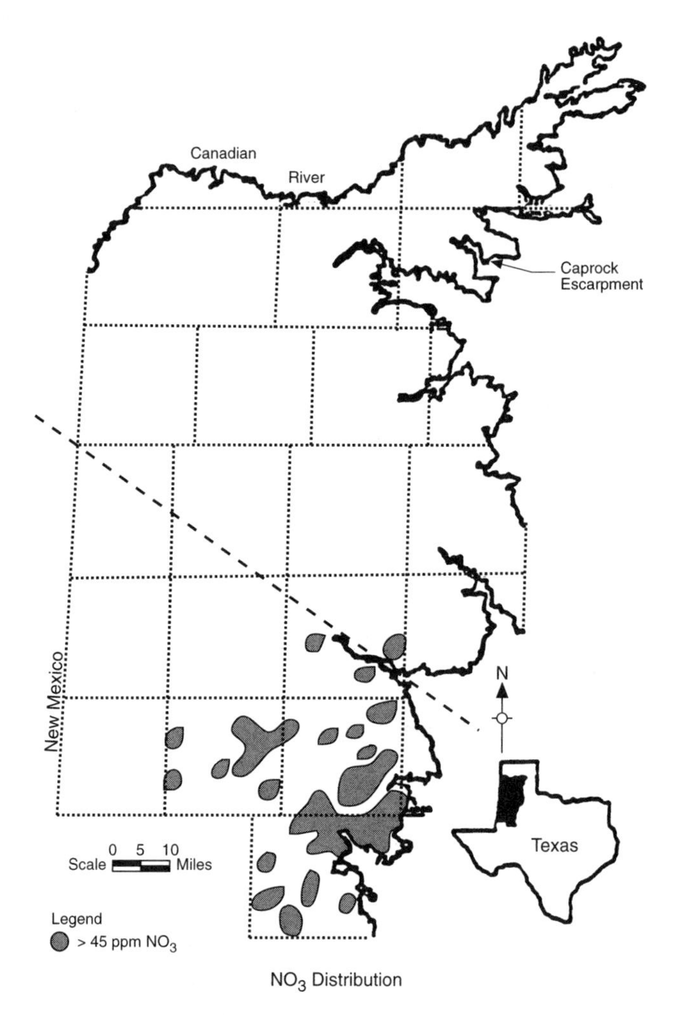

Figure 3.1 Distribution of nitrate in groundwater on Southern High Plains, Texas. (From Reeves, C. C., Jr. and Miller, W. D., *Ground Water*, Vol. 16, No. 3, 1978, pp. 167–173. With permission.)

Reeves and Miller (1978) noted that a multiplicity of sources of local increases in nitrate, chloride, and dissolved solids have been identified in the southern high plains; however, the regional pattern of poor quality water identified in this study is probably due to regional rather than local causes. The most obvious sources for high nitrate values in Ogallala groundwater are the high nitrate

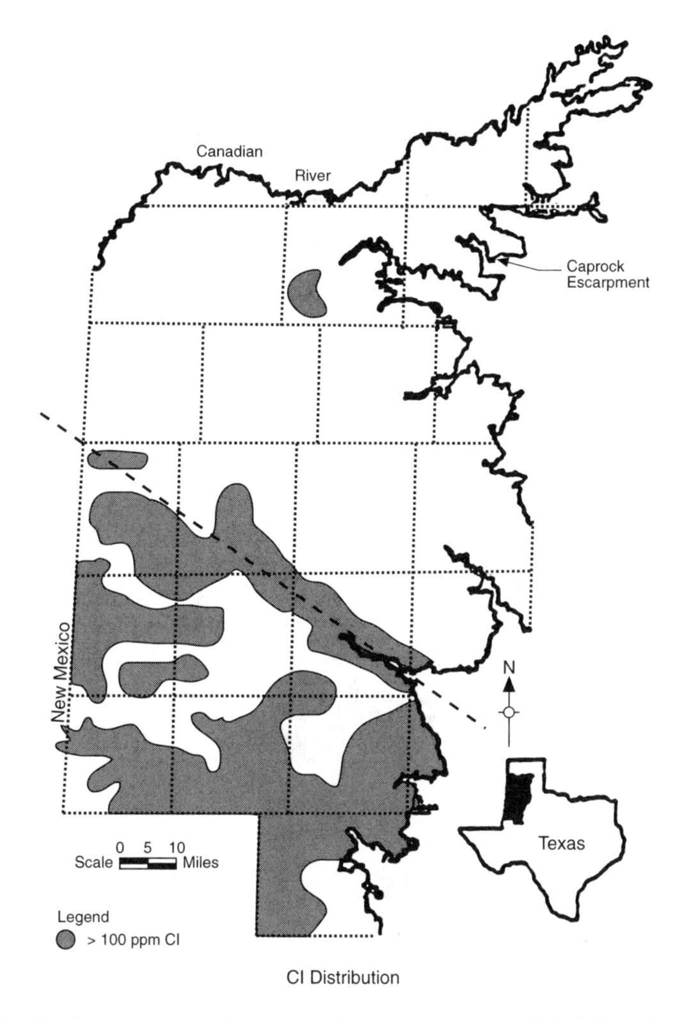

Figure 3.2 Distribution of chloride in groundwater on Southern High Plains, Texas. (From Reeves, C. C., Jr. and Miller, W. D., *Ground Water*, Vol. 16, No. 3, 1978, pp. 167–173. With permission.)

fertilizers used during the 1960s and 1970s. Most of the high nitrate values (>45 mg/l) occur in areas having sandy soils that have been intensively cultivated; thus, leaching of nitrogen-based fertilizers is suspect. However, the high chloride and dissolved solids, which exist in essentially the same geographic area, probably represent vertical to local lateral seepage of saline water from large alkali lake basins and local vertical migration from saline Cretaceous aquifers.

The major sources of nitrate in California groundwaters used as water supplies include leachate produced from agricultural activities and animal and poultry wastes, individual septic tank systems, effluent from wastewater treatment plants, and municipal and industrial runoff (Anton, Barnickol, and Schnaible, 1988). The largest sources are those related to agricultural activities; in particular,

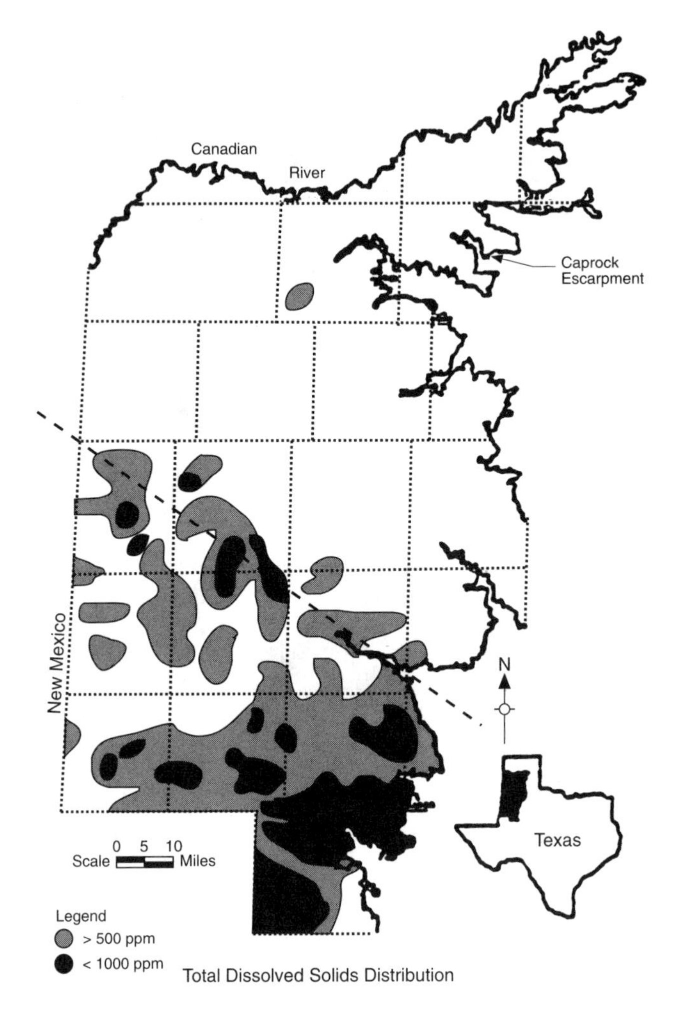

Figure 3.3 Distribution of dissolved solids in groundwater on Southern High Plains, Texas. (From Reeves, C. C., Jr. and Miller, W. D., *Ground Water*, Vol. 16, No. 3, 1978, pp. 167–173. With permission.)

those that utilize the application of nitrogen fertilizers in one form or another. Several case histories have indicated the geographical extent and seriousness of the problems, and two examples will be noted. First, the Metropolitan Water District of Southern California has indicated that it annually loses 4% of its drinking water supply to total dissolved solids and nitrate (primarily nitrate), compared to less than 1/2% from toxic organic chemicals. About 12% of the wells sampled in the service area exceeded the state MCL (maximum contaminant level) for nitrate. In the second example from the Salinas Valley, it is estimated that by the year 2000, the groundwater in most of the water table aquifers will exceed the state drinking water standard for nitrate of 45 mg/l (Anton, Barnickol, and Schnaible, 1988).

Over 9000 domestic wells in North Carolina were sampled for nitrate, chloride, conductivity, and pH in 1989. The results of the survey indicated that 3.2% of the wells sampled contained nitrate-nitrogen at concentrations exceeding the drinking water standard of 10 mg/l (Jennings, Sneed, Huffman, Humenik, and Smolen, 1991). Regional differences in the results were noted, with the Coastal Plain having the highest incidence of contamination and the Blue Ridge having the lowest. The predominant factor related to the nitrate level was well depth. Shallow, poorly constructed wells in close proximity to croplands or animal operations were typically found to be high in nitrate.

Multistate groundwater quality testing has been undertaken for several years at Heidelberg College in Tiffin, Ohio; Table 3.2 summarizes the resultant nitrate-nitrogen data (Swanson, 1992). The data indicate higher nitrate-nitrogen levels in the sampled groundwater from Illinois, Virginia, New Jersey, and Kentucky.

Table 3.2 Summary of Nitrate Data by State

State	Counties tested	Number of samples	Average nitrate concentration (mg/l)	Percent over 10 mg/l
Illinois	8	286	5.76	19.9
Indiana	33	5,685	0.92	3.5
Kentucky	90	4,559	2.50	4.6
Louisiana	23	997	1.19	0.8
New Jersey	5	1,108	2.60	6.8
Ohio	80	18,202	1.32	3.0
Virginia	24	1,054	2.92	7.1
West Virginia	13	1,288	0.83	0.8

From Swanson, G.J., *Water Well Journal*, Vol. 46, 1992, 39–41. With permission.

A 1991 study of herbicides and nitrates in near-surface (within 50 ft of the land surface) aquifers in 12 states (Illinois, Indiana, Iowa, Kansas, Michigan, Minnesota, Missouri, Nebraska, North Dakota, Ohio, South Dakota, and Wisconsin) in the corn and soybean-producing region of the U.S. has been conducted. The study was based on groundwater samples collected during the spring and summer from 303 wells completed in near-surface unconsolidated and near-surface bedrock aquifers (Kolpin, Burkart, and Thurman, 1994). Nitrate concentrations equal to or greater than 3.0 mg/l (termed "excess nitrate") were detected in 29% of the 599 nitrate analyses, and nitrate-nitrogen concentrations equal to or greater than the MCL for drinking water of 10 mg/l were found in 6% of the samples.

Hydrogeologic factors, land use, agricultural practices, local features, and water chemistry were analyzed for their possible relation to excess nitrate detections. Excess nitrate was detected more frequently in near-surface unconsolidated aquifers than in near-surface bedrock aquifers. The depth to the top of the aquifer was inversely related to the frequency of detection of excess nitrate, and no significant seasonal differences were determined for the frequency of excess nitrate (Kolpin, Burkart, and Thurman, 1994).

Factors and Practices Influencing Nitrates in Groundwater

A number of factors and practices can influence the nitrate concentration in groundwater. Pertinent factors include precipitation/runoff, soil type, denitrification, and nonagricultural sources of nitrogen in the subsurface environment; and pertinent practices include fertilizing intensity, and crop type and land usage. Additional illustrations of these factors and practices from case studies in Europe and elsewhere will be described later.

The influence of land usage (urban vs. agricultural) has been addressed by DeRoo (1980). The study was conducted to determine if fertilizers contribute to nitrate pollution of groundwater in the Connecticut River Valley. To study agricultural use of fertilizer, a shade-grown wrapper-tobacco, which has been grown in the Connecticut River valley for many years, was chosen. It is heavily fertilized, mostly with natural nitrogenous organics; and in combination with cover cropping can, in time, build high soil organic nitrogen levels. This study was at the farm of the Valley Laboratory of the Connecticut Agricultural Experiment Station in Windsor, and at a commercial shade tobacco farm in Suffield. To typify nonagricultural use of fertilizer, turf plots at the Valley Laboratory farm were chosen. A network of sampling wells at the Valley Laboratory allowed sampling of the groundwater flowing into, under, and out of the farm. A similar, but smaller, network of wells was installed at the commercial farm in Suffield. Both test sites were on nearly level to gently sloping terraces. The deep, well-drained, rapidly permeable soils at both locations are underlain by impervious glacial-lake deposits of silt and clay, over which perched groundwater flows and fluctuates. These conditions allowed sampling of the groundwater with relatively shallow wells.

The general finding from the study reported by DeRoo (1980) was that groundwater under the Valley Laboratory farm averaged 3 mg/l nitrate-nitrogen over 3 years. The experimental farm operations increased the 2.5 mg/l concentration of the groundwater flowing into the farm to an average concentration of 4 mg/l nitrate-nitrogen leaving the farm. Concentrations of over 10 mg/l nitrate-nitrogen were observed after heavy rainfall. These temporary increases usually occurred in the fall, downgradient from areas treated with generous amounts of fertilizer. In the fall of 1977, surface water entering the northwest corner of the farm showed a sudden increase in NO_3–N concentration (up to 18 mg/l), apparently from an adjacent, upstream residential area. However, groundwater under the turf plots was not significantly influenced by moderate fertilization treatments during the $2^1/_2$ years of observations. These observations suggest that as long as a reasonable nitrogen fertilizer program on lawns is followed, the potential for nitrate leaching under turf is not significant.

On the Merrell Farm in Suffield, in a rural environment with comparable soil conditions, a different situation was noted (DeRoo, 1980). Long-term effects of fertilizations with predominantly organic materials and intensive shade tobacco management sustained year-round high levels of nitrate-nitrogen concentrations in the groundwater. The nitrate-nitrogen concentrations averaged 20 mg/l, except in the upstream corner of the field, possibly due to dilution with incoming water with a very low nitrate concentration and/or by denitrification under the less than well-drained soil conditions in this area. Under these conditions of high soil nitrogen levels, decreasing the amount of fertilizer nitrogen applied or increasing the number and total amount of post-planting applications did not significantly affect nitrate leaching.

The influence of various land uses on nitrates in groundwater was also examined by Bachman (1984) in a study of the Columbia aquifer in the central Delmarva Peninsula in Maryland. Six categories of land use and their associated sources of nitrogen input to the subsurface environment are shown in Table 3.3 (Bachman, 1984). Some of the land uses, such as field crops and field crops with residence, have at least one common source of nitrogen. The relationships between nitrate concentrations in groundwater and land usage are shown in Figure 3.4 (Bachman, 1984). The difference between groundwater concentration medians is statistically significant.

Table 3.3 Land-Use Categories and Sources of Nitrogen in Maryland Study

Land use	Sources of nitrogen
Woods and wetlands	Natural fixation of atmospheric nitrogen and decay of organic matter
Woods with residence	Fixation of atmospheric nitrogen; decay of organic matter; and infiltration of septic-tank effluent, lawn and garden fertilizer
Field crops	Inorganic nitrogen fertilizer; nonpoint application of manure; natural nitrogen fixation and decay of organic matter
Field crops with residence	Inorganic nitrogen fertilizer; nonpoint application of manure; natural nitrogen fixation and decay of organic matter; infiltration of septic tank effluent
Chickenhouses	Leaching from chicken manure
Urban	Infiltration of septic tank effluent; leachate from sanitary landfills; infiltration of sewage from sewage lines; lawn and garden fertilizer; dog and cat feces; natural nitrogen fixation

From Bachman, L. J., "Nitrate in the Columbia Aquifer, Central Delmarva Peninsula, Maryland," Water Resources Investigation Report No. 84-4322, 1984, U.S. Geological Survey, Towson, MD.

Groundwater at sites with agricultural land uses, including chicken houses, had the highest median nitrate values. Water from wells drilled near chicken houses had the highest median concentration, 9.7 mg/l. Sites with field crops and residences were next, followed by sites with field crops only, urban sites, and wooded sites. Wooded and field crop sites with residences (and presumably on-site sewage disposal systems) had water with higher median nitrate values than sites without residences, although the Kruskal-Wallis test indicated a nonsignificant difference between residential and nonresidential median concentrations for sites with a woodland land use (Bachman, 1984).

In addition to land usage, geological factors can also influence the groundwater concentrations of nitrate. For example, Tryon (1976) conducted a study of 675 water wells in and around Phelps County, Missouri. This county encompasses an area of 677 square miles in the Ozarks region of south-central Missouri. The county is underlain primarily by cherty dolomite bedrock of Ordovician age, but some quartzitic sandstone inclusions are also present and widespread.

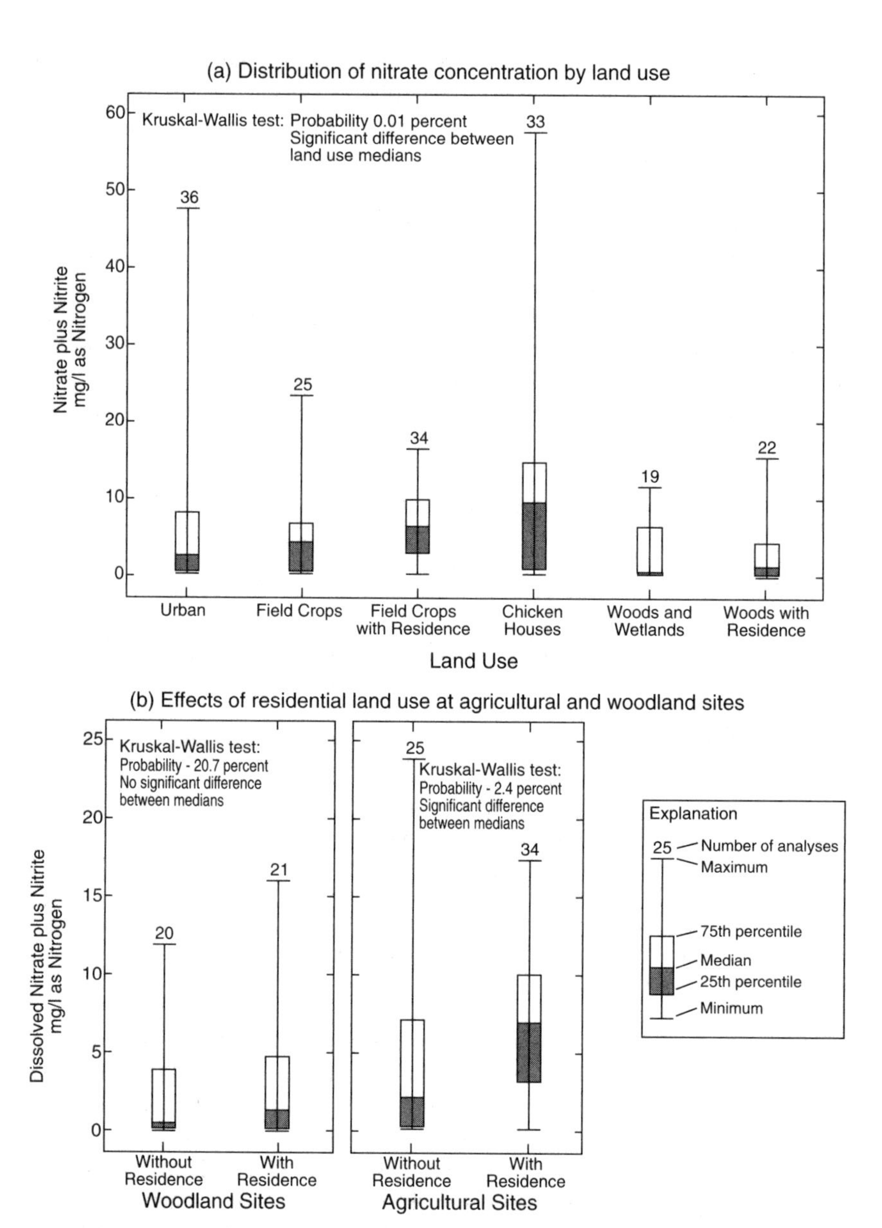

Figure 3.4 Relation between land use and nitrate concentration in groundwater of Columbia Aquifer in the Central Delmarva Peninsula, Maryland. (From Bachman, L. J., "Nitrate in the Columbia Aquifer, Central Delmarva Peninsula, Maryland," Water Resources Investigation Report No. 84-4322, 1984, U.S. Geological Survey, Towson, MD.)

Topographic features range from local areas of gently rolling land to large areas of steep and highly dissected land, with local relief up to 350 ft. Karst features such as caves, springs, sinkholes, influent streams that lose their surface flow to the subsurface groundwater reservoir, and solutionally enlarged water transport conduits in the bedrock occur throughout the county, although their abundance and areal extent of subsurface integration vary widely.

Based on the data analyses, Tryon (1976) indicated that discrete areas of differing groundwater quality can be identified and mapped. The best quality groundwater, as judged by its low nitrate content and coliform bacterial density, was found in areas of relatively little agricultural (i.e., pasture and livestock) land use. The poorest quality was found in areas of intensely developed karst and greater agricultural land use. The adverse effect of agricultural land use on groundwater quality was more severe in the intensely developed karst than in the less intensely developed. Rural population density and soil association variations had no readily discernible effects on groundwater quality.

Irrigation and fertilization practices can also influence nitrate concentrations in groundwater. For example, Saffigna and Keeney (1977) documented the nitrate and chloride concentrations in groundwater associated with irrigated agricultural land in a central Wisconsin sand plain study area. The study area was incapable of supporting intensive agriculture without fertilizer and water additions; the principle crops were potatoes, corn, and vegetables. The study was conducted at three levels of intensity; a small potato field during one growing season, a University Experimental Farm over several years, and many local farms during part of a growing season. Differences in concentrations of nitrate-nitrogen and chloride in the top zone of the groundwater underlying the moderately fertilized potato field closely reflected the fertilizer and irrigation management practices on different parts of the field during that season. In general, increased irrigation and fertilization caused increased nitrates and chlorides in the groundwater.

Differences in concentrations of nitrate-nitrogen and chloride between irrigation wells on the Experimental Farm also closely reflected the irrigation and fertilizer practices on surrounding fields (Saffigna and Keeney, 1977). Nitrate-nitrogen and chloride concentrations in groundwater were lower in areas where the land had been idle for many years and only recently cropped. However, levels were high where adjacent fields were in continual use and had received large fertilizer and irrigation inputs.

In general, the results of the Saffigna and Keeney (1977) study indicated that the nitrate and chloride concentrations in the groundwater of the central Wisconsin sand plains are significantly above background, and that the main source is the irrigated agricultural practices in the region. Nitrate-nitrogen concentrations ranged from nil to 56 mg/l, and chloride from nil to 68 mg/l. Nitrate-nitrogen concentrations exceeded 10 mg/l in 15 of 33 irrigation wells and in two of three domestic wells. Nitrate-nitrogen and chloride concentrations varied widely between adjacent wells, but the NO_3–N:Cl ratio was much less variable. This suggested similar relative inputs of nitrogen and chloride, presumably from nitrogen and potassium (potassium chloride) fertilizers.

The influence of fertilization intensity on nitrates in the groundwater of Suffolk County (Long Island), New York, was examined by Baier and Rykbost (1976). The primary crop in the study area was potatoes, and previous work had shown that fertilization practices could result in substantial nitrogen losses; in many cases, excessive use of nitrogen could reduce crop yields. Annual nitrogen losses of 50 lb/acre (55.5 kg N/ha) had been shown sufficient to cause a concentration in the aquifer's uppermost layer of 10 mg/l nitrate-nitrogen (New York State Drinking Water Standard). The average potato grower on Long Island applied 200 to 250 lb N/ac (222 to 278 kg N/ha) at planting time; and, depending upon a number of factors, the N recovered in harvested tubers varies from 75 to 150 lb N/ac (83 to 167 kg N/ha). Losses to the groundwater could vary from 50 to 175 lb N/ac (55.5 to 194 kg N/ha).

The study by Baier and Rykbost (1976) demonstrated that the application of 150 lb N/ac (167 kg N/ha) could still maintain maximum potato yields and keep the nitrogen loss to groundwater below 50 lb N/ac (55.5 kg N/ha) by improving nitrogen-use efficiency. This was done by splitting nitrogen applications so that one third to one half was applied at planting and the remainder was applied prior to the period of rapid crop growth and nutrient uptake. These findings can aid in reducing crop production costs and protecting groundwater quality.

Exner and Spalding (1990) reported on the results of nitrate analyses in 5826 groundwater samples in Nebraska collected from 1984 through 1988. The large number of data points (an average of one sample every 13 square miles) reflects the importance with which nitrate contamination of the groundwater is viewed.

Nitrate-nitrogen concentrations measured in the 5826 sampled wells ranged from undetectable levels to 343 mg/l (Exner and Spalding, 1990). Because all areas of known or suspected nitrate contamination were sampled more heavily, the bar graphs depicted in Figure 3.5 are not representative of the overall condition of the groundwater quality with respect to nitrate contamination (Exner and Spalding, 1990). About 80% of the wells comprising this assessment had nitrate-nitrogen concentrations below 10 mg/l and more than 60% had concentrations below 5.0 mg/l. Public supply, domestic, and stock wells were less likely to exceed 10 mg/l than were irrigation wells.

With the exception of stock wells, the percentage of wells with concentrations of 10 mg/l or more decreased within each progressively higher concentration range up to 50 mg/l (see Figure 3.5). The occurrence of concentrations ≥50 mg/l was much higher in domestic (1.3%) and stock (3.7%) wells than in irrigation (0.4%) wells. Some 14 domestic wells had concentrations above 86 mg/l, the maximum concentration detected in the irrigation wells. Generally, these excessive concentrations are symptomatic of groundwater contaminated by point sources, and the presence of these concentrations in domestic wells supports the premise that domestic wells are more likely to be highly contaminated by nitrate from concentrated point sources than are irrigation wells (Exner and Spalding, 1990). Point-source contamination, which from a health perspective is at least as important as nonpoint or dispersed source contamination, occurs most frequently in eastern Nebraska.

In Nebraska, irrigated small grain crops require the use of almost twice as much nitrogen fertilizer as dryland crops, and irrigated corn requires more nitrogen fertilizer than other irrigated crops. There are a large number of wells with elevated nitrate concentrations in irrigated crop-producing areas hydrogeologically

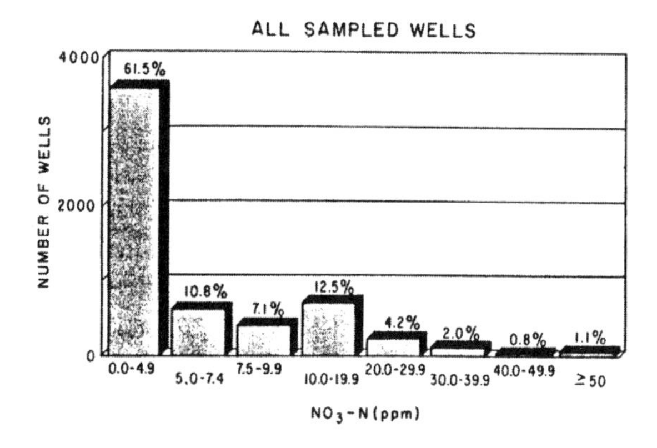

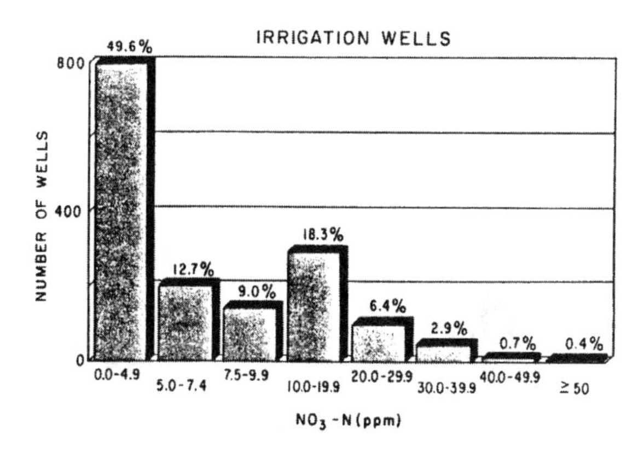

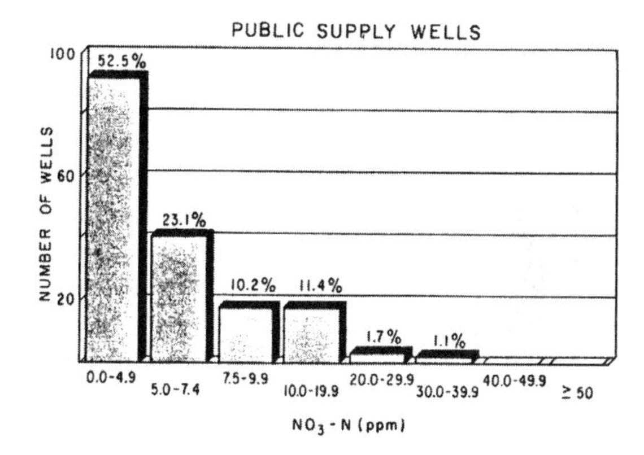

Figure 3.5 Distribution of nitrate-nitrogen concentrations by well type in Nebraska. (From Exner, M. E. and Spalding, R. F., "Occurrence of Pesticides and Nitrate in Nebraska's Groundwater," Report WC1, 1990, Water Center, Institute of Agriculture and Natural Resources, University of Nebraska, Lincoln, pp. 3–30.)

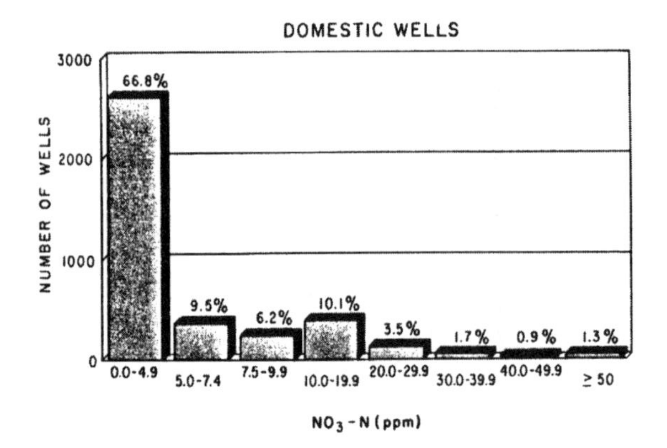

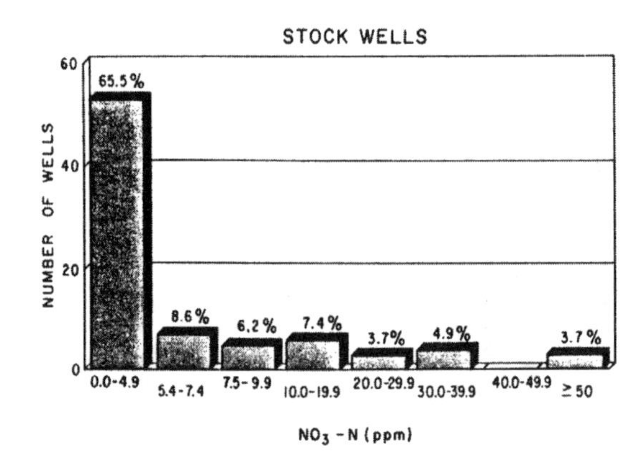

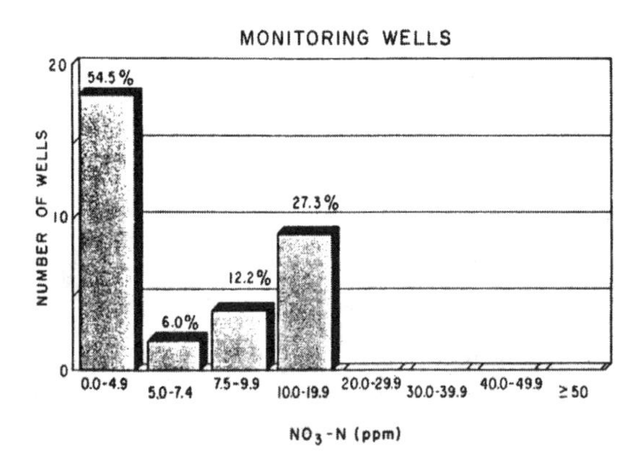

Figure 3.5 (continued)

vulnerable to contamination. About 51% of the wells with concentrations of 10 mg/l or more are in these areas. Comparisons between previous surveys and this assessment indicate that the concentrations of nitrate in the irrigated corn-producing areas are increasing (Exner and Spalding, 1990).

A comprehensive review of factors and practices influencing nitrates in groundwater has been assembled by Hallberg (1986). It was noted that numerous studies, in all parts of the world, and ranging from controlled plot studies to basin-size inventories, have shown that nitrate concentrations in groundwater (in shallow, freshwater aquifers) can be directly related to agricultural land use. Nitrate concentrations in groundwater under forest, unfertilized (or low-level fertilization) pastures, meadows, and grasslands are cited generally as less than 2 mg/l nitrate-nitrogen and often less than 1 mg/l, whereas nitrate concentrations under fertilized crops and intensive animal production areas are commonly greater than 5 mg/l, and may range to over 100 mg/l nitrate-nitrogen.

Hallberg (1986) noted that numerous studies show a direct relationship between nitrate concentrations in groundwater and nitrogen fertilization rates and/or fertilization history. For example, during the 1950s and 1960s, the nitrate concentration in groundwater in the Big Spring Basin in Iowa averaged about 3 mg/l nitrate-nitrogen (12 to 14 mg/l expressed as NO_3). By the 1980s, the nitrate concentration had increased about threefold; for example, to an annual average of 9 mg/l nitrate-nitrogen (39 mg/l as NO_3) in water-year 1982 and 10 mg/l nitrate-nitrogen (46 mg/l NO_3) in 1983. Data from a variety of groundwater wells in the area revealed the same trend. The primary sources of nitrogen in this Basin were manure and fertilizer applications. Manure nitrogen loadings had increased only about 30%, while fertilizer nitrogen (FN) applied increased 2.5 to 3 times as a function of the increasing rate of nitrogen application and the increase in corn acreage. As shown in Figure 3.6, the increase in nitrate concentration in groundwater directly paralleled the increase in the amount of FN applied in the basin. In a more recent study, nitrate analyses on 44,000 groundwater samples in Iowa indicated that 18% exceeded the federal maximum contaminant level (MCL) for nitrates in drinking water (Walker and Hoehn, 1989).

Hallberg (1986) summarized various studies by noting that they typically demonstrated a range from a 3- to 60-fold increase in nitrate concentrations in groundwater between forested-pasture-grassland areas and nearby intensively cultivated and fertilized areas. Many variables affect the resultant concentrations of nitrate that reach groundwater, but most studies have indicated that, over the long term, there are three primary controlling factors: the amount of nitrogen source available, the amount of infiltrating or percolating water (and the hydraulic conductivity of the surface and subsurface material), and the potential for nitrate reduction and/or denitrification.

Nitrogen Budgets

A nitrogen budget for a 7.4-km^2 study area in the Susquehanna River Basin in Pennsylvania was developed by Pionke and Urban (1985). The nutrient budget

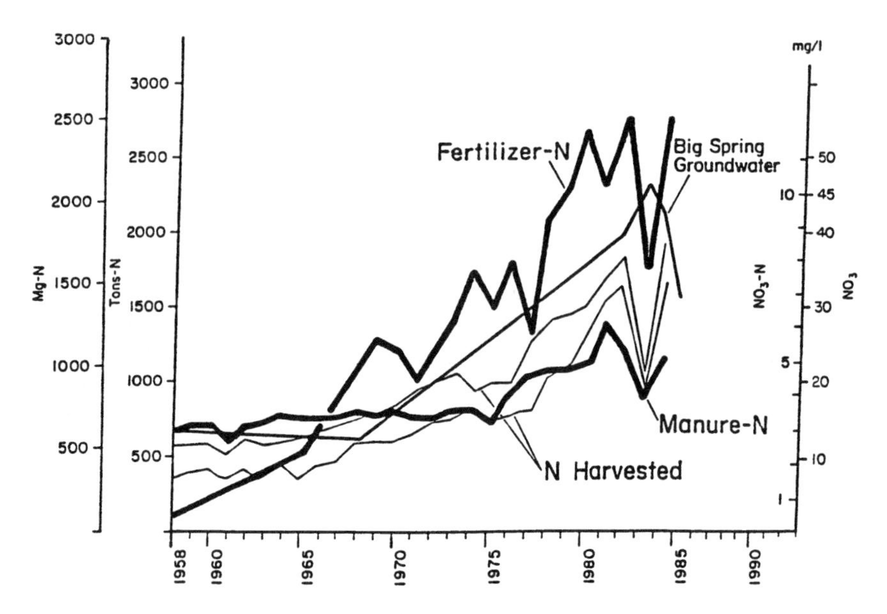

Figure 3.6 Mass of fertilizer-N and manure-N applied in the Big Spring Basin, annual average NO_3 concentration in groundwater at Big Spring, and mass of N harvested in corn grain. (From Hallberg, G. R., *Proceedings of the Conference on Agricultural Impacts on Groundwater*, August 1986, National Water Well Association, Dublin, OH. pp. 1–63. With permission.)

for cropland added nutrient gains from fertilizer, manure, and precipitation and subtracted nutrient losses by crop removal and gaseous nitrogen losses. The computed nutrient difference in the root zone is assumed to be a measure of potential impact on the underlying groundwater quality. The results of the budget calculations are shown in Table 3.4 (Pionke and Urban, 1985).

Table 3.4 Nutrient Budget of Cropland, Based Primarily on Nutrient Added in Commercial Fertilizer, Manure, and Precipitation (N), and Subtracted by Crop Harvest and Gaseous Losses (N)

	Applied[a]					
Nutrient	Fertilizer (kg)	Manure (kg)	Removed by harvest[a] (kg)	Computed gaseous N loss[b] (kg)	Difference	Nutrient gain[c] (kg/ha/yr)
N	36,770	17,860	33,000	14,445	7,185	+22.3
P	10,320	2,640	6,980	0	5,980	+14
K	18,100	9,860	21,880	0	6,080	+14

[a] Average annual nutrient application and removal determined by survey (1970–77).
[b] Gaseous N losses of 50 and 15% for added manure and fertilizer, respectively.
[c] Difference ÷ 421 ha cropland, except for N where 5.3 kg/ha/year was measured from precipitation is added to Difference (17 kg/ha/year).

From Pionke, H. B. and Urban, J. B., *Ground Water*, Vol. 23, No. 1, 1985, pp. 68–80. With permission.

Farm management records from farmers in the study watershed were used to establish the cropland nutrient budget. These records provided fertilizer and manure use, and crop production from 1970 to 1977 on 42 ha of cropland that comprised the 740-ha watershed. Because animal types and numbers changed on each farm over the 8-year survey period due to shifts in crop type, crop yield, animal price, and farmer preference, the N, P, and K contents of manure were computed on the basis of the overall hog, poultry, and cattle mix to be 7.5 kg N, 1.1 kg P, and 4.2 kg K per metric ton wet manure produced. Nutrient removal by crop harvesting was computed by multiplying the appropriate nutrient content values by the crop yield. From the farm survey, the annual yields in metric tons for the watershed were: hay, 294; corn grain, 1073; small grain, 472; straw, 708; and potato, 170 (Pionke and Urban, 1985).

Nutrient removal by each specific crop was computed as follows: hay, mostly a 50:50 red clover-grass mix, removed 7 kg N, 2 kg P, and 10 kg K/t dry matter yield. Only grass was assumed to take up applied nitrogen, which accounted for the low kg/t value. The nitrogen source for red clover, a legume, was assumed to be nitrogen fixation. Correspondingly, the nitrogen fixed by the legume was not added to the nitrogen budget when the crop was terminated. Small grain uptake was computed as a total, where small grains were considered approximately 50:50 oats:wheat, or barley. Removal was based on 19 kg N, 5 kg P, and 3 kg K/t grain yield. It was assumed that 1.5 t of straw was harvested for each metric ton of grain with 6 kg N, 1 kg P, and 19 kg K/t straw. Corn yield was mostly corn grain with very little silage being harvested. The removal was based on 16 kg N, 3 kg P, and 3 kg K/t corn grain yield. The potato yield was 34 t/ha with removal of 3.3 kg N, 0.6 kg P, and 5.0 kg K/t of potato yield (Pionke and Urban, 1985).

Gaseous nitrogen losses by denitrification plus ammonia volatilization were assumed to be 15% of the commercial nitrogen fertilizer application and 50% of the manure nitrogen application. Such nitrogen losses are highly variable depending on the nitrogen source and environmental conditions. The estimated 15% nitrogen fertilizer loss was based on a review which noted that about 10-30% of the total mineral input nitrogen in the U.S. was lost by denitrification, and on a field lysimeter study where the denitrification loss of ^{15}N-labeled fertilizer was found to be 17% under similar climatic, hydrologic, and geologic conditions (Pionke and Urban, 1985). The gaseous nitrogen loss from manure can range from very low to nearly complete, depending on environmental conditions and how the manure is handled. The predominant handling method was either to temporarily stack and store, or to apply immediately to the soil surface. In both cases, the applied manure was exposed at the soil surface from several days to weeks. Based on a nitrogen budget study in New York, a 50% loss appeared reasonable (Pionke and Urban, 1985).

Tools for Groundwater Nitrate Surveys

Several tools are available to aid in the conduction and interpretation of nitrate data in groundwater; two examples will be cited: (1) nitrogen-isotope analyses and (2) computerized mapping techniques. Kreitler and Browning (1983) described the use of the natural variation of the nitrogen isotopes, ^{14}N and ^{15}N, to identify potential sources of nitrates in groundwater. The nitrogen isotopes were measured through the use of mass spectrometry.

Kreitler and Browning (1983) also noted that measuring nitrogen isotope ratios (^{15}N:^{14}N) of nitrate from different sources and soil environments allows the comparison of these ratios with the nitrogen isotope ratios of nitrate in groundwater. This technique has been shown to be an effective method for identifying sources of nitrate that may be contaminating an aquifer. The $\delta^{15}N$ (the ratio of ^{15}N to ^{14}N) of nitrate in soils cultivated without fertilizer addition ranged from +2 to 8‰, whereas the $\delta^{15}N$ of nitrate resulting from the decomposition of animal waste material ranged from +10 to +22‰. Using this technique, it was determined that high concentrations of nitrate in the groundwater in southern Runnels County, Texas, were the result of the oxidation of organic nitrogen in the soil. Cultivation without nitrogen fertilizers was the dominant land use. Conversely, in Macon County, Missouri, where dairy farming is the dominant land use, the nitrogen isotope analyses indicated that animal wastes were the dominant source of nitrate in groundwater. Further usage of this tool suggested that groundwater nitrate in Suffolk County, New York, was related to the effects of agriculture, whereas the groundwater nitrate in Nassau and Queens counties probably resulted from septic tank system drainage and sewer leakage.

Spalding, et al. (1978) described the use of SYMAP (gray-scale mapping) in a study of nonpoint nitrate contamination of groundwater in Merrick County, Nebraska. The SYMAP system is a geographically based, computerized information system that has been used in numerous environmental studies. Environmental data on several factors can be overlapped on a composite map. Spalding, et al. (1978) found that the areal distribution of 293 samples from the groundwater of Merrick County had definite patterns of high (>20 mg/l), intermediate (10–20 mg/l), and low (<10 mg/l) nitrate-nitrogen concentrations. Where contamination was present, the nitrate-nitrogen concentrations were relatively homogeneous, indicating large diffuse nonpoint sources. The use of SYMAP indicated exceptionally good correlation between the irrigated coarse-textured soils and the higher nitrate-nitrogen levels. The obvious implication is that the nitrate-nitrogen levels are directly dependent on the leaching of nitrogenous material dispersed in or on the coarser textured soils. Additional mapping methods are described in Chapter 4.

ILLUSTRATIONS FROM EUROPE AND OTHER COUNTRIES

Table 3.5 contains summary comments on 27 references dealing with case studies of nitrate groundwater pollution. A total of 22 references address European studies, with the balance highlighting work from Canada (Hill, 1982), India (Handa, 1983; Jacks and Sharma, 1983), Israel (Ronen and Margaritz, 1985), and Chile (Schalscha, et al., 1979). One reference (Anonymous, 1975) is a conference proceedings with 52 papers addressing nitrogen as a surface and groundwater pollutant; and Tessendorff (1985) and Kraus (1993) provide a general discussion

Table 3.5 References Related to European and Other Non-U.S. Case Studies

Author(s) year	Comments
Andersen and Kristiansen (1984)	Depth profiles of nitrate in groundwater in agricultural areas of the Karup Basin in Denmark
Anonymous (1975)	Proceedings of a conference including 52 papers addressing nitrogen as a pollutant in surface and groundwater
Anonymous (1983)	Nitrates in water supplies in France
Conway and Pretty (1991)	Discussion of groundwater contamination by nitrates in the U.K.; including information on nitrate profiles in the unsaturated zones of aquifers in relation to the history of land use in the area
Custodio (1982)	Agricultural nitrate pollution of groundwater in irrigated areas of Spain
Duyvenbooden and Loch (1983)	Agricultural inputs to nitrate concentrations in Netherland groundwaters
Forslund (1986)	Summary information on nitrate pollution of groundwater in agricultural areas in Denmark
Foster and Bath (1983)	Depth profiles of nitrates in the unsaturated zone of the Chalk aquifer in Britain
Greene (1978)	Nitrates in groundwater in the Anglian region of Britain
Gustafson (1983)	Agricultural influences on groundwater quality in Sweden
Handa (1983)	Discussion of the effect of fertilizer use on the concentrations of nitrates, potassium, and phosphates in shallow unconfined and deeper semi-confined to confined aquifers in different parts of India
Hill (1982)	Nitrates in a shallow water table aquifer underlying an agricultural area near Alliston, Ontario, Canada
Jacks and Sharma (1983)	Agricultural and domestic contributions to nitrates in groundwater in an area in southern India
Khanif, Van Cleemput, and Baert (1984)	Influence of fertilization and rainfall on groundwater pollution in a sandy area of Belgium
Kraus (1993)	Summary information on nitrate groundwater pollution in the member countries of the European Community
Overgaard (1984)	Trends in nitrate pollution of groundwater in 184 waterworks in Denmark
Parker, Booth, and Foster (1987)	Nitrate leaching from agricultural soils and penetration into the Norfolk Chalk in Britain
Pekny, Skorepa, and Vrba (1989)	Study of the effects of nitrogen fertilizers on the quality of shallow aquifers in Czechoslovakia
Ronen and Margaritz (1985)	Influence of sewage irrigated agricultural land in Israel on solute concentrations in groundwater
Schalscha, et al. (1979)	Influences of sewage irrigated agricultural land in Chile on nitrate concentrations in groundwater
Skorepa, Vcislova, and Vrba (1982)	Agricultural causes of nitrates in groundwater in the vicinity of the Middle Elbe River in Czechoslovakia
Sontheimer and Rohmann (1984)	Agricultural causes of nitrate pollution of groundwater in Germany
Stibral (1982)	Nitrate leaching as a function of soil type and crop in Czechoslovakia
Tessendorff (1985)	Discussion of groundwater nitrate problems in the European Community countries
Tester and Carey (1985)	Treatment of nitrates in groundwater in the Anglian region of Britain
White (1983)	Review of the nitrate problem in British surface and groundwaters
Young (1983)	Agricultural impacts on groundwater quality in the U.K.

of groundwater nitrate problems in the European community countries. Studies from seven countries will be noted but not discussed in detail; these include: (1) Czechoslovakia (Skorepa, Vcislova, and Vrba, 1982; Stibral, 1982; Pekny, Skorepa, and Vrba, 1989); (2) Germany (Sontheimer and Rohmann, 1984); (3) The Netherlands (Duyvenbooden and Loch, 1983); (4) the U.K. (Tester and Carey, 1985; White, 1983; Young, 1983; Conway and Pretty, 1991; Parker, Booth, and Foster, 1987); (5) Denmark (Forslund, 1986); (6) Israel (Ronen and Margaritz, 1985); and (7) Chile (Schalscha, et al., 1979). Information from the remaining references will be presented in terms of four topical issues: (1) problem assessment surveys, (2) factors and practices influencing nitrates in groundwater, (3) nitrate variations with soil depth, and (4) the calculation of nitrogen budgets.

Problem Assessment Surveys

Examples of problem assessment surveys conducted in four countries will be highlighted; these include Denmark (Overgaard, 1984), France (Anonymous, 1983), Spain (Custodio, 1982), and the U.K. (Greene, 1978). Overgaard (1984) reported on a nationwide investigation of nitrate concentrations in groundwater in Denmark based on analyses of samples from approximately 11,000 wells and drinking water from 2800 groundwater works. The trends in nitrate concentrations were investigated on the basis of time-series of nitrate concentrations for 184 water works. The investigation indicated that: the overall mean level of nitrate concentrations in the groundwaters had trebled within the last 20 to 30 years; there are pronounced regional differences in the nitrate concentrations of the groundwaters and in the trends of the nitrate concentrations; 36% of the investigated time-series of nitrate concentrations in drinking water show a rising trend and there is no sign of leveling off; and drinking water from 8% of the water works exceed the European Economic Community (EEC) maximum admissible concentration. The impact of agricultural activities and the role of the geological conditions in Denmark are considered as causative factors related to the results of this investigation.

A survey performed in France revealed that 81% of the population had nitrate levels less than 25 mg/l, and 96 to 98% had levels less than 50 mg/l in their drinking water supplies (Anonymous, 1983). Out of 53 million people accounted for in the survey, 280,000 at most had a water supply exceeding 100 mg/l of nitrate at least once over a 3-year period. Out of the 20,000 distribution units surveyed, about 1000 had nitrate above 50 mg/l; however, only 61 units were above 100 mg/l. Most of the high nitrates were found in groundwater supplies and most of the wells affected were in highly developed agricultural areas.

Custodio (1982) reported that agricultural nitrate pollution is a widespread problem in irrigated areas in Spain and is accompanied by an increase in sulfate and bicarbonate, but not necessarily potassium contents. Nitrate in groundwater frequently exceeds 50 mg/l and sometimes reaches 500 mg/l.

A regional survey in the Anglian area of the U.K. in 1975 indicated that 50 public supply boreholes, wells, and springs accounting for about 160 tcmd had recorded nitrate levels in excess of 11.3 mg/l (as nitrogen); 12 of these sources had recorded levels in excess of 15 mg/l nitrate-nitrogen (Greene, 1978). The major "problem" areas were identified in the East Anglian chalk aquifers and the Lincolnshire limestone. Lack of data on many groundwater sources prevented a fully comprehensive survey of all public supply groundwater systems.

Finally, it should be noted that groundwater supplies 75% of drinking water in the European community as a whole, rising to 88% in Italy and 98% in Denmark (Kraus, 1993). The pollution of groundwater from industry and waste dumps is a serious problem, particularly in the more developed countries in the community. High concentrations of nitrates are the other main cause of groundwater pollution. The major source of nitrates is the use of inorganic fertilizers and the disposal of farmyard manure. Concentrations are high in many areas of the community. For example, in Denmark, the average nitrate level of groundwater has tripled over the last 30 years and is increasing at a rate of about 3.3 mg/l per year, with the result that 8% of the water produced in Danish waterworks now has a nitrate concentration above the admissible community limit of 50 mg/l (Kraus, 1993). It is estimated that 800,000 people in France, 850,000 in the U.K. and 2.5 million in Germany are drinking water with nitrate concentrations above the permitted community limit.

Given that many of the pollutants washed out of the soil over the past decade have not reached the water table, it will take between 25 and 50 years for nitrate levels in groundwater in the shallow wells of the Netherlands, Belgium, Denmark, and Germany to decline to an acceptable concentration, in accordance with the Community Directive on drinking water, despite recent cuts in the use of fertilizers in some member states (Kraus, 1993).

Factors and Practices Influencing Nitrates in Groundwater

Several factors and practices influence nitrates in groundwater and nitrate variations with soil depth. Examples of factors include precipitation/runoff, soil type, denitrification, and nonagricultural sources of nitrogen in the subsurface environment. Practices influencing nitrates in groundwater include fertilizing intensity, crop type, and land usage.

A 2-year field study on the influence of nitrogen fertilization and rainfall on nitrates in groundwater was conducted in a sandy area of Belgium (Khanif, Van Cleemput, and Baert, 1984). The fertilizer and manure application schedule is presented in Table 3.6, and the monthly rainfall during the study is given in Table 3.7 (Khanif, Van Cleemput, and Baert, 1984). The depth of the water table fluctuated between 0.4 and 2.0 meters, depending on the rainfall and time of year. The measured nitrate content in the groundwater at different depths is in Figure 3.7; Figure 3.8 displays the nitrate:chloride ratio in the groundwater (Khanif, Van Cleemput, and Baert, 1984). The movement of nitrate and chloride in soil occurs in a similar way; therefore, because chloride is not affected by biological processes, the nitrate:chloride ratio can indicate the mechanism by which nitrate is lost from the soil profile. A constant nitrate:chloride ratio would suggest leaching, whereas a declining ratio would indicate denitrification.

Table 3.6 Fertilizer and Manure Application Schedule in Belgium Study

Crops	Fertilizer/manure application	
	Date	Amount (type)
Barley	17/03/80	50 kg N/ha (NH_4NO_3)
	19/04/80	50 kg N/ha (NH_4NO_3)
Turnips	End July	30,000 dm³/ha (slurry) (3‰N)
	30/07/80	153 kg N/ha (NH_4NO_3)
Maize	Feb. 1980	25 ton/ha (manure) (5‰N)
	Feb. 1981	78,000 dm³/ha (slurry) (3‰N)
	16/04/81	105 kg N/ha (NH_4NO_3)
	10/06/81	35 kg N/ha (NH_4NO_3)

From Khanif, Y. M., Van Cleemput, O., and Baert, L., *Water, Air and Soil Pollution*, Vol. 22, No. 4, 1984, pp. 447–452. With permission of Kluwer Academic.

Table 3.7 Monthly Rainfall (mm) During the Belgium Study

Year	Months												Total
	J	F	M	A	M	J	J	A	S	O	N	D	
1980	73	64	71	32	20	93	161	55	14	86	52	102	823
1981	90	32	95	40	93	71	25	32	51	145	70	52	796

From Khanif, Y. M., Van Cleemput, O., and Baert, L., *Water, Air and Soil Pollution*, Vol. 22, No. 4, 1984, pp. 447–452. With permission of Kluwer Academic.

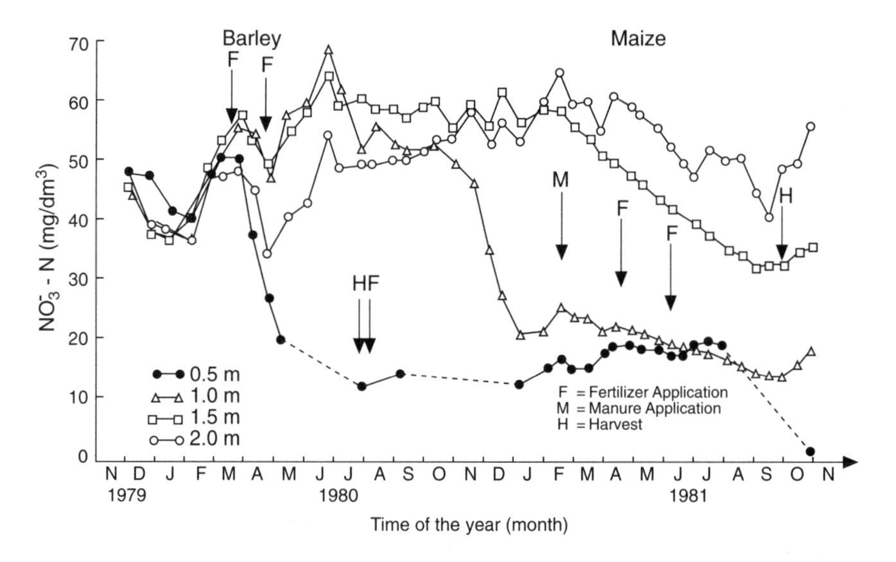

Figure 3.7 Nitrate concentration of the groundwater in the Belgium study. (From Khanif, Y. M., Van Cleemput, O., and Baert, L., *Water, Air and Soil Pollution*, Vol. 22, No. 4, 1984, pp. 447–452. With permission of Kluwer Academic.)

Khanif, Van Cleemput, and Baert (1984) applied 50 kg/ha fertilizer nitrogen to barley in March, 1980. However, in the next weeks, the plants started to develop nitrogen deficiency symptoms. As the nitrogen fertilization was not reflected in

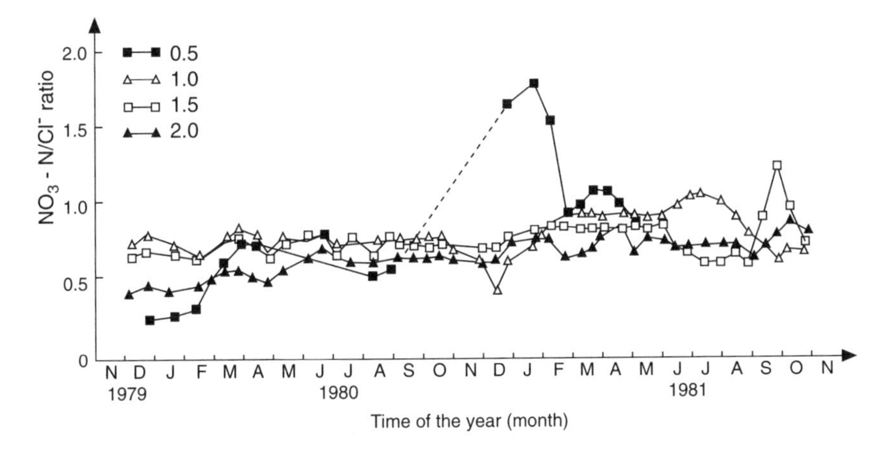

Figure 3.8 Nitrate:chloride ratio of groundwater from November 1979 to November 1981 in Belgium. (From Khanif, Y. M., et al., *Water, Air and Soil Pollution*, Vol. 22, No. 4, 1984, pp. 447–452. With permission of Kluwer Academic.)

the groundwater at 0.5 m (see Figure 3.7) and because of the unchanged ratio of NO_3^-–N:Cl^- (see Figure 3.8), it was suggested that due to the high rainfall in March, the applied nitrogen was denitrified in the top zone of the soil profile. The high rainfall in March not only stimulated denitrification, but caused dilution of the nitrogen content in the groundwater at lower depths after the relatively dry February month. Following the deficiency symptoms, a second nitrogen application of 50 kg N/ha was made in April. Again, there was no increase in the nitrate–nitrogen content at 0.5-m depth. At this time, rainfall was limited and no downward movement at 0.5-m occurred. Also, most of the applied nitrogen was taken up by the plants. As a result of the leaching out of the 0.5-m zone, toward harvesting, there was some increase of the nitrate concentrations at 1.0-, 1.5-, and 2.0-m depths. During the 1980 winter, the nitrate-nitrogen content at the 1.0-m depth decreased, most probably due to leaching. The results show that the turnips sown in July had no influence on the nitrate–nitrogen content at the different depths. In the second growing season, maize was grown on the same field. Addition of manure and fertilizer nitrogen at the beginning of the growing season did not affect the nitrate–nitrogen concentration of the groundwater. During this period, there was no excess rainfall. Thus, most of the applied nitrogen remained in the top soil for efficient use by the crop. The concentration at all depths declined during the growing season (Khanif, Van Cleemput, and Baert, 1984).

As seen in Figure 3.8, during the first growing season, the NO_3^-–N:Cl^- ratio at all depths did not fluctuate widely, thus suggesting that denitrification was not so prominent. Due to the slurry and manure application, the NO_3^-–N:Cl^- ratio at 0.5 m was increased at the beginning of 1981. From February on, however, the ratio decreased, indicating some nitrate–nitrogen loss by denitrification. At lower depths, there was no significant change in the ratio, implying that there was no important loss of nitrate–nitrogen by denitrification. Significant denitrification at these depths cannot be expected because of the low organic matter content (%C < 0.1) (Khanif, Van Cleemput, and Baert, 1984).

The type of soil can also influence nitrate in groundwater. For example, Gustafson (1983) studied the influence of soil type on nitrogen leaching. The study area was in southern Sweden. The sandy soils lost more than twice as much nitrogen as compared with the clay soils. The root depth in the sandy soils rarely exceeds 40 to 60 cm, which is one explanation why the losses were so great. Nitrogen below this level is naturally available for the crop and is exposed to leaching. For a clay soil with good structure, the situation is different. The root penetration can easily reach 1 m or more, which results in a more stable uptake of nitrogen by the crop. Some of the nitrogen in a well-aggregated clay soil (i.e., nitrogen found inside aggregates and micropores) is effectively protected against leaching. Most of the percolating water occurs in root canals and macropores. This physical state does not exist in a sandy soil, in which the water percolates through pores (Gustafson, 1983).

The influence of fertilizing intensity was also examined by Gustafson (1983). Greater leaching losses, on a mass balance basis, can be expected for higher fertilizer application rates. In addition, and depending on crop type, if more fertilization results in higher amounts of crop residues, this can also increase nitrogen losses to the subsurface due to mineralization after harvest. The nitrogen in the soil to a depth of 3 m in a cropping system with three different fertilization levels is shown in Figure 3.9 (Gustafson, 1983). The soil samples were collected during December, 1981. The highest amount of nitrogen used causes a substantial accumulation of nitrate in the soil profile. This accumulation is obvious throughout the profile (Figure 3.9). There is a difference of 89 kg N/ha between the 100-N treatment and the treatment with no nitrogen. Therefore, it is evident that crop production at a normal fertilization level also causes a minor accumulation of nitrate in the soil profile (Gustafson, 1983).

The type of crop can also influence the level of nitrates in groundwater. For example, Gustafson (1983) noted that crops harvested late cause less mineralization of crop residues, since the temperature normally drops steeply as winter approaches. Winter wheat and other crops sown during the autumn can absorb nitrogen late in the season. A grass ley of several years' standing would provide optimum conditions. The following crop succession was started in 1973: winter wheat, spring rape, winter wheat, barley, oats with re-seed, and, finally, 3 years of ley. The ley was plowed in November, 1980, with no effects on the nitrogen losses the following winter, presumably because of the late date of the plowing. The mean contents of nitrate found at a depth of 1.7 m from a similar study area are shown in Table 3.8 (Gustafson, 1983). Of the three crops shown, potatoes yielded the highest nitrate concentrations in the subsurface.

The influence of crop type and land use was also examined in a study conducted of a shallow water table aquifer underlying a sand plain near Alliston, Ontario, Canada (Hill, 1982). Ground water nitrate and chloride concentrations were related to mean percent areas in the various land use categories and average fertilizer application rates based on the 4-year period, 1977–1980, under the

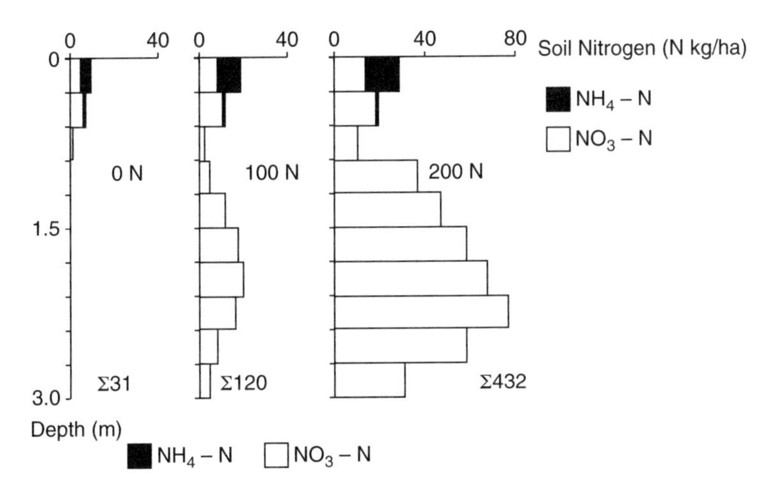

Figure 3.9 Influence of fertilization levels on nitrogen in soil in Sweden study. (From Gustafson, A., *Environmental Geology*, Vol. 5, No. 2, 1983, pp. 65–71. With permission.)

Table 3.8 Influence of Crop Type on Nitrate in Groundwater

Crop	Potatoes	Grass ley	Plowing up and oats
Nitrate (N mg/1)[a]	24	11	20
Time of observation (months)	5	28	19

[a] Mean concentration at a depth of 1.7 m.

From Gustafson, A., *Environmental Geology*, Vol. 5, No. 2, 1983, pp. 65–71. With permission.

assumption that the element concentration in the shallow aquifer may reflect inputs over several years rather than in the year immediately preceding sample collection. Where necessary, variables were log transformed to increase the normality of the data prior to regression analysis. Correlations between nitrate-N and chloride concentrations in groundwater and land use are shown in Table 3.9 for 93 and 83 well sites, respectively. Nitrate-nitrogen concentration exhibited a weak but significant positive correlation with percent area of potatoes and heavily fertilized crops (potatoes, corn, sod, and asparagus). Forest and scrub and permanent pasture had a negative association with nitrate-nitrogen concentrations. A significant positive correlation of $r = 0.76$ was also found between groundwater nitrate-nitrogen concentration and the average rate of fertilizer nitrogen application (Table 3.9). The correlations between groundwater chloride concentration and land use variables were very similar to those found for nitrate-nitrogen (Hill, 1982).

Fertilizer nitrogen may not be the only source of nitrate in the groundwater of agricultural areas. As noted in Chapter 1, other potential sources include natural organic nitrogen, livestock, septic tank systems, and atmospheric inputs (Hill, 1982). Point pollution sources identified in the Alliston, Ontario, Canada, study area included septic tank systems and several cattle barns.

Table 3.9 Correlation Coefficients for Groundwater Nitrate–N and Chloride Concentrations Regressed Against Average Percent Areas of Land Use and Fertilizer Application Rates for 1977 to 1980

Land use variable	Nitrate-N (mg/l) (N = 93)	Chloride (mg/l) (N = 83)
Potatoes	0.71[a]	0.74[a]
Grain	0.20	0.19
Log forest and scrub	–0.51[a]	–0.57[a]
Log pasture	–0.32[a]	–0.22
Heavily fertilized crops	0.72[a]	0.68[a]
Fertilizer N (lb/acre)	0.76[a]	
Fertilizer KCl (1b/acre)		0.79[a]

[a] Significant at the 0.001 level.

From Hill, A. R., *Ground Water*, Vol. 20, No. 6, 1982, pp. 696–702. With permission.

Nitrate Variations with Soil Depth

Several studies have documented that nitrates and other inorganic constituents exhibit variations with increasing depth from the land surface. Examples of such variations were described in a study of groundwater and soils in the Karup Basin in Denmark (Andersen and Kristiansen, 1984). It was found that the nitrate concentration from wells located in cultivated land was high, being from 5 to 25 mg/l in the upper parts of the groundwater zone, while the iron in solution was low (<0.5 mg/l Fe). However, at a given depth, within a few decimeters, the nitrate content fell to near-zero and the iron content increased. In forested areas, the nitrate-nitrogen content was very low, being below 0.1 mg/l, even in the upper part of the groundwater zone where oxidizing conditions existed and little or no iron was found in solution. Deeper in the aquifer, the nitrate content approached zero and the iron content increased, thus indicating reducing conditions. In cultivated land, the oxidation zone, with its rather high content of nitrate, was found at depths of 8 to 15 m below the groundwater table. At a forested site, the oxidized nitrate zone was thinner, approximately 5 to 7 m, and exhibited a low nitrate content. These findings indicated that agricultural activity results in a recharge with a high content of nitrate, whereas the groundwater in forested areas contains very little nitrate. The thickness of the oxidized zone below the farm land, in comparison to that below forest, may also indicate the effect of nitrate runoff from agriculture, as the nitrate zone below plowed land is roughly twice as thick as the nitrate zone below uncultivated land (Andersen and Kristiansen, 1984).

Foster and Bath (1983) reported on the results of unsaturated zone investigations at a site on the Chalk aquifer area near Cambridge, England. The site was 60 m on a side and part of a large flat field, fairly typical of agricultural land on the aquifer outcrop in eastern England, which has been in arable farming for at least 40 years. Fifteen cereal crops were cultivated during the period 1960 to

1980 with inorganic fertilizer applications mainly in the range of 60 to 100 kg N/ha per year and 40 to 55 kg/ha per year of both P_2O_5 and K_2O, higher rates being applied in split application to winter-sown cereals. Sugar beets and peas were the only other crops. Fertilizer application rates are believed to have increased greatly during the period 1950 to 1960 with increasingly frequent cereal cropping and, at this site, have been reduced marginally since 1975. Cambridge is one of the driest parts of eastern England and the site was estimated to have a long-term average excess rainfall of about 170 mm/year. Prior to the study years, the climatic conditions were extreme; for example, the prolonged drought that ended in the autumn of 1976 was followed by three unusually wet winters and springs, with, for example, major infiltration occurring as late as May, 1978. Hydrogeologically, the site is on the Middle Chalk with a groundwater table not thought ever to rise above a 15-m depth below the land surface. The Chalk is a highly porous limestone, but at the study site, as in much of eastern England, it is slightly more marly and less homogeneous than the Upper Chalk of southern England.

Foster and Bath (1983) developed six nitrate profiles during initial investigations in May, 1979. The shape of each profile was remarkably similar but laterally, at the same depth, concentrations varied quite widely, as shown in Figure 3.10 (Foster and Bath, 1983). The profiles revealed high nitrate concentrations with a "front" of major proportions at 4- to 8-m depths. Over the depth range to the base of the front (at 8 m), the profiles contain more than 1100 kg nitrate–nitrogen per hectare. Comparable profiles, but with generally lower concentrations and deeper fronts, have been observed in numerous arable fields in two other regions of eastern England. Chloride and sulfate profiles with depth, exhibit similar patterns to those for nitrates (Foster and Bath, 1983).

Based upon the results shown in Figures 3.10 and related information, Foster and Bath (1983) concluded that large quantities of solutes leached from arable land are present in the unsaturated zone of the Chalk system. The rates of leaching increased substantially with more intensive cultivation from about 1960 to 1980. The pore water profiles presented also demonstrate the complexity of solute transport mechanisms (from the soil through the unsaturated zone) in a fissured microporous limestone, such as the Chalk. From the numerous lines of investigation pursued in detail at the Cambridge site, the following summary observations were suggested (Foster and Bath, 1983).

1. In the uppermost few meters, the precise depth being a function of the hydrogeological properties of the weathered profile, solute movement is strongly dispersive, with intermittent rapid downward fluxes occurring in summer.
2. Solutes are eluted from this upper zone and transported downward by infiltrating excess rainfall, the extent of nonequilibration, or dispersion, being essentially a function of the intensity of infiltration, the matrix hydraulic properties, the fissure geometry, and the appropriate diffusion coefficient of the solute concerned.

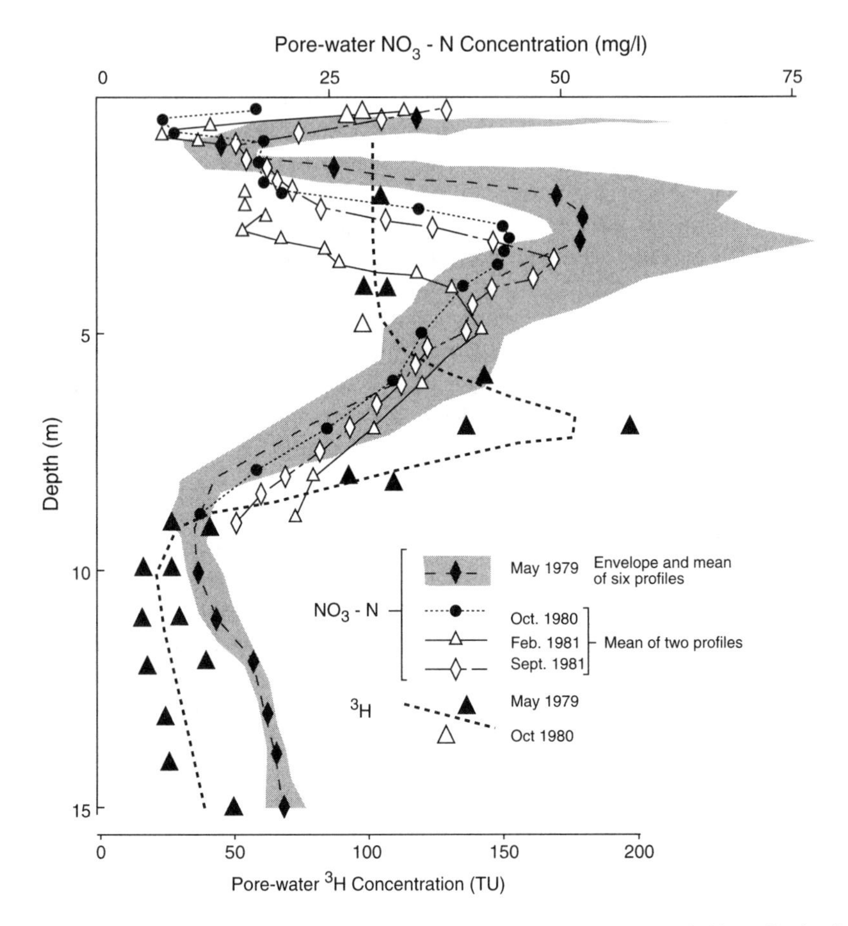

Figure 3.10 Nitrate-nitrogen profiles of chalk unsaturated zone at Cambridge, England. (From Foster, S. S. D. and Bath, A. H., *Environmental Geology,* Vol. 5, No. 2, 1983, pp. 53–59. With permission.)

Nitrogen Budgets

Jacks and Sharma (1983) described the development of a nitrogen budget for the Noyil River basin in southern India. Nitrogen fluxes were estimated for nitrogen in precipitation, nitrogen fixation, nitrogen excretion by humans and animals, leaching into the groundwater, and loss from the area due to runoff water. The nitrate in precipitation ranged from 1 to 3 mg/l, and with 600 mm of precipitation per year, the deposition would range from 135 to 405 kg/km²/year. The final balance used a nitrogen input rate of 250 to 500 kg/km²/year.

Nitrogen fixation by plants was accounted for by considering that: (1) a leguminous crop may fix 23 to 45 kg N per acre and growing season; (2) the area with nitrogen-fixing crops; and (3) the number of crops per year. Based on the relevant data, the nitrogen fixation ranged from about 600 to 1200 kg/km²/year.

In addition to nitrogen fixation by plants, there is a nonsymbiotic fixation by bacteria such as Azotobacter and Clostridium. Assuming a rate of 10 to 25 kg/ha/year and applying it to the rest of the cropped area not covered by leguminous species, a flux of 440 to 1100 kg/km^2/year was obtained. Therefore, the total nitrogen fixation was considered to range from about 1000 to 2000 kg/km^2/year (Jacks and Sharma, 1983).

Nitrogen excretion by humans and animals totaled 760 and 3200 kg/km^2/year, respectively. These materials are typically placed on the soil in southern India. The factor for humans was based on an average daily intake of 6.7 g N, and a population density of 310 persons per square kilometer (Jacks and Sharma, 1983). Nitrogen excretion from large animals averaged 90 g/day, with an animal density of 90/km^2. For large animals, the total excretion was 2960 kg/km^2/year. For small animals, the excretion rate for humans was used along with an animal density of 90/km^2; therefore, for small animals, the total excretion was 220 kg/km^2/year.

Fertilizer usage in the study area averaged 30 kg/ha/year; thus, the nitrogen input from this source was 1900 kg/km^2/year (Jacks and Sharma, 1983). Some nitrogen losses from the soil to the atmosphere occur; however, these were not quantified in the nitrogen budget.

The level of nitrogen exported from the basin by runoff water was 80 kg/km^2/year. The level for the dry part of the basin was 40 kg/km^2/year for the corresponding period. The weighted mean of the concentration of nitrate in runoff water was 9 mg/l, considerably below the mean content in groundwater. This may be due to several factors, such as dilution of groundwater inflow by precipitation falling on the river bed and river banks, resorption of nitrogen in discharge areas into the ecosystem, and the binding of nitrogen in plankton in runoff water. In view of the runoff pattern, characterized by floods of short duration, it is likely that the first effect was dominating (Jacks and Sharma, 1983).

The soil pool of nitrogen (total nitrogen) was of the order of 2×10^5 kg/km^2. The content of total nitrogen in the soils ranged from 0.04 to 0.1% in the upper 20 cm of soil. The loss through deep leaching into the groundwater can be calculated from net infiltration and the mean content of nitrate in groundwater. The net infiltration in the study area was 29 mm/year (Jacks and Sharma, 1983). The areal mean concentration of nitrate in the study area groundwater was 40 mg/l; thus, the loss of nitrate–nitrogen from the soil zone to the groundwater amounted to about 260 kg/km^2/year.

The groundwater pool was also of interest for the assessment of eventual changes in nitrate content over time. It is the active parts of the aquifers that are involved in mixing with the recent percolation. The size of the active part of the groundwater depends on the hydraulics of the aquifer and the depth of wells along with the rate of extraction. If the depth is tentatively set as 50 m and the fracture volume is 2%, the groundwater pool would contain 4×10^4 kg/km^2. Based on the overall nitrogen budget, it is obvious that the losses in deep leaching are small in comparison with the accounted yearly cycling of nitrogen in the ecosystem (Jacks and Sharma, 1983).

SUMMARY

This chapter has provided a review of 56 references related to illustrations of nitrate pollution in groundwater. Case studies are briefly described from the U.S. of America, several European countries, Canada, India, Israel, and Chile. Thus, it can be stated that groundwater pollution by nitrates is a widespread problem in many locations in the world. Man-made or man-caused sources of nitrogen introduction into the subsurface environment include agricultural fertilizers, septic tank systems, and animal waste disposal. Agricultural sources received the primary emphasis in this chapter. A number of hydrogeological factors and agricultural practices influence the concentration of nitrates in groundwater at specific locations. Examples of such factors and practices cited in this chapter include precipitation/runoff, irrigation, soil type and depth, geological features such as karst areas, denitrification, fertilizing intensity, and crop types and land usage.

SELECTED REFERENCES

Andersen, L.J. and Kristiansen, H., "Nitrate in Groundwater and Surface Water Related to Land Use in the Karup Basin, Denmark," *Environmental Geology*, Vol. 5, No. 4, 1984, pp. 207–212.

Anonymous, *Proceedings of Conference on Nitrogen as a Water Pollutant*, International Association on Water Pollution Research, August, 1975, Copenhagen, Denmark.

Anonymous, "Nitrate in Waters for Human Consumption: The Situation in France (1979-1981)," *Aqua*, Vol. 2, April, 1983, pp. 74–78.

Anton, E.C., Barnickol, J.L., and Schnaible, D.R., "Nitrate in Drinking Water — Report to the Legislature," Report No. 88-11 WQ, October, 1988, Division of Water Quality, State Water Resources Control Board, State of California, Sacramento, California, pp. 35–38.

Bachman, L.J., "Nitrate in the Columbia Aquifer, Central Delmarva Peninsula, Maryland," Water Resources Investigation Report No. 84-4322, 1984, U.S. Geological Survey, Towson, Maryland.

Baier, J.H. and Rykbost, K.A., "The Contribution of Fertilizer to the Ground Water of Long Island," *Ground Water*, Vol. 14, No. 6, November-December, 1976, pp. 439–448.

Buller, W., Nichols, W.J., and Harsch, J.F., "Quality and Movement of Ground Water in Otter Creek-Dry Creek Basin, Cortland County, New York," Water Resources Investigation Report No. 78-3, 1978, U.S. Geological Survey, Albany, New York.

Conway, G.R. and Pretty, J.N., *Unwelcome Harvest — Agriculture and Pollution*, Earthscan Publications, Ltd., London, England, 1991, pp. 183–190.

Cummings, T.R., Twenter, F.R., and Holtschlag, D.J., "Hydrology and Land Use in Van Buren County, Michigan," Water Resources Investigation Report No. 84-4112, 1984, U.S. Geological Survey, Lansing, Michigan.

Custodio, E., "Nitrate Build-up in Catalonia Coastal Aquifers," *Memoires, International Association of Hydrogeologists*, Vol. 16, No. 1, 1982, pp. 171–181.

DeRoo, H.C., "Nitrate Fluctuations in Ground Water as Influenced by Use of Fertilizer," Bulletin No. 779, June, 1980, Connecticut Agricultural Experiment Station, New Haven, Connecticut.

Duyvenbooden, W.V. and Loch, J.P.G., "Nitrate in the Netherlands—A Serious Threat to Groundwater," *Aqua*, Vol. 2, April, 1983, pp. 59–60.

Exner, M.E. and Spalding, R.F., "Evolution of Contaminated Groundwater in Holt County, Nebraska," *Water Resources Research*, Vol. 15, No. 1, February, 1979, pp. 139–147.

Exner, M.E. and Spalding, R.F., "Occurrence of Pesticides and Nitrate in Nebraska's Ground Water," Report WC1, 1990, Water Center, Institute of Agriculture and Natural Resources, The University of Nebraska, Lincoln, Nebraska, pp. 3–30.

Forsland, J., "Groundwater Quality Today and Tomorrow," *World Health Statistics Quarterly*, Vol. 39, No. 1, 1986, pp. 81–92.

Foster, S.S.D. and Bath, A.H., "Distribution of Agricultural Soil Leachates in the Unsaturated Zone of the British Chalk," *Environmental Geology*, Vol. 5, No. 2, 1983, pp. 53–59.

Greene, L.A., "Nitrates in Water Supply Abstractions in the Anglian Region: Current Trends and Remedies Under Investigation," *Water Pollution Control*, Vol. 77, No. 4, 1978, pp. 478–491.

Gustafson, A., "Leaching of Nitrate from Arable Land into Groundwater in Sweden," *Environmental Geology*, Vol. 5, No. 2, 1983, pp. 65–71.

Hall, D.W., "Effects of Nutrient Management on Nitrate Levels in Ground Water Near Ephrata, Pennsylvania," *Ground Water*, Vol. 30, No. 5, September-October, 1992, pp. 720–730.

Hallberg, G.R., "Overview of Agricultural Chemicals in Ground Water," *Proceedings of the Conference on Agricultural Impacts on Ground Water*, August 1986, National Water Well Association, Dublin, Ohio, pp. 1–63.

Hallberg, G.R., Libra, R.D., and Hoyer, B.E., "Nonpoint Source Contamination of Ground Water in Karst-Carbonate Aquifers in Iowa," *Proceedings of National Conference on Nonpoint Source Pollution*, May, 1985, U.S. Environmental Protection Agency, Kansas City, Missouri, pp. 109–114.

Hamilton, P.A. and Shedlock, R.J., "Are Fertilizers and Pesticides in the Groundwater? A Case Study of the Delmarva Peninsula, Delaware, Maryland, and Virginia," U.S. Geological Survey Circular 1080, 1992, U.S. Geological Survey, Denver, Colorado, pp. 4–9.

Handa, B.K., "Effect of Fertilizer Use on Ground Water Quality in India," *Proceedings of a Symposium on Ground Water in Water Resources Planning*, Vol. II, IAHS Publication No. 142, 1983, pp. 1105–1119.

Hergert, G.W., "Distribution of Mineral Nitrogen Under Native Range and Cultivated Fields in Nebraska Sandhills," July, 1982, Water Resources Research Center, University of Nebraska, North Platte, Nebraska.

Hill, A.R., "Nitrate Distribution in the Ground Water of the Alliston Region of Ontario, Canada," *Ground Water*, Vol. 20, No. 6, November-December, 1982, pp. 696–702.

Jacks, G. and Sharma, V.P., "Nitrogen Circulation and Nitrate in Ground Water in an Agricultural Catchment in Southern India," *Environmental Geology*, Vol. 5, No. 2, 1983, pp. 61–64.

Jennings, G.D., Sneed, R.E., Huffman, R.H., Humenik, F.J., and Smolen, M.D., "Nitrate and Pesticide Occurrence in North Carolina Wells," Paper No. 912107, *1991 International Summer Meeting of the American Society of Agricultural Engineers*, June, 1991, St. Joseph, Michigan.

Jensen, R., "Agriculture's Role in Nonpoint Source Pollution," *Texas Water Resources*, Vol. 17, No. 1, Spring, 1991, Texas Water Resources Institute, College Station, Texas, pp. 1–6.

Khanif, Y.M., Van Cleemput, O., and Baert, L., "Interaction Between Nitrogen Fertilization, Rainfall and Groundwater Pollution in Sandy Soil," *Water, Air and Soil Pollution*, Vol. 22, No. 4, May, 1984, pp. 447–452.

Klein, J.M. and Bradford, W.L., "Distribution of Nitrate and Related Nitrogen Species in the Unsaturated Zone, Redlands and Vicinity, San Bernardino, California," Water Resources Investigation Report No. 79-60, July, 1979, U.S. Geological Survey, Menlo Park, California.

Kolpin, D.W., Burkart, M.R., and Thurman, E.M., "Herbicides and Nitrate in Near-Surface Aquifers in the Midcontinental United States, 1991," U.S. Geological Survey Water-Supply Paper 2413, 1994, U.S. Geological Survey, Denver, Colorado, pp. 1–5, and 15–33.

Kraus, H.H., Ed., "The European Parliament and EC Environment Policy," Working Paper W-2, April, 1993, European Parliament, Luxembourg, p. 12.

Kreitler, C.W. and Browning, L.A., "Nitrogen Isotope Analysis of Ground Water Nitrate in Carbonate Aquifers: Natural Sources Versus Human Pollution," *Journal of Hydrology*, Vol. 61, 1983, pp. 285–301.

Lawrence, A.E., et al., "Distribution of Nitrate in Ground Water, Redlands, California," USGS/WRI-76-117, March, 1977, U.S. Geological Survey, Menlo Park, California.

Overgaard, K., "Trends in Nitrate Pollution of Groundwater in Denmark," *Nordic Hydrology*, Vol. 15, No. 4-5, 1984, pp. 177–184.

Parker, J.M., Booth, S.K., and Foster, S.S.D., "Penetration of Nitrate from Agricultural Soils into the Groundwater of the Norfolk Chalk," *Proceedings of the Institution of Civil Engineers*, Vol. 83, Part 2, March, 1987, pp. 15–32.

Pekny, V., Skorepa, J., and Vrba, J., "Impact of Nitrogen Fertilizers on Groundwater Quality — Some Examples from Czechoslovakia," *Journal of Contaminant Hydrology*, Vol. 4, 1989, pp. 51–67.

Pionke, H.B. and Urban, J.B., "Effect of Agricultural Land Use on Groundwater Quality in a Small Pennsylvania Watershed," *Ground Water*, Vol. 23, No. 1, 1985, pp. 68–80.

Reeves, C.C., Jr. and Miller, W.D., "Nitrate, Chloride and Dissolved Solids, Ogallala Aquifer, West Texas," *Ground Water*, Vol. 16, No. 3, May-June, 1978, pp. 167–173.

Ronen, D. and Margaritz, M.,"High Concentration of Solutes at the Upper Part of the Unsaturated Zone (water table) of a Deep Aquifer Under Sewage-Irrigated Land," *Journal of Hydrology*, Vol. 80, No. 3/4, 1985, pp. 311–323.

Saffigna, P.G. and Keeney, D.R., "Nitrate and Chloride in Ground Water Under Irrigated Agriculture in Central Wisconsin," *Ground Water*, Vol. 15, No. 2, March-April, 1977, pp. 170–177.

Schalscha, E.B., et al., "Nitrate Movement in a Chilean Agricultural Area Irrigated with Untreated Sewage Water," *Journal of Environmental Quality*, Vol. 8, No. 1, 1979, pp. 27–30.

Sievers, D.M. and Fulhage, C.D., "Survey of Rural Wells in Missouri for Pesticides and Nitrate," *Ground Water Monitoring Review*, Vol. 12, No. 4, Fall, 1992, pp. 142–150.

Skorepa, J., Vcislova, B., and Vrba, J., "Occurrence of Nitrogen Materials in Ground Water in the Territory of the Middle Elbe River in Bohemia," *Memoires, International Association of Hydrogeologists*, Vol. 16, No. 1, 1982, pp. 313–322.

Sontheimer, H. and Rohmann, U., "Groundwater Pollution by Nitrates: Causes, Significance and Solutions," *Gas and Wasserfach, Wasser Abwasser*, Vol. 125, No. 12, December, 1984, pp. 599–608.

Spalding, R.F. et al., "Nonpoint Nitrate Contamination of Ground Water in Merrick County, Nebraska," *Ground Water*, Vol. 16, No. 2, March-April, 1978, pp. 86–94.

Spruill, T.B., "Nitrate Nitrogen Concentrations in Ground Water from Three Selected Areas in Kansas," April, 1982, U.S. Geological Survey, Lawrence, Kansas.

Spruill, T.B., "Relationship of Nitrate Concentrations to Distance of Well Screen Openings Below Casing Water Levels," *Water Resources Bulletin*, Vol. 19, No. 6, December, 1983, pp. 977–981.

Stibral, J., "Biological Protection of Groundwater from Nitrate Pollution," *Memoires, International Association of Hydrogeologists*, Vol. 16, No. 1, 1982, pp. 335–346.

Swanson, G.J., "Well Testing Program is Cooperative Effort," *Water Well Journal*, Vol. 46, No. 8, August, 1992, pp. 39–41.

Taylor, R.G. and Bigbee, P.D., "Fluctuations in Nitrate Concentrations Utilized as an Assessment of Agricultural Contamination to an Aquifer of a Semiarid Climatic Region," *Water Research*, Vol. 7, No. 8, August, 1973, pp. 1155–1161.

Tessendorff, H., "Nitrates in Groundwater: A European Problem of Growing Concern," *Aqua*, Vol. 4, 1985, pp. 192–193.

Tester, D.J. and Carey, M.A., "Prediction of Nitrate Pollution in Chalk Groundwater-Derived Potable Supplies and Renovation of an Affected Source," *Water Pollution Control*, Vol. 84, No. 3, 1985, pp. 366–379.

Tryon, C.P., "Ground Water Quality Variation in Phelps County, Missouri," *Ground Water*, Vol. 14, No. 4, July-August, 1976, pp. 214–223.

Walker, D.R. and Hoehn, J.P., "The Economic Costs of Nitrate Contamination in Rural Groundwater Supplies," Agricultural Economics Report No. 525, June, 1989, Michigan State University, East Lansing, Michigan.

White, R.J., "Nitrate in British Waters," *Aqua*, Vol. 2, April, 1983, pp. 51–57.

Young, C.P., "Data Acquisition and Evaluation of Groundwater Pollution by Nitrate, Pesticides, and Disease-Producing Bacteria," *Environmental Geology*, Vol. 5, No. 1, 1983, pp. 11–18.

4 VULNERABILITY MAPPING OF GROUNDWATER RESOURCES

INTRODUCTION

The potential for groundwater contamination depends upon the natural attenuation that may take place between the sources of pollution and the aquifer system. During infiltration through soils and transport in aquifers, many contaminants may be naturally attenuated; however, not all subsurface environments have equally effective purifying capacities. The degree of attenuation generally varies with different aquifer materials and with the distance a pollutant must travel through unsaturated materials to reach groundwater. Several hydrogeologically based susceptibility ranking systems have been developed as aids for the evaluation of the capacity of the subsurface environment to attenuate pollutants; many such systems involve mapping of resultant zones of vulnerability to pollution. These classifications can be used as rapid and cost-saving tools for preliminary screening related to differing land use choices, fertilizers application rates, monitoring needs, and development of groundwater protection strategies.

Groundwater vulnerability to contamination can be defined as the tendency or likelihood for contaminants to reach a specified position in the groundwater system after introduction at some location above the uppermost aquifer (Committee on Techniques for Assessing Ground Water Vulnerability, 1993). It should be noted that all groundwater is vulnerable to contamination inasmuch as vulnerability is not an absolute property, but a relative indication of where contamination is likely to occur; no groundwater, with possible exceptions such as deep sedimentary basin brines, is invulnerable (Committee on Techniques for Assessing Ground Water Vulnerability, 1993).

Methods for predicting groundwater vulnerability can typically be considered in three major classes: (1) overlay and index methods that combine specific physical characteristics affecting vulnerability, often giving a numerical score; (2) process-based methods consisting of mathematical models that approximate the behavior of substances in the subsurface environment; and (3) statistical methods that draw associations with areas where contamination is known to have occurred (Committee on Techniques for Assessing Ground Water Vulnerability, 1993). This chapter addresses class (1), Chapter 5 provides an example of the development of an index, and Chapter 6 highlights mathematical models (class 2) and statistical methods (class 3).

This chapter begins with some information on the context of vulnerability mapping. It then compares ten techniques that can be used to develop numerical indices or classifications of geographical areas with regard to the vulnerability of their groundwater resources to man-made pollution. Some information on six of the ten techniques is included in this chapter, with further information on a nitrate pollution index provided in Chapter 5 (Ramolino, 1988). The ten mapping techniques were selected based upon a review of extant literature; there are probably more than 50 such techniques in worldwide use (Ramolino and Canter, 1990). Finally, the advantages and limitations of vulnerability mapping techniques are summarized.

CONTEXT OF VULNERABILITY MAPPING

The susceptibility of a groundwater resource to pollution from the usage of agricultural chemicals (fertilizers or pesticides) is related to several local hydrogeological factors. Mapping of aquifer vulnerability to such pollution usually entails the composite consideration of several factors descriptive of the depth and permeability of the unsaturated zone and the local area hydrological balance. The primary technical issue addressed in vulnerability mapping is the subsurface transport and fate of potential pollutant chemicals such as nitrates (Canter, 1987). While not limited to the unsaturated or vadose zone, the general emphasis of most vulnerability mapping techniques is on transport through this media as opposed to transport within the saturated zone.

Vulnerability mapping techniques typically lead to either numerical indices for, or classifications of, geographical areas relative to their susceptibility to groundwater contamination. In order to develop these indices or classifications, the typical approach involves consideration of multiple physical and chemical factors along with their relative importance weighting. In addition, information is typically provided on measurements for the factors and their evaluation, generally through the use of a numerical approach. The final index or classification for a given geographical area will be based upon the summation of the factor scores and/or the summation of the products of the factor scores multiplied by their relative importance weights.

Vulnerability mapping has some particularly important uses or purposes relative to agricultural activities and groundwater pollution concerns. One of the most important uses is to identify geographical locations with high susceptibility to groundwater pollution, and thus plan monitoring programs in these "hot spots." This is particularly important in light of the expenditures necessary for appropriate planning and implementation of monitoring programs. It is more expensive to conduct a groundwater monitoring program than a comparable program for surface water resources; accordingly, any tool that can aid in identifying those areas of greatest concern is valuable. Another potential use of vulnerability mapping would be as a basis for determining appropriate fertilizer application rates or for permitting various agricultural chemicals for usage in particular geographical locations. For example, it might be desirable to limit the usage of mobile and persistent pesticides in geographical areas that have high susceptibility to groundwater pollution. In addition, vulnerability mapping could be the basis for delineating time periods and quantities of chemical usage as well as quantities of irrigation water allowed in particular geographical locations.

In general, groundwater vulnerability mapping techniques take into account processes affecting pollutants and the protective properties of soil and aquifer materials. During infiltration through soils and transport in aquifers, many contaminants are naturally attenuated. Figure 4.1 illustrates processes causing pollutant attenuation in soil and aquifer materials (Foster, 1987). These processes occurring in the subsurface environment are complex and may work individually or in combination to provide varying degrees of attenuation, depending on the individual pollutants in the system and on site-specific soil and aquifer characteristics. Therefore, not all subsurfaces have the same capacity to protect groundwater; as well, knowledge of soils and aquifer materials is essential in evaluating the attenuation capacity of the environment.

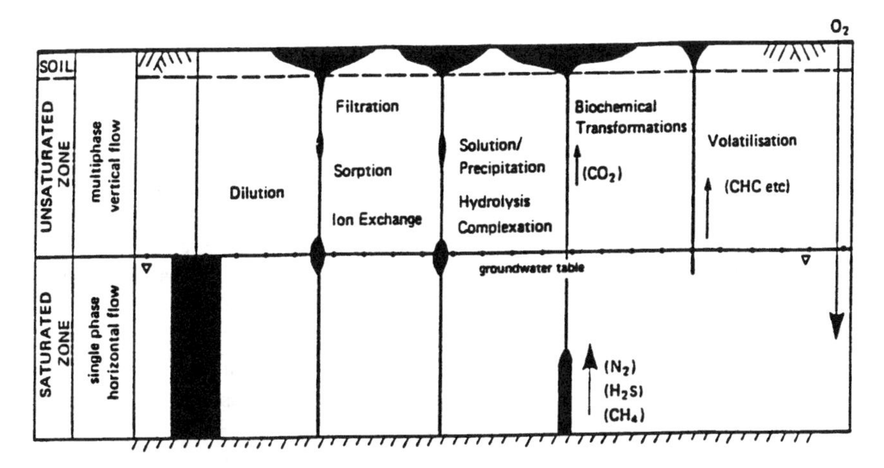

Figure 4.1 Processes causing pollutant attenuation in groundwater systems and their relative importance in the soil, above and below the water tables. (From Foster, S. S. D., in *Vulnerability of Soil and Ground Water to Pollutants*, Van Duijvenbooden, W. and Van Waegeningh, H. G., Eds., National Institute of Public Health and Environmental Hygiene, The Hague, The Netherlands, 1987, pp. 69–86).

Soil is the first layer encountered by infiltrating water or entering pollutants. Among the most significant factors typically used to evaluate the degree of protection of soil against pollution are: (1) texture, (2) permeability, (3) thickness, (4) pH, (5) soil organic matter content, (6) soil sorption, (7) soil cation exchange capability, and (8) soil acid neutralizing capacity.

1. *Soil texture* is measured by the percentage of sand-, silt-, and clay-sized particles present in representative soil samples. Fine-grained soils are, in gen-

eral, more conducive for attenuation processes than coarse grained. Clay is effective in removing pollutants because its pores are small and its particles have a high capacity for adsorption and ion exchange.

2. *Permeability* is the measure of the rate at which water moves through soil. Generally, water moves through large pores at a relatively rapid rate and this decreases the chance for a pollutant moving with water to be attenuated. On the other hand, water moves through the smaller soil pores at a slower rate, which increases the potential for pollutant attenuation.
3. *Soil thickness* is an important factor because the effectiveness of the soil in attenuating pollutants depends on the amount of contact time that transported pollutants have with the mineral and organic constituents of the soil layers.
4. *Organic matter content* of soil is an important factor because it increases the ability of soil to hold nutrients, water, and complex heavy metals, and to adsorb organic materials. It is a valuable energy source for soil microorganisms that play an important part in the breakdown of organic wastes and pesticides.
5. The *pH* level of a soil is an important factor because the breakdown processes function better at a pH between 6 and 7.
6. *Sorption* refers to the physical (absorption) or chemical (adsorption) attraction of individual molecules by soil particles. The rate of sorption will vary depending on the contaminant and the properties of the soil material. It will also vary with the pH and the oxidation potential.
7. *Cation exchange capacity* refers to the chemical exchange capability of soil. Clay and humus colloids, present in both weathering rocks and soils, contain negative electric charges at their particle surfaces; these charges are neutralized by electrostatically absorbed cations, which are subject to relatively rapid exchange reactions with other salts (or pollutants) that may percolate through the soil or weathering rocks.
8. The *acid neutralization capacity* (ANC) of soil or rock material is defined as the sum of cations minus the sum of strong acid anions, expressed as their potential to consume or produce protons above pH 3.

Groundwater is found in saturated rock and soil formations. These formations may be consolidated bedrock or unconsolidated deposits. Unconsolidated aquifers appear as formations of sand and gravel. Geologic materials with large openings such as coarse sand and gravel permit pollutants to move through aquifers with relative rapidity without reducing their concentrations. Consolidated aquifers generally occur in sedimentary rock formations of sandstone and limestone or dolomite. Two characteristics of consolidated aquifers determine the permeability and therefore the susceptibility to pollution: (1) the pore spaces within a consolidated formation, and (2) the presence of cracked fissures and solution cavities within a limestone formation. Such features are formed when infiltrating water dissolves the limestone, enlarging existing features and joints in the material. Solution cavities make aquifers highly vulnerable to pollution. If the soil layer is thin or absent, offering little attenuation, karst aquifers may be readily contaminated by pollutants introduced to the ground surface above them. However, it should be noted that most existing vulnerability mapping techniques address only transport that occurs by simple percolation and ignore preferential flow paths such as biochannels, cracks, joints, and solution channels in the vadose zone (Committee on Techniques for Assessing Ground Water Vulnerability, 1993).

Other hydrogeological factors typically used in assessing the vulnerability of groundwater to contamination include:

1. Net recharge from irrigation and/or rainfall that contributes to the downward movement of pollutants in soil. The recharge water may be the principal medium by which contaminants are dissolved and consequently transported.
2. The depth to the water table is an expression of the thickness of the vadose zone and is related to the time to reach the groundwater table. In general, the longer the time and the greater the distance traveled, the greater the potential for attenuation, because dilution increases with distance traveled and sorption becomes more complete with distance traveled.
3. Water table gradients, both the direction and the rate of flow of groundwater, are important in evaluating contamination potential. Movement of water from a disposal site is much less favorable than movement toward it.
4. Infiltration describes the tendency of water to enter the surface of the soil or a waste disposal site. A site with a large infiltration will thus be more susceptible to contamination of groundwater.
5. A large distance between a contamination source and water supply is generally considered to be a favorable factor, especially where the movement is through loose granular material with some sorption capacity.

Generally, the choice of factors in developing vulnerability mapping techniques is influenced by the availability of published data. As noted in the above discussion, different types of data are needed to characterize and understand contaminant behavior in the subsurface environment (Jaffe and Dinovo, 1987).

SUMMARY OF TEN MAPPING TECHNIQUES

The ten vulnerability mapping techniques summarized herein are compared in Table 4.1. They are all similar in that mapped areas are divided into categories representing more or less vulnerable conditions. These mapping techniques can be categorized relative to whether they have been developed to address point or nonpoint (area) sources of groundwater pollution. Point source methodologies include examples for landfills and hazardous waste sites. Nonpoint sources are related to agriculture, acid deposition, and septic tank systems (Canter and Sabatini, 1990). Six of the mapping techniques listed in Table 4.1 that are more related to nitrate pollution of groundwater will be highlighted in order to illustrate the details of the techniques.

VULNERABILITY MAPPING IN BELGIUM

In the Flanders area of Belgium, the quality of groundwater is threatened by possible pollution from either point sources (waste disposal sites, sewage pits, etc.) or diffused sources (fertilizers and acid rain). The Minister for Water and

Table 4.1 Comparison of Groundwater Vulnerability Mapping Techniques

Methodology (Ref.)	Pollution sources addressed	Factors in mapping	When developed and by whom	Actual/potential uses
Belgium vulnerability mapping (DeSmedt et al., 1987)	Point sources, area sources	Aquifer material (composition, hydraulic conductivity); covering layer (thickness and characteristics); unsaturated zone (thickness)	Developed in 1987 by Minister for Water and Environment of the Flemish Region of Belgium	Used to assess the vulnerability of groundwater in Flanders, Belgium Can be used to define the degree of the risk for the contamination of groundwater in the upper aquifers by contaminants entering the soil from the surface Provides a comprehensive view of conditions mainly for planning of regional scale Serves as a guideline for proper site investigations
Vulnerability of U.K. groundwater to agricultural nitrate pollution (Carter, Palmer, and Monkhouse 1987)	Area sources (nitrate in agricultural areas)	Aquifer (permeability, thickness of the bedrock cover); soil (texture, organic carbon content, permeability and wetness)	Developed in 1987 by Soil Survey of England and Wales and the British Geological Survey	Used to investigate the vulnerability of the Stafford Triassiac sandstone aquifer to nitrate pollution
Assessing aquifer sensitivity to acid deposition in Europe (Holmberg, Johnston, and Maxe, 1987)	Area sources (acid deposition)	Soil (texture, depth, soil-water contact, base cation content); aquifer (size, potential recharge, mineral composition, residence time)	Developed in 1987 by the International Institute for Applied Systems Analysis (IIASA)	Used to estimate the sensitivity of aquifers to acid deposition and the associated risk that geochemical changes may occur as a result of acid deposition Can be used as screening tool to locate aquifers at risk to experience changes first
Terrain suitability rating for septic tank systems in Canada (Sauriol, 1982)	Area sources (septic tank effluents)	Terrain unit (texture, lithology, stratigraphy, thickness, topography, drainage, depth to groundwater)	Developed in 1982 by Geo-Analysis (Canadian Engineering Geology firm)	Used to assess soil attenuation properties for septic tank effluent May be used at planning stage to encourage development in suitable areas and alternate land usage in less suitable areas

Table 4.1 Comparison of Groundwater Vulnerability Mapping Techniques (Continued)

Methodology (Ref.)	Pollution sources addressed	Factors in mapping	When developed and by whom	Actual/potential uses
Groundwater contamination susceptibility in Minnesota (Porcher, 1988)	Point sources, area sources	Aquifer (grain size, percentage of solution openings); recharge potential; soil (texture, clay content); vadose zone (texture, rock fracture and solution features)	Developed in 1988 by Minnesota Pollution Control Agency	Used to assess groundwater contamination susceptibility in Minnesota Can be used to provide information needed for local planning/protection, management, and education projects
Standardized system for evaluating waste disposal sites in the U.S. (LeGrand, 1983)	Point sources (waste disposal)	Depth to groundwater; water table gradient; permeability; sorption characteristics; contaminant severity (toxicity, concentration and volume, mobility in the water, and persistence); aquifer areal extent of potential importance as a groundwater source	Developed in 1983 by National Water Well Association	Standardized system for describing and rating waste disposal and other land surface sites in terms of their potential for contamination of groundwater Can be used by regulatory agencies and owners/operators of waste sites during planning and operational stages of contaminant handling
DRASTIC: A standardized system for evaluating groundwater pollution potential in the U.S. (Aller et al., 1987)	None specifically; pesticides in Pesticide DRASTIC	Depth to water; net recharge; aquifer media; soil media; topography; impact of the vadose zone media; and conductivity	Developed in 1985 by National Water Well Association	Used to evaluate groundwater pollution potential using hydrogeologic settings Can be used to assist planners, managers and administrators in the task of evaluating the relative vulnerability of an area to groundwater contamination from various sources of contamination

Table 4.1 Comparison of Groundwater Vulnerability Mapping Techniques (Continued)

Methodology (Ref.)	Pollution sources addressed	Factors in mapping	When developed and by whom	Actual/potential uses
Soil/aquifer field evaluation (SAFE) in the U.S. (Roux, DeMartinis, and Dickson, 1986)	Area sources (pesticides)	Aquifer (recharge characteristics, yield, water quality, degree of aquifer protection, recharge area, flow direction); soil (major soil unit, soil layers, layer permeability, soil association, soil pH, soil temperature)	Developed in 1986 by Roux Associates Inc. and CIBA Geigy	Used for determining the vulnerability of groundwater to pesticide contamination Used to plan monitoring program for pesticides
Nitrate pollution index for groundwater in the U.S. (Ramolino, 1988)	Area sources (nitrates applied as fertilizers)	Nitrogen fertilization intensity; soil texture; net recharge; depth to groundwater	Developed in 1988 at the Environmental and Ground Water Institute, University of Oklahoma	Used to assess the potential risk of groundwater contamination from nitrate applied as fertilizers in agricultural areas
Susceptibility of U.K. groundwaters to acid deposition (Edmunds and Kinniburgh, 1986)	Area sources (acid deposition)	Neutralization capability of bedrock geology (rock type, carbonate mineral content, and resistance to acidification)	Developed in 1986 by Hydrogeology Research Group of the British Geological Survey	Used to assess the susceptibility of U.K. groundwater to acidification

Environment has established groundwater vulnerability maps at a scale of 1:100,000 for each Flemish province (DeSmedt, et al., 1987). The maps provide a comprehensive view of the conditions, and are useful for planning at the regional scale. They can be used in analyzing potential and actual problems associated with the nitrate pollution of groundwater. In mapping the vulnerability of groundwater in Flanders, three categories of information were taken into consideration: (1) the aquifer material, (2) the covering formation, and (3) the unsaturated zone (DeSmedt, et al., 1987).

The determining factors used to assess the potential of aquifer material to attenuate pollutants included the type of aquifer formation, the permeability of the aquifer, and the potential behavior of contaminants in the aquifer. Aquifers were classified into four groups as follows:

A. Chalk, limestone, sandstone, and marl
B. Gravel
C. Sand
D. Loam sand and argillaceous (clayey) sand

The factors used to assess the potential of the cover layer to attenuate pollutants or prevent their entrance in groundwater include the type of formation, the thickness, and the hydraulic resistance. A cover layer less than 5 m in thickness was regarded as a nonexistent layer; also, a cover layer of sand was not considered to be a protective cover. The cover layers were divided into three classes as follows:

a. No cover (less than 5 m and/or sandy)
b. Loamy cover
c. Clayey cover

The unsaturated zone was classified into two groups on the basis of thickness. In the absence of a covering formation the unsaturated zone, if sufficiently thick, may act as a protective barrier against contamination of groundwater; therefore, a distinction was made as follows:

1. Thickness of 10 m or less
2. Thickness of more than 10 m

According to the information collected on the three categories of factors (aquifer, covering layer, and unsaturated zone), a vulnerability index was developed as shown in Table 4.2 (DeSmedt, et al., 1987). The index was subdivided into five classes; each specific class was a result of the combination of information on the three categories of factors. The vulnerability index was developed by assigning a numerical value varying from 1 to 5 to each class. A "1" rating was assigned to the extremely vulnerable class; "2" to the highly vulnerable; "3" to the vulnerable; "4" to the moderately vulnerable; and "5" to the least vulnerable.

Table 4.2 Classification of the Vulnerability Index Used in Belgium

Degree	Indices	Aquifer	Covering layer	Unsaturated zone (m)
1. Extremely vulnerable	Aa1	Chalk, limestone, sandstone, marl	<5 m or sandy	<10
	Ba1	Gravel	<5 m or sandy	<10
2. Very vulnerable	Aa2	Chalk, limestone, sandstone, marl	<5 m or sandy	>10
	Ba2	Gravel	<5 m or sandy	>10
	Ca1	Sand	<5 m or sandy	<10
3. Vulnerable	Ab	Chalk, limestone, sandstone, marl	Loamy	
	Bb	Gravel	Loamy	
	Ca2	Sand	<5 m or sandy	>10
4. Moderately vulnerable	Ac	Chalk, limestone, sandstone, marl	Clayey	
	Bc	Gravel	Clayey	
	Cb	Sand	Loamy	
	Da1	Loamy or clayey sand	<5 m or sandy	<10
	Da2	Loamy or clayey sand	<5 m or sandy	>10
5. Less vulnerable	Cc	Sand	Clayey	
	Db	Loamy or clayey sand	Loamy	
	Dc	Loamy or clayey sand	Clayey	

After DeSmedt, P., et al., *Aqua*, 5, 264–267, 1987.

GROUNDWATER VULNERABILITY TO AGRICULTURAL NITRATE POLLUTION IN THE U.K.

The Soil Survey of England and Wales (SSEW) and the British Geological Survey (BGS) have developed maps to investigate the vulnerability of the Stafford Triassic sandstone aquifer to nitrate pollution from agricultural practices (Carter, Palmer, and Monkhouse, 1987). This study was undertaken for the Severn Trent Water Authority (STWA). The maps were developed based on using published information on soil and geology. They can be used as aids for management decisions in the water and agricultural industries (Carter, Palmer, and Monkhouse, 1987).

Soil physical properties that influence nitrate leaching were assessed for pertinent soil types and used to classify soil leaching in the vulnerability scale. Four leaching classes, ranging from extreme to low, were determined as shown in Table 4.3 (Carter, Palmer, and Monkhouse, 1987). Topsoil and subsoil texture and organic carbon content were chosen as indicators of soil water retention. Porosity data were taken to represent soil permeability, and wetness classes gave an indication of the soil water regime and the influence of drainage treatments. A score was assigned to each of these parameters for individual soil type; and by adding each parameter score, an overall risk rating was given as shown in Table 4.3.

Table 4.3 Soil Leaching Classes and Their Characteristics

Score[a]	Rating	Class	Soil characteristics
21–30	Extreme	1	Deep permeable sandy soils, some affected by fluctuating groundwater in winter; shallow loamy and sandy soils over sandstone; 1.0–2.5% organic carbon[b]
16–20	High	2	Deep permeable light loamy soils; 1.5–2.5% organic carbon[b]
11–15	Moderate	3	Deep moderately permeable medium loamy soils; moderately permeable, medium loamy soils with dense slowly permeable subsoils; 2.0–4.0% organic carbon[b]
0–15	Low	4	Slowly permeable loamy, loamy over clayey, or clayey soils; deep moderately permeable loamy alluvial soils; 2.0–4.0% organic carbon[b]

[a] Based on considering soil texture, organic carbon content, porosity, and wetness.
[b] Relates to arable soils.

From Carter, A. D., Palmer, R. C., and Monkhouse, R. A., in *Vulnerability of Soil and Ground Water to Pollutants*, Van Duijvenbooden, W. and Van Waegeningh, H. G., Eds., National Institute of Public Health and Environmental Hygiene, The Hague, The Netherlands, 1987, pp. 333–342.

The determining factors used in assessing the ability of aquifer materials to control the entrance of nitrate into groundwater were the permeability and thickness of the bedrock cover. Three types of aquifers were identified by the degree of protection afforded to the aquifer by the bedrock cover; these categories are shown in Table 4.4 (Carter, Palmer, and Monkhouse, 1987).

Table 4.4 Aquifer Classification and Characteristics

Aquifer type	Aquifer characteristics
1	Aquifer outcrop without drift cover or with cover of permeable drift (sands and gravels)
2	Aquifer outcrop covered by less permeable drift (thin lacustrine clays, peat), or by thin and/or patchy boulder clay
3	Aquifer outcrop covered by impermeable drift (boulder clay), or by impermeable bedrock cover (Mercia Mudstone Group), or no aquifer present

From Carter, A. D., Palmer, R. C., and Monkhouse, R. A., in *Vulnerability of Soil and Ground Water to Pollutants*, Van Duijvenbooden, W. and Van Waegeningh, H. G., Eds., National Institute of Public Health and Environmental Hygiene, The Hague, The Netherlands, 1987, pp. 333–342.

The vulnerability of groundwater to nitrate pollution was determined using a matrix that combined the assessments of soil leaching and aquifer classification. The resultant groundwater vulnerability categories are shown in Table 4.5 (Carter, Palmer, and Monkhouse, 1987). The maps derived from the soil and geological assessments were overlain and all the land overlying an aquifer was assigned particular vulnerability categories. This system of assessing the vulnerability of aquifers to pollution was strictly designed for the leaching of agricultural nitrate through soils and rocks to groundwater. The behavior of other substances such as pesticides and simple anions (e.g., sulfate and chloride) may differ significantly

Table 4.5 Groundwater Vulnerability Categories for Agricultural Nitrate Pollution in the U.K.

Aquifer classification (type)	Soil leaching classification (class)			
	1	2	3	4
1	Extreme	High	Moderate	Low
2	High	Moderate	Low	Low
3	Low	Low	Low	Low

Note: Extreme, type 1 aquifer with shallow or sandy very permeable soils; High, type 1 aquifer with moderately deep or deep permeable loamy and/or peaty soils; or type 2 aquifer with shallow or sandy very permeable soils; Moderate, type 1 aquifer with moderately permeable loamy over clayey and some clayey soils; or type 2 aquifer with deep or moderately deep permeable loamy soils; and Low, type 1 aquifer with slowly permeable loamy or clayey soils; or type 2 aquifer with moderately permeable loamy over clayey and some clayey soils; or type 3 aquifer with any soils.

From Carter, A. D., Palmer, R. C., and Monkhouse, R. A., in *Vulnerability of Soil and Ground Water to Pollutants*, Van Duijvenbooden, W. and Van Waegeningh, H. G., Eds., National Institute of Public Health and Environmental Hygiene, The Hague, The Netherlands, 1987, pp. 333–342. With permission.

as a result of their particular adsorption coefficients within soil rock, and also their solubility and mobility (Carter, Palmer, and Monkhouse, 1987).

TERRAIN SUITABILITY RATING FOR SEPTIC TANK SYSTEMS IN CANADA

A feasibility study was performed to determine suitability ratings of a proposed suburban district of the township of Oxford-on-Rideau in Canada for a development utilizing private sewerage services. The study area was located in the vicinity of the town of Kemptville, which is 50 km from Ottawa, Ontario, Canada (Sauriol, 1982). The suitability of the terrain to sustain a conventional septic tank system was evaluated on the basis of the following factors: (1) capacity of the natural soil material to attenuate septic tank system effluent; (2) depth to groundwater table; (3) depth to impervious material; (4) drainage; and (5) surface topography. This study was undertaken to ensure that proper distances are used to separate wells, houses, and drain tile fields, and that proper attenuation takes place to preserve the integrity of the groundwater supply from nitrate and bacterial pollution (Sauriol, 1982).

Photogeologic analysis on a small scale was used to identify different terrain units in the study area. Five terrain units were defined based on lithology, thickness of overburden, topography, drainage, and depth to groundwater (Sauriol, 1982). An integration of these characteristics permitted the determination of the capacity of these terrain units to attenuate septic tank system effluents. Table 4.6 summarizes the individual terrain characteristics and their adequacy to attenuate

Table 4.6 Terrain Suitability Evaluation for Septic Tank Systems

Terrain unit	Alluvium or organic	Sand	Clay	Sand/till	Till
Lithology	Silt, sand, clay, peat	Medium- to very fine-grained, silty	Silty clay upper 0.5 m weathered	Medium- to fine-grained sand with granules over sandy silt till with angular clasts	Sandy silt till with angular cobbly clasts
Stratigraphy thickness	Layered, 1- to 2-m thick	Layered, 1- to 3-m thick	Dense, massive blue gray clay, 3- to 6-m thick	Shallow sand, 1- to 2-m thick over till up to 10-m thick	Dense, compacted 3- to 10-m thick
Topography	Stream valley, low local relief plain	Low local relief, undulating plain	Low local relief, undulating plain	Moderate local relief ridge, undulating	Moderate to low local relief, undulating
Origin	Fluvial, flood plain and channel deposits	Marine, shallow water deposits	Marine deep water deposits	Marine beach over glacial till knob	Glacial
Drainage	Poor, high water table, wet	Moderately to poorly drained, high water table in places	Poorly drained wet, high water table	Good drainage dry, low water table	Good drainage, low water table, dry
Suitability for septic systems	Erosion and flood hazards, not suitable	Marginal suitability	Marginal suitability	Good suitability	Very best suitability
Minimum lot sizes on private services	Not applicable	0.3-ha lot	0.4-ha lot	0.2-ha lot	0.2-ha lot
Development recommendations	None, low-bearing capacity	Grouted wells, raised tile beds	Grouted wells, raised tile beds, because of high water table	Grouted wells, aggregate potential drainage problems, high estimated percolation time	Grouted wells

From Sauriol, J. J., *Proceedings of the Sixth National Ground Water Quality Symposium*, National Water Well Association, Dublin, OH, 1982, pp. 270–275. With permission.

effluent from septic tank systems (Sauriol, 1982). Four suitability grades for development were defined: (1) very best area; (2) good area; (3) marginal area; and (4) area unsuitable for development.

Terrain suitability ratings for the attenuation of septic tank system effluents, when used at the planning stage, may help to reduce contaminant loadings and avoid groundwater contamination. This information could then be used to encourage development in suitable areas and alternate land uses in less suitable zones (Sauriol, 1982).

GROUNDWATER CONTAMINATION SUSCEPTIBILITY IN MINNESOTA

A method for assessing relative susceptibility to groundwater contamination in Minnesota was designed by Porcher (1988). The resultant information on groundwater susceptibility to contamination is given in map form and should be useful in targeting state regulatory monitoring, planning, preventive, and public information efforts in those areas determined to be most critical (Porcher, 1988). The method could be used in nitrate groundwater pollution studies. The parameters chosen for assessing groundwater vulnerability were those that were readily available using published maps; they included (1) aquifer materials, (2) recharge potential, (3) soil materials, and (4) vadose zone materials (Porcher, 1988).

Aquifer materials refer to the aquifer nearest the land surface — which is composed of consolidated or unconsolidated material such as sand, gravel, and porous or fractured bedrock, and which yields sufficient quantities of water for use. Each unit was evaluated on the basis of its generalized hydraulic conductivity. Larger grain sizes and higher percentages of solution openings within an aquifer were taken as indicative of greater permeability and greater groundwater contamination susceptibility. From these considerations, each unit was assigned a rating value that reflected its potential ability to prevent the entrance of pollutants into groundwater; these ratings are shown in Table 4.7 (Porcher, 1988).

Table 4.7 Ratings for Aquifer Media in Minnesota

Media	Rating
Nonaquifer	1
Shaly sandstone	3
Metamorphic/igneous	4
Bedded limestone, sandstone, and shale sequences	5
Sandstone	6
Limestone/dolomite	7
Sand and gravel	9
Karstic limestone/dolomite	10

From Porcher, E., "Ground Water Contamination Susceptibility in Minnesota," 1988, Minnesota Pollution Control Agency, St. Paul, MN.

Recharge potential refers to the amount of water that may penetrate the ground surface relative to other areas in Minnesota. The recharging water is considered the principal medium by which contaminants are dissolved and consequently transported to and within the aquifer. Table 4.8 summarizes the numerical ratings assigned for recharge potential in Minnesota (Porcher, 1988).

Table 4.8 Ratings for Recharge Potential in Minnesota

Recharge potential	Rating
No recharge	0
Very low	1
Low	3
Moderate	6
High	9

From Porcher, E., "Ground Water Contamination Susceptibility in Minnesota," 1988, Minnesota Pollution Control Agency, St. Paul, MN.

Generalized soil materials refers to the soils in the uppermost six ft of the earth's surface. In this assessment of groundwater contamination susceptibility, soil was evaluated on the basis of texture and the amount of clay content. The presence of smaller grain sizes (and therefore smaller permeabilities) results in a decreased groundwater contamination susceptibility. The soil materials component was classified into seven different soil types, and a rating value (as shown in Table 4.9) was assigned to each soil type (Porcher, 1988).

Table 4.9 Ratings for Generalized Soil Materials in Minnesota

Soil material	Rating
No rating	0
Clay	2
Clay loam	3
Silt loam	4
Loam	5
Sandy loam	6
Peat	8
Sand	9
Thin or absent	10

From Porcher, E., "Ground Water Contamination Susceptibility in Minnesota," 1988, Minnesota Pollution Control Agency, St. Paul, MN.

Vadose zone materials refer to the materials in the zone above the water table that are unsaturated. The ability of vadose zone materials to control the downward migration of contaminants was evaluated by either their texture or rock fracture and solution features. Ratings for each of the selected vadose zone materials are shown in Table 4.10 (Porcher, 1988).

The four groups of parameters evaluated above were then combined into a single computer file in order that relative values could be assigned to each of the unique hydrogeological areas in Minnesota. This computer file contains a list of 4- or 5-digit numbers that represent the various parameters and their respective

Table 4.10 Ratings for Vadose-Zone Materials in Minnesota

Material	Rating
No rating	0
Clay	2
Clay loam	3
Precambriam igneous and metamorphic	3
Bedrock aquifer confining units	4
Silt	4
Loam	5
Shaly sandstone	5
Other metamorphic/igneous	5
Sandy loam	5
Sandstone	6
Sand	6
Limestone/dolomite	7
Sand and gravel	8
Peat (no vadose zone)	10
Karstic limestone/dolomite	10

From Porcher, E., "Ground Water Contamination Susceptibility in Minnesota," 1988, Minnesota Pollution Control Agency, St. Paul, MN.

factors. Each of the numbers (ones, tens, hundreds, thousands, and ten thousands columns) is representative of a specific physical characteristic and its appropriate numerical rating. The rating values were summed via the computer to give a total value. The range of actual total values (lowest = 6; highest = 37) is 31; this range was divided by five to obtain five nearly equal classes as shown in Table 4.11 (Porcher, 1988). To serve as an example, suppose the number 8399 is determined from the data file for a site. This number is interpreted as follows:

Ones column — high recharge potential = 9 points (Table 4.8)
Tens column — sandy soil materials = 9 points (Table 4.9)
Hundreds column — shaly sandstone aquifer material = 3 points (Table 4.7)
Thousands column — sand and gravel vadose zone materials = 8 points (Table 4.10)

The total score for this unique set of characteristics is 29 out of 37 possible points. In Minnesota, well over 1000 unique combinations of physical characteristics exist, thus indicating a variety of susceptible regions across the State of Minnesota (Porcher, 1988).

DRASTIC: A STANDARDIZED SYSTEM FOR EVALUATING GROUNDWATER POLLUTION POTENTIAL IN THE U.S.

A numerical rating scheme, called DRASTIC, has been developed for evaluating the potential for groundwater pollution in given areas based on their hydrogeological setting (Aller, et al., 1987). This rating scheme is based on seven

Table 4.11 Groundwater Contamination Susceptibility Classes in Minnesota

Class	Summation of respective ratings
Lowest groundwater contamination susceptibility	6–11
—	12–17
—	18–23
—	24–29
Highest groundwater contamination susceptibility	30–37

From Porcher, E., "Ground Water Contamination Susceptibility in Minnesota," 1988, Minnesota Pollution Control Agency, St. Paul, MN.

factors chosen by over 35 groundwater scientists from throughout the U.S. Information on these factors is presumed to exist for all locations in the U.S. In addition, the scientists also established relative importance weights and a point rating scale for each factor. The acronym DRASTIC is derived from the seven factors in the rating scheme:

D = Depth to groundwater
R = Recharge rate (net)
A = Aquifer media
S = Soil media
T = Topography (slope)
I = Impact of the vadose zone
C = Conductivity (hydraulic) of the aquifer

Determination of the DRASTIC index for a given area (the smallest applicable size is about 100 acres) involves multiplying each factor weight by its point rating and summing the total. The higher sum values represent greater potential for groundwater pollution, or greater aquifer vulnerability. For a given area being evaluated, each factor is rated on a scale of 1 to 10 that indicates the relative pollution potential of the given factor for that area. Once all factors have been assigned a rating, each rating is multiplied by the assigned weight, and the resultant numbers are summed as follows:

$$D_rD_w + R_rR_w + A_rA_w + S_rS_w + T_rT_w + I_rI_w + C_rC_w = \text{Pollution Potential}$$

where:

r = rating for the area being evaluated
w = importance weight for the factor

Weights from 1 to 5 determine the relative importance of the factors with respect to each other. Ratings are obtained from tables for each factor, while the weights are determined from tables representing either general applicability (Table 4.12), or for the potential pollution from pesticide applications (Table 4.13) (Aller, et al., 1987). The importance weights from Table 4.12 are for the generic DRASTIC, whereas the weights from Table 4.13 are for the modified version

Table 4.12 Assigned Importance Weights for Factors in Generic DRASTIC

Factor	Importance weight
Depth to groundwater (D)	5
Net recharge (R)	4
Aquifer media (A)	3
Soil media (S)	2
Topography (T)	1
Impact of the vadose-zone media (I)	5
Hydraulic conductivity of the aquifer (C)	3

From Aller, L., Bennett, T., Lehr, J. H., Petty, R. J., and Hackett, G., "DRASTIC: A Standardized System for Evaluating Ground Water Pollution Potential Using Hydrogeologic Settings," EPA/600/2-87/035, 1987, U.S. Environmental Protection Agency, Ada, OK, pp. 11–25, 68, and 101.

Table 4.13 Assigned Importance Weights for Factors in Pesticide DRASTIC

Factor	Importance weight
Depth to groundwater (D)	5
Net recharge (R)	4
Aquifer media (A)	3
Soil media (S)	5
Topography (T)	3
Impact of the vadose-zone media (I)	4
Hydraulic conductivity of the aquifer (C)	2

From Aller, L., Bennett, T., Lehr, J. H., Petty, R. J., and Hackett, G., "DRASTIC: A Standardized System for Evaluating Ground Water Pollution Potential Using Hydrogeologic Settings," EPA/600/2-87/035, 1987, U.S. Environmental Protection Agency, Ada, OK, pp. 11–25, 68, and 101.

called "Pesticide DRASTIC." The Pesticide DRASTIC importance weights would be more appropriate for studies related to nitrate groundwater pollution.

The depth to the water table is an important factor, primarily because it determines the depth of material through which a contaminant must travel before reaching the aquifer; this factor can also be used to determine the contact time with the surrounding materials. Table 4.14 contains the ranges and ratings for the depth to groundwater (Aller, et al., 1987). The ranges were determined based on what the study group of groundwater professionals considered to be depths where the potential for groundwater pollution significantly changed.

Net recharge refers to the total quantity of water that infiltrates from the ground surface to reach the aquifer. Net recharge includes the average annual amount of infiltration and does not take into consideration the distribution, intensity, or duration of recharge events. The ranges and ratings for net recharge are given in Table 4.15 (Aller, et al., 1987). The attenuation capacity of the aquifer media is evaluated on the basis of grain sizes, fractures, and solution openings. Ranges and ratings for the aquifer media factor are illustrated in Figure 4.2 and

Table 4.14 Evaluation of the Depth to Groundwater Factor in DRASTIC

Depth to groundwater	
Range (ft)	**Rating**
0–5	10
5–15	9
15–30	7
30–50	5
50–75	3
75–100	2
100+	1

From Aller, L., Bennett, T., Lehr, J. H., Petty, R. J., and Hackett, G., "DRASTIC: A Standardized System for Evaluating Ground Water Pollution Potential Using Hydrogeologic Settings," EPA/600/2-87/035, 1987, U.S. Environmental Protection Agency, Ada, OK, pp. 11–25, 68, and 101.

Table 4.15 Evaluation of the Net Recharge Factor in DRASTIC

Net annual recharge	
Range (in.)	**Rating**
0–2	1
2–4	3
4–7	6
7–10	8
10+	9

From Aller, L., Bennett, T., Lehr, J. H., Petty, R. J., and Hackett, G., "DRASTIC: A Standardized System for Evaluating Ground Water Pollution Potential Using Hydrogeologic Settings," EPA/600/2-87/035, 1987, U.S. Environmental Protection Agency, Ada, OK, pp. 11–25, 68, and 101.

Table 4.16 Evaluation of the Aquifer Media Factor in DRASTIC

Aquifer media		
Range	**Rating**	**Typical rating**
Massive shale	1–3	2
Metamorphic/igneous	2–5	3
Weathered metamorphic/igneous	3-5	4
Glacial till	4–6	5
Bedded sandstone, limestone, and shale sequences	5–9	6
Massive sandstone	4–9	6
Massive limestone	4–9	6
Sand and gravel	4–9	8
Basalt	2–10	9
Karst limestone	9–10	10

From Aller, L., Bennett, T., Lehr, J. H., Petty, R. J., and Hackett, G., "DRASTIC: A Standardized System for Evaluating Ground Water Pollution Potential Using Hydrogeologic Settings," EPA/600/2-87/035, 1987, U.S. Environmental Protection Agency, Ada, OK, pp. 11–25, 68, and 101.

Table 4.16 (Aller, et al., 1987). The soil media is considered as the upper weathered zone of the earth that averages a depth of 6 ft or less from the ground surface. The soil media is evaluated on the basis of the type of clay present, the

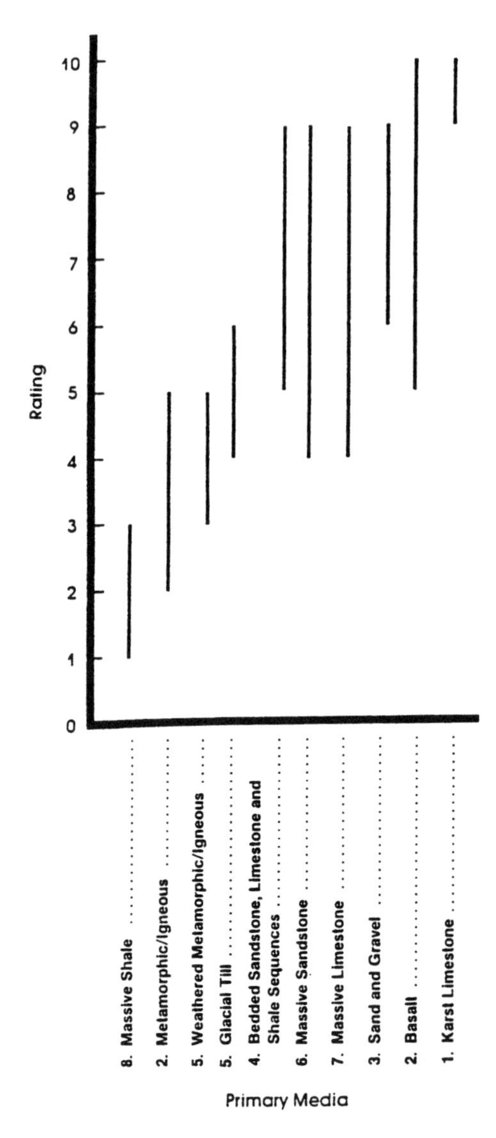

Figure 4.2 Ranges and ratings for the aquifer media factor in DRASTIC. Relative ranges of ease of pollution for the principal aquifer types. Ranges are based upon consideration of route length and tortuosity, potential for consumptive sorption, dispersion, reactivity, and degree of fracturing. The primary factors controlling the rating of each media are (1) reactivity (solubility and fracturing); (2) fracturing; (3) route length and tortuosity, sorption, dispersion — all essentially determined by grain size, sorting, and packing; (4) route length and tortuosity as determined by bedding and fracturing; (5) sorption and dispersion; (6) fracturing, route length, and tortuosity — influenced by intergranular relationships; (7) reactivity (solubility) and fracturing; (8) fracturing and sorption. (From Aller, L., Bennett, T., Lehr, J. H., Petty, R. J., and Hackett, G., "DRASTIC: A Standardized System for Evaluating Ground Water Pollution Potential Using Hydrogeologic Settings," EPA/600/2-87/035, 1987, U.S. Environmental Protection Agency, Ada, OK, pp. 11–25, 68, and 101.)

Table 4.17 Evaluation of the Soil Media Factor in DRASTIC

Soil media	
Range	**Rating**
Thin or absent	10
Gravel	10
Sand	9
Shrinking and/or aggregated clay	7
Sandy loam	6
Loam	5
Silty loam	4
Clay loam	3
Muck	2
Nonshrinking and nonaggregated clay	1

From Aller, L., Bennett, T., Lehr, J. H., Petty, R. J., and Hackett, G., "DRASTIC: A Standardized System for Evaluating Ground Water Pollution Potential Using Hydrogeologic Settings," EPA/600/2-87/035, 1987, U.S. Environmental Protection Agency, Ada, OK, pp. 11–25, 68, and 101.

Table 4.18 Evaluation of the Topography Factor in DRASTIC

Topography	
Range (% slope)	**Rating**
0–2	10
2–6	9
6–12	5
12–18	3
18+	1

From Aller, L., Bennett, T., Lehr, J. H., Petty, R. J., and Hackett, G., "DRASTIC: A Standardized System for Evaluating Ground Water Pollution Potential Using Hydrogeologic Settings," EPA/600/2-87/035, 1987, U.S. Environmental Protection Agency, Ada, OK, pp. 11–25, 68, and 101.

shrink/swell potential of that clay, and the grain size of the soil. The ranges and ratings for the soil media factor are shown in Table 4.17 (Aller, et al., 1987).

As used in the DRASTIC methodology, topography refers to the slope and slope variability of the land surface. Table 4.18 contains the slope ranges chosen as significant relative to groundwater pollution potential (Aller, et al., 1987). The vadose zone is defined as that zone above the water table which is unsaturated or discontinuously saturated. The vadose zone is evaluated on the basis of grain sizes, fracturing and solution openings, and sorption potential. Figure 4.3 and Table 4.19 display the ranges and ratings for the impact of the vadose zone media factor (Aller, et al., 1987). Finally, values for hydraulic conductivity are calculated from aquifer pumping tests. Information on hydraulic conductivity is typically available from published hydrogeological reports in given geographical areas. The ranges and ratings of the hydraulic conductivity factor are given in Table 4.20 (Aller, et al., 1987).

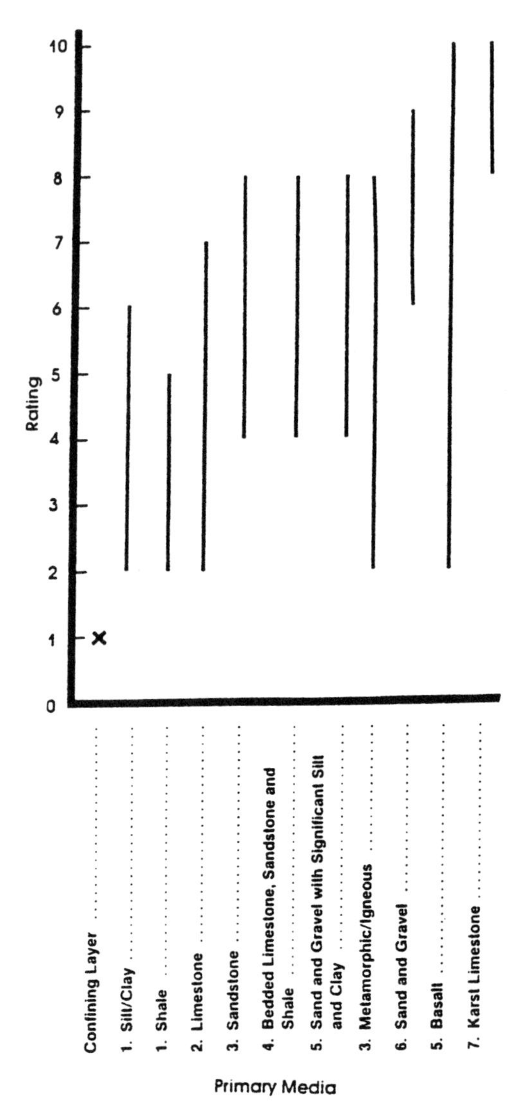

Figure 4.3 Ranges and ratings for the impact of the vadose-zone media factor in DRASTIC. Relative impact of the principal vadose-zone media types. Range based upon path length and tortuosity, potential for dispersion and consequent dilution, reactivity (solubility), consumptive sorption, and fracturing. Primary factors affecting the rating: (1) consumptive sorption and fracturing; (2) fracturing and reactivity; (3) fracturing—path length as influenced by intergranular relationships; (4) fracturing—path length and tortuosity as influenced by bedding planes, sorption, and reactivity; (5) path length and tortuosity as impacted by bedding, grain size, sorting and packing, and sorption; (6) path length and tortuosity as influenced by grain size, sorting, and packing; (7) reactivity and fracturing. (From Aller, L., Bennett, T., Lehr, J. H., Petty, R. J., and Hackett, G., "DRASTIC: A Standardized System for Evaluating Ground Water Pollution Potential Using Hydrogeologic Settings," EPA/600/2-87/035, 1987, U.S. Environmental Protection Agency, Ada, OK, pp. 11–25, 68, and 101.)

Table 4.19 Evaluation of the Impact of the Vadose Zone Media Factor in DRASTIC

Impact of the vadose-zone media		
Range	Rating	Typical rating
Confining layer	1	1
Silt/clay	2–6	3
Shale	2–5	3
Limestone	2–7	6
Sandstone	4–8	6
Bedded limestone, sandstone, shale	4–8	6
Sand and gravel with significant silt and clay	4–8	6
Metamorphic/igneous	2–8	4
Sand and gravel	6–9	8
Basalt	2–10	9
Karst limestone	8–10	10

From Aller, L., Bennett, T., Lehr, J. H., Petty, R. J., and Hackett, G., "DRASTIC: A Standardized System for Evaluating Ground Water Pollution Potential Using Hydrogeologic Settings," EPA/600/2-87/035, 1987, U.S. Environmental Protection Agency, Ada, OK, pp. 11–25, 68, and 101.

Table 4.20 Evaluation of the Hydraulic Conductivity Factor in DRASTIC

Hydraulic conductivity	
Range (gpd/ft^2)	Rating
1–100	1
100–300	2
300–700	4
700–1000	6
1000–2000	8
2000+	10

From Aller, L., Bennett, T., Lehr, J. H., Petty, R. J., and Hackett, G., "DRASTIC: A Standardized System for Evaluating Ground Water Pollution Potential Using Hydrogeologic Settings," EPA/600/2-87/035, 1987, U.S. Environmental Protection Agency, Ada, OK, pp. 11–25, 68, and 101.

The DRASTIC methodology was developed to allow for the systematic evaluation of the groundwater in the U.S. This system has been prepared to assist planners, managers, and administrators in the task of evaluating the relative vulnerability of areas to groundwater contamination from various sources of pollution (Aller, et al., 1987). While not developed specifically for nitrate pollution, it could be used in such assessments. The methodology has also been used in Sweden (Swanson, 1990).

One presumption basic to the DRASTIC method is that data or information are available on each of the seven factors. Table 4.21 summarizes various sources of hydrogeological information required in the use of DRASTIC (Aller, et al., 1987). The most common sources for each factor are (Aller, et al., 1987):

1. Depth to groundwater: well logs or hydrogeologic reports
2. Net recharge: water resource reports combined with data on precipitation from the National Oceanic and Atmospheric Administration

Table 4.21 Sources of Hydrogeological Information for Use in DRASTIC

Source	Depth to water	Net recharge	Aquifer media	Soil media	Topography	Impact of the vadose media	Hydraulic conductivity of the aquifer
U.S. Geological Survey	X	X	X		X	X	X
State Geological Survey	X	X	X			X	X
State Department of Natural/Water Resources	X	X	X			X	X
U.S. Department of Agriculture-Soil Conservation Service		X		X	X		
State Department of Environmental Protection	X	X	X			X	X
Clean Water Act "208" and other Regional Planning Authorities	X	X	X			X	X
County and regional water supply agencies and companies (private water suppliers)	X		X			X	X
Private consulting firms (hydrogeologic, engineering)	X		X			X	X
Related industry studies (mining, well drilling, quarrying, etc.)	X		X			X	
Professional associations (Geological Society of America, National Water Well Association, American Geophysical Union)	X	X	X			X	X
Local colleges and universities (Department of Geology, Earth Sciences, and Civil Engineering)	X	X	X			X	X
Other federal/state agencies (Army Corps of Engineers, National Oceanic and Atmospheric Administration)	X	X	X			X	

From Aller, L., Bennett, T., Lehr, J. H., Petty, R. J., and Hackett, G., "DRASTIC: A Standardized System for Evaluating Ground Water Pollution Potential Using Hydrogeologic Settings," EPA/600/2-87/035, 1987, U.S. Environmental Protection Agency, Ada, OK, pp. 11–25, 68, and 101.

3. Aquifer media: published geologic and hydrogeologic reports
4. Soil media: published soil survey reports or local mapping projects conducted by the U.S. Soil Conservation Service
5. Topography: published U.S. Geological Survey topographic maps (various scales)
6. Impact of the vadose zone media: published geologic reports
7. Hydraulic conductivity of the aquifer: published hydrogeologic reports

As summarized in Chapter 1, the U.S. EPA's Office of Pesticide Programs and Office of Drinking Water have sponsored a national survey of pesticides and nitrates in groundwater (Alexander and Liddle, 1986). One of the goals of this survey was to assess the relationship between the agricultural usage of pesticides and nitrates, the measured distribution of pesticide residues and nitrates in groundwater, and the hydrogeologic factors that influence groundwater contamination. One aspect of this survey included a county-level classification of groundwater vulnerability for all 3144 counties in the U.S. The agricultural (pesticide) DRASTIC system was utilized in a modified form to rank the vulnerability of groundwater to pollution for each county. This cursory classification survey was accomplished in a 3-month period; the form used in the survey is presented in Table 4.22 (Alexander and Liddle, 1986). As shown in Table 4.22, information from each county segment was aggregated into a county level by considering the percentage of the county characterized by the various ranges in the DRASTIC factors. The result of this study was a nationwide categorization of counties into those with high, medium, or low vulnerability to groundwater pollution.

Based upon the importance weights and typical ratings as described above, the range of possible pollution potential scores for the generic DRASTIC is 26 to 226; whereas, for the pesticide DRASTIC, it is 29 to 256. Conceptually, the greater the score for an area, the greater the potential for groundwater pollution. There are no standard subdivisions of the scores; however, Table 4.23 contains suggested divisions and associated map color codes (Aller, et al., 1987). The colors of the spectrum were chosen to show relative vulnerability to pollution. The warm colors (red, orange, and yellow) indicate areas with the potentially greatest vulnerability; the cool colors (blue, indigo, and violet) indicate areas of lower vulnerability to pollution. Two varying shades of green depict the middle ranges.

The developers of DRASTIC noted the following cautions regarding its usage (Aller, et al., 1987):

1. The user must remember that the methodology is neither designed nor intended to replace on-site investigations or to specifically site any type of facility or practice.
2. DRASTIC provides the user with a measure of relative groundwater vulnerability to pollution and, therefore, may be one of many criteria used in siting or other decisions, but should not be the sole criteria.
3. DRASTIC may be used for preventive purposes through the prioritization of areas where groundwater protection is critical. The system may also be used to identify areas where special attention or protection efforts are warranted.

Table 4.22 Data Collection Form for Agricultural (Pesticide) DRASTIC

State	County		FIPS Coding
Ground Water Region ____		Estimated % of County ____	
Hydrogeologic Setting ____ ____		Estimated % of County ____	

Ranges in DRASTIC Factors	Agricultural Weighted Ratings	Estimated % of County	Ranges in DRASTIC Factors	Agricultural Weighted Ratings	Estimated % of County
Depth to Water (ft)			**Topography (% slope)**		
0-5	50	____	0-2	30	____
5-10	45	____	2-6	27	____
10-15	40	____	6-12	15	____
15-30	35	____	12-18	9	____
30-50	25	____	18+	3	____
50-75	15	____	**Impact of Vadose Zone Media**		
75-100	10	____	SI/CL	4	____
100+	5	____	SH	12	____
Net Recharge (in)			LS	24	____
0-2	4	____	SS	24	____
2-4	12	____	LS, SS, SH (bedded)	24	____
4-7	24	____	SA & GVL w/SI & CL	24	____
7-10	32	____	META/IGN	16	____
10+	36	____	SA & GVL	32	____
Aquifer Media			BASALT	36	____
SH (massive)	6	____	LS (karst)	40	____
META/IGN	9	____	**Hydraulic Conductivity (gpd/sq. ft.)**		
META/IGN (wthd)	12	____	1-100	2	____
SS, LS, SH (in bed)	18	____	100-300	4	____
SS (massive)	18	____	300-700	8	____
LS (massive)	18	____	700-1000	12	____
SA & GVL	24	____	1000-2000	16	____
BASALT	27	____	2000+	20	____
LS (karst)	30	____	**Agricultural DRASTIC Score** ________		
Soil Media					
THIN OR ABSENT	50	____	**Index Variability** + ____ - ____		
GVL	50	____			
SA	45	____	**Data Confidence** ____		
CL (shrinking)	35	____	(3 = High, 2 = Medium, 1 = Low)		
LOAM (SA)	30	____			
LOAM	25	____	**By** ________		
LOAM (SI)	20	____	**Firm**	**Date** ____ 85	
LOAM (CL)	15	____			
CL (nonshrinking)	5	____	**Data Sources or Comments on back** –		

From Alexander, W. J. and Liddle, S. K., *Proceedings of the Conference on Agricultural Impacts on Groundwater*, August 1986, National Water Well Association, Dublin, OH, pp. 77–87. With permission.

DRASTIC, coupled with other factors such as application methods, may help delineate areas where pesticides or nitrates may pose a greater threat to groundwater.

4. DRASTIC may be used to identify data gaps that affect pollution potential assessment. For example, justification could be provided for further reconnaissance of the hydrogeologic parameter that would subsequently form a better database for future resource assessments or another DRASTIC analysis.
5. As with any model or classification scheme, it is possible to enter inaccurate or erroneous data that affect the reliability of the results. It is also possible to

Table 4.23 Suggested National Color Code for DRASTIC Index Ranges

DRASTIC index range	Color
<79	Violet
80–99	Indigo
100–119	Blue
120–139	Dark green
140–159	Light green
160–179	Yellow
180–199	Orange
>200	Red

From Aller, L., Bennett, T., Lehr, J. H., Petty, R. J., and Hackett, G., "DRASTIC: A Standardized System for Evaluating Ground Water Pollution Potential Using Hydrogeologic Settings," EPA/600/2-87/035, 1987, U.S. Environmental Protection Agency, Ada, OK, pp. 11–25, 68, and 101.

extend the system beyond the intended use of the methodology. For example, the use of DRASTIC to determine the vulnerability of groundwater to pollution by an injection well is an inappropriate use of the methodology. By directly injecting the contaminant into the aquifer, the opportunity for the pollutant to be attenuated by the physical factors included in DRASTIC are removed.

6. DRASTIC cannot be used to replace site-specific investigations or to preclude the consideration of particular factors that may be deemed important by a professional hydrogeologist at a site.

SOIL/AQUIFER FIELD EVALUATION (SAFE)

Another technique that can be used to determine the vulnerability of a groundwater system to pesticide contamination is referred to as the soil/aquifer field evaluation (SAFE) methodology (Roux, DeMartinis, and Dickson, 1986). The technique is somewhat more simple than DRASTIC, and is based on characterization of the soil layer and underlying unsaturated (vadose) and saturated zones in a given geographical area. The technique can also be used in nitrate groundwater pollution studies.

To define areas of potentially vulnerable aquifers, subsurface geology is evaluated on a state or county scale using published information. The general flow sequence of the sensitive aquifer determination is displayed in Figure 4.4 (Roux, DeMartinis, and Dickson, 1986). For those areas where pesticides (or nitrates) are used, the first step is to assemble appropriate hydrogeologic data and identify aquifers in terms of whether or not they are present and developable as a major water supply. As suggested in Figure 4.4, if the groundwater system is one that yields less than 50 gallons per minute, it is eliminated from review. For those groundwater systems that are potentially developable as a major water supply (greater than 50 gallons per minute yield), the next question is related to the existing water quality in the groundwater system. If the groundwater quality is not suitable for public supply based on the application of typical water quality

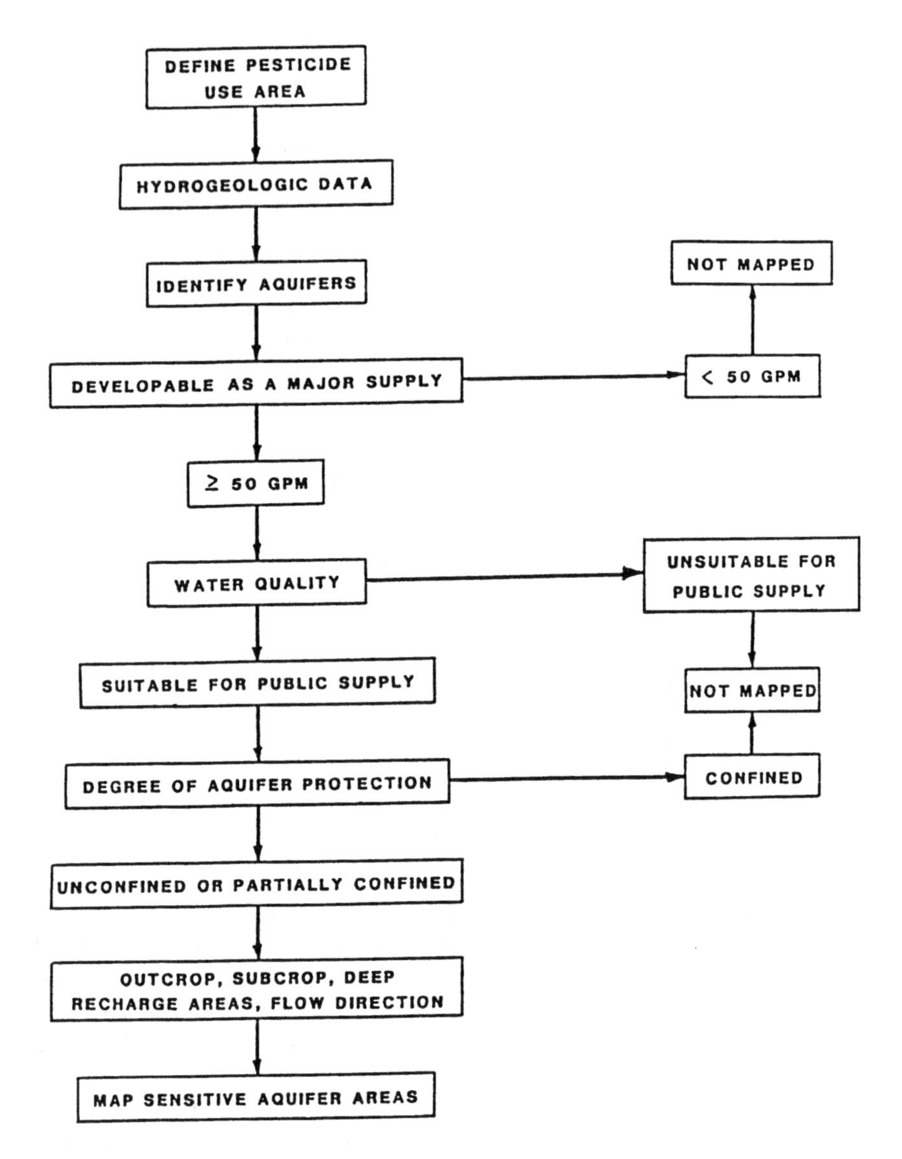

Figure 4.4 Sensitive aquifer determination in SAFE. (From Roux, P., DeMartinis, J., and Dickson, G., *Proceedings of the Conference on Agricultural Impacts on Ground Water,* National Water Well Association, Dublin, OH, 1986, pp. 145–158. With permission.)

criteria, then this system would also be eliminated from further consideration. However, if it is suitable for public supply, the next question is associated with the degree of aquifer (groundwater) protection. If the aquifer is confined, it would not be mapped as a sensitive aquifer in the geographical area. For those aquifer systems that are unconfined or only partially confined, the next question is related to recharge characteristics. Aquifers can be directly recharged from precipitation on their outcrop/subcrop areas, downward leakage from an overlying aquifer or surface water body, and upward leakage from a deeper, artesian aquifer. Principle routes for surface-supplied chemicals such as pesticides or nitrates to reach an aquifer are precipitation (or irrigation) from the outcrop area and flow from the shallow contaminated water table aquifer into the subcrop. Thus, the outcrop areas (and subcrop areas where they intersect shallow water table aquifers) of major aquifers are mapped in order to define sensitive aquifer areas (Roux, DeMartinis, and Dickson, 1986).

The second emphasis in this technique is related to the evaluation of sensitive (susceptible) soils. The important soil characteristics that may affect pesticide transport through the subsurface environment include permeability, thickness, pH, and temperature (Roux, DeMartinis, and Dickson, 1986). Permeability, because of the extremes in range, appears to be the most critical factor. Other important factors such as clay content, organic content, and moisture are generally related to permeability. Since permeability is a factor that can be mapped on a county scale from existing information, it is considered as an appropriate indicator in delineating sensitive soils.

The general approach for this sensitive soils evaluation is displayed in Figure 4.5 (Roux, DeMartinis, and Dickson, 1986). The determination of sensitive soils requires a step-wise procedure initiated by considering the pesticide (or fertilizer) use area and the associated soil data. If less than 20% of the soil associations are farmed, then this would not be considered within this technique. For those soil associations where greater than 20% is farmed, then major soil units and soil layers are identified along with the associated permeabilities. Sensitive farmed soil associations for pesticide use areas are ranked from 1 (least sensitive) to 5 (most sensitive) according to their comparison with Long Island, New York, soil types. Long Island soil types were used because detectable concentrations of agricultural chemicals have been found in the underlying groundwater. Table 4.24 delineates the sensitivity rankings of the soils: the higher the sensitivity rank, the greater the likelihood for groundwater contamination (Roux, DeMartinis, and Dickson, 1986).

Each soil layer within the soil unit is ranked according to its sensitivity, as delineated in Table 4.24. When each soil layer rank has been assigned, the sensitivity rank of the overall soil (typically composed of three layers) is calculated. For example, if the permeability ranks of all three layers of soil are 4, then the soil is ranked 4. However, if the individual layers have different sensitivity rankings, then the system takes the thicknesses and relative positions of the various layers into account (Roux, DeMartinis, and Dickson, 1986).

After each soil layer is ranked and an overall unit permeability is determined based on an analysis of the dominance of the rank assignment and the thickness of the individual layers, the overall soil associations in the study area are evaluated and one numerical rank is developed. The numerical rank for the overall soil associations is based on a geographically weighted average of the ranks for the individual soil units. At that point, soil associations that have final rankings between 1 and 3 are not considered as sensitive soils. Sensitive soils are those having a final ranking scale of 4 or 5.

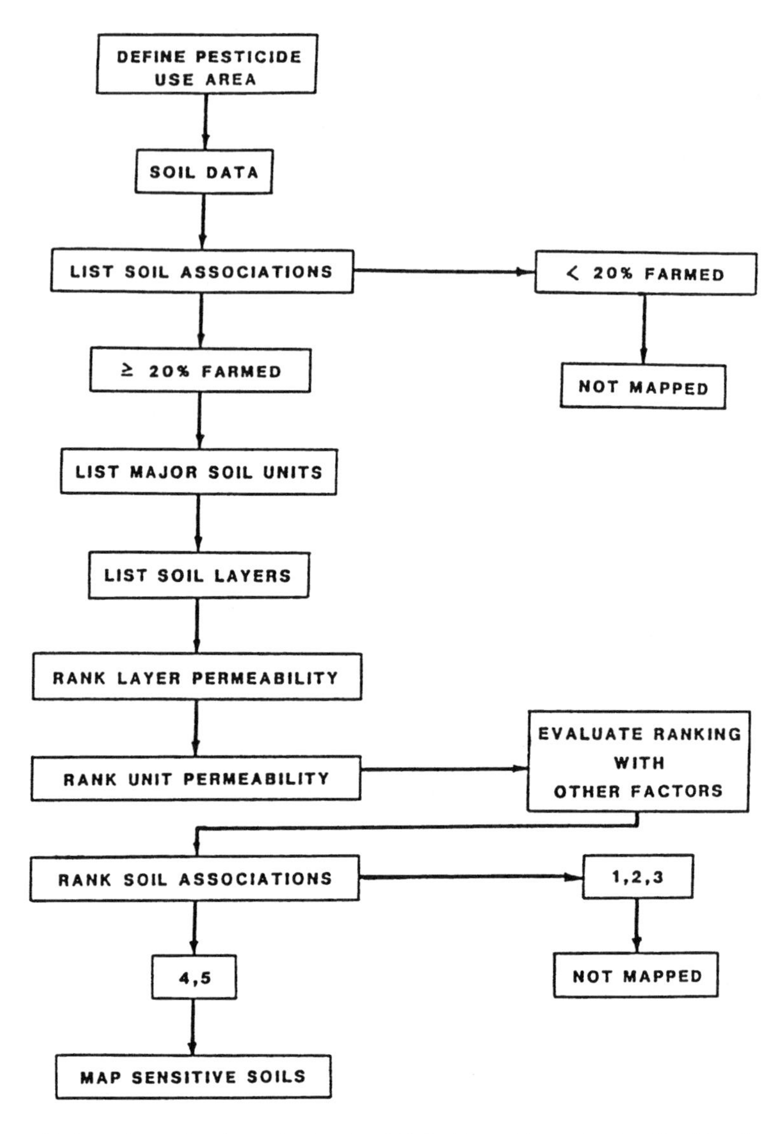

Figure 4.5 Sensitive soils determination in SAFE. (From Roux, P., De Martinis, J., and Dickson, G., *Proceedings of the Conference on Agricultural Impacts on Groundwater,* National Water Well Association, Dublin, OH, 1986, pp. 145–158. With permission.)

The end results of the aquifer and soil ranking processes are maps showing sensitive areas for the aquifers and sensitive areas for soils in the region being evaluated. The two maps can then be combined and the sensitive or vulnerable areas determined based on the overlap of areas of sensitive aquifers and soil types. This final vulnerability map can then be used to plan monitoring programs for pesticides or nitrates, either by assisting in the placement of new monitoring wells or in the selection of existing wells for monitoring in both vulnerable and non-

Table 4.24 Permeability and SAFE Sensitivity Ranks of Long Island Soils Used as a Standard for This Method

Soil unit	Class	Permeability (in./h)	Sensitivity rank
Atison	Medium sand	>6.3	5
Carver	Medium coarse sand	>6.3	5
Riverhead	Sandy loam (over sand)	2.0–6.3	4
Walpole	Loamy sand (over sand)	2.0–6.3	4
Bridgehampton	Fine sandy loam	0.63–2.0	3
Haven	Fine sandy loam within clay content <40%	0.63–2.0	3
Montauk Variant	Silty sandy loam	0.2–2.0	2
Raynham	Silty sandy loam with clay content >40%	0.2–2.0	2
Whitman	Light loam over fill	<0.63	1
Candice	Silty clayey loam	<0.20	1

Note: A rank of 5 is the most sensitive for pesticide migration to groundwater.

From Roux, P., DeMartinis, J., and Dickson, G., *Proceedings of the Conference on Agricultural Impacts on Ground Water*, National Water Well Association, Dublin, OH, 1986, pp. 145–158. With permission.

vulnerable areas (Roux, DeMartinis, and Dickson, 1986). Other factors, such as depth to groundwater, vertical hydraulic conductivity within the unsaturated zone, precipitation, irrigation, and so forth, could also be considered in selecting specific monitoring well locations.

ADVANTAGES AND LIMITATIONS RELATIVE TO VULNERABILITY MAPPING

Groundwater vulnerability to pollution may be defined as the sensitivity of its quality to anthropogenic activities that may prove detrimental to the present and/or intended usage value of the resource (Bachmat and Collin, 1987). Several limitations as well as advantages and needs of vulnerability mapping techniques can be delineated. Perhaps the primary advantage of vulnerability mapping is as a tool in screening, in a geographical context, those areas most susceptible to groundwater contamination. Screening can be used as an aid in the planning of monitoring programs directed toward determining worst-case conditions in a geographical study area. This is particularly important in developing information on nitrate groundwater quality in agricultural areas.

Ideally, vulnerability assessments should be tested against field observations of groundwater contamination. However, it is not possible to test vulnerability assessments in the same way that a field-scale simulation model can be tested. One difficulty is that vulnerability assessment methods typically yield an index value rather than a concentration. Thus, vulnerability, as treated in many methods, is not a property that can be directly measured in the field (Committee on Techniques for Assessing Ground Water Vulnerability, 1993). To compare predicted values of vulnerability with observed constituent concentrations, one must either know the history of contaminant loading to the subsurface or assume that the contaminant loading has been spatially and temporally uniform. One approach for evaluating a vulnerability assessment method is to compare the concentrations or the percent detections of one or more contaminants among different vulnerability classes predicted by the method.

The use of groundwater quality data to examine differences among vulnerability classes should be done with considerable caution for a number of reasons. Some of these reasons are (Committee on Techniques for Assessing Ground Water Vulnerability, 1993):

1. The production zone of the well may be quite different from the reference location of the vulnerability method.
2. Differences in groundwater quality observed among vulnerability classes may be an artifact of spatial and temporal variations in chemical loadings.
3. Short-circuiting of natural flow paths by movement down wells or their annuli can cause misleading results.
4. Contaminants introduced at or near the land surface may not have had sufficient time to reach the water table but may do so at a future date if they are sufficiently persistent and mobile.
5. Information on well construction features, condition of the well, and location of the sampling point relative to water distribution, storage, or treatment are needed to evaluate the suitability of the well for sampling the constituents of concern. This information is incomplete for many wells. Information on the location of open interval(s) and the hydrogeologic unit(s) to which the well is open also may be lacking for many wells.
6. Temporal variations in water quality may be a complicating factor, particularly for wells in shallow aquifers or with significant variations in pumping.
7. Limitations in the protocols used for water quality monitoring may cause considerable uncertainty in the measured concentrations of constituents that are to be compared to model predictions.

To serve as an example of field study difficulties, some comparisons of groundwater quality data with DRASTIC scores have been made using data from two national monitoring programs — the National Pesticide Survey and the National Alachlor Well Water Survey. In both studies, little association between contamination by agricultural chemicals and DRASTIC scores was found. The lack of association between contamination and the DRASTIC scores could reflect significant limitations in DRASTIC as a vulnerability assessment method; however, the lack of association may be related to the problems listed above in relating groundwater quality to vulnerability indices (Committee on Techniques for Assessing Ground Water Vulnerability, 1993).

Another issue of concern relative to vulnerability mapping is the maximum or minimum geographical area that should be mapped. Should this geographical area be defined based on average hydrogeological conditions for a small area such as several hundred acres, or can geographical areas as large as counties be mapped? For example, as noted earlier, the DRASTIC technique was developed as a screening tool for county-size areas down to a resolution of approximately 100 acres (Alexander and Liddle, 1986). In contrast, regional mapping of aquifer vulnerability and sensitivity has been accomplished for the conterminous U.S.

(Pettyjohn, Savoca, and Self, 1991). Additional consideration needs to be given to geographical limitations in the development of specific numerical indices or classifications.

It is important to realize that vulnerability mapping techniques assume uniform conditions within the subsurface environment, at least for the geographical area to be addressed. However, the subsurface environment is not uniform relative to hydrogeological factors. Pesticides and fertilizers are also not uniformly applied over time and space in a given geographical area. In addition, the subsurface transport and fate of nitrate can differ considerably from location to location. The nonuniformity of conditions does not mean that vulnerability mapping should not be used; however, it does suggest that very careful interpretation is needed for developed numerical indices and classification systems.

Another issue is related to the divisions between classifications used to identify areas highly susceptible to groundwater pollution, in contrast with those that are moderately or minimally susceptible to groundwater pollution. Again, it should be recognized that these arbitrary divisions are based on the collective judgment of the individuals developing the technique.

Despite these limitations, vulnerability maps can be used for regional planning in relation to land usage, planned developments, and applications of fertilizers and/or pesticides. Such maps should always be used together with an explanatory guide (Fobe and Goossens, 1990). Furthermore, the dynamic nature of groundwater depths and flow characteristics should be recognized, particularly if influenced by pumping activities.

Finally, uncertainty is inherent in all methods of mapping groundwater vulnerability. Uncertainties arise from errors in obtaining data, natural spatial and temporal variability of data, and in the mapping of data. Table 4.25 summarizes sources of errors into six classes (Committee on Techniques for Assessing Ground Water Vulnerability, 1993). Results of vulnerability assessments are usually displayed on a regional map depicting various subareas having different levels of vulnerability. The distinction between each level is usually arbitrary. Further, the estimates of vulnerability are associated with a level of uncertainty. The inability to distinguish differences between adjacent cells with differing vulnerability scores increases with increasing uncertainties in methods and data. As a result of the numerous uncertainties, Table 4.26 contains a series of research recommendations aimed at reducing uncertainty in vulnerability assessments and improving opportunities to use them effectively (Committee on Techniques for Assessing Ground Water Vulnerability, 1993).

SUMMARY

The natural subsurface environment can restrain the introduction of pollutants into groundwater; thus, numerous vulnerability mapping techniques described herein were developed to evaluate the capacity of the subsurface environment to attenuate pollutants. The parameters used to assess groundwater susceptibility to contamination are quite varied and overlapping. They include

Table 4.25 Sources of Errors in Groundwater Vulnerability Assessments

Sources of Errors
Errors in obtaining data
Accuracy in locating sites
Sample collection and handling
Laboratory preparation and analysis
Interpretation
Errors due to natural spatial and temporal variability
Random sampling error
Bias
Regionalization, extrapolation, interpolation
Scale effects, changes in variance due to averaging
Interpretation
Errors in computerization (digitizing) and storage of data
Data entry
Data age
Changes in storage format
Errors in programs to access data
Use of surrogate data and procedures
Adjustments in scale
Determining boundaries
Changes in representation of data
Interpretation
Data processing errors
Numerical, truncation, and round-off errors
Discretization errors
Problems in solution convergence
Interpretation
Modeling and conceptual errors
Process representation and coupling
Parameter identification
Scale effects
Interpretation
Output and visualization errors
Determination of boundaries
Classification into vulnerability categories
Interpretation

From Committee on Techniques for Assessing Ground Water Vulnerability, *Ground Water Vulnerability Assessment — Predicting Contamination Potential Under Conditions of Uncertainty,* National Research Council, National Academy Press, Washington, D.C., 1993, pp. 1–11 and 42–86. With permission.

but are not limited to soil media, topography, depth to groundwater, aquifer media, vadose zone materials, net recharge, hydraulic conductivity of aquifer, hydraulic gradient, distance to nearest drinking water supply, depth to bedrock, unsaturated and saturated zone permeability and thickness, net precipitation, sorption, soil organic content, and acid neutralization capacity. The selection of factors used in developing vulnerability mapping techniques is often limited by the availability of published data.

Several limitations and needs of vulnerability mapping techniques can be delineated. One of the primary limitations of vulnerability mapping techniques is based on the appraisal of soil and geological characteristics and the typical absence of verification based on results from field measurements. Vulnerability

Table 4.26 Research Recommendations Related to Reducing Uncertainty in Groundwater Vulnerability Assessments

- Develop a better understanding of all processes that affect the transport and fate of contaminants.
- Establish simple, practical, and reliable methods for measuring *in situ* hydraulic conductivities of the soil and the unsaturated and saturated zones. Develop methods for scaling measurements that sample different volumes of porous materials to provide equivalent measures. Develop simple, practical, and reliable methods for measuring *in situ* degradation rates (e.g., hydrolysis, methylation, biodegradation), and develop methods for characterizing changes in degradation rate as a function of other physical parameters (e.g., depth in soil).
- Develop improved approaches to obtaining information on the residence time of water along flow paths and identifying recharge and discharge areas.
- Develop unified ways to combine soils and geologic information in vulnerability assessments.
- Improve the chemical databases, currently the source of much uncertainty in vulnerability assessments.
- Detemine the circumstances in which the properties of the intermediate vadose zone are critical to vulnerability assessments and develop methods for characterizing the zone for assessments.
- Establish in the soil mapping standards of USDA's Soil Conservation Service an efficient soil sampling scheme for acquiring accurate soil attribute data in soil mapping unit polygons and documenting the uncertainty in these data. A need exists to better characterize the inclusions of other soil types in soil mapping units, including fractional area of included soil and distribution of inclusions.
- Establish reliable transfer functions for estimating *in situ* hydraulic properties using available soil attribute data (e.g., bulk densities, particle-size distributions, etc.). Develop ways to determine the additional uncertainty arising from the use of transfer functions in groundwater vulnerability assessments.
- Develop methods for merging data obtained at different spatial and temporal scales into a common scale for vulnerability assessment.
- Improve analytical tools in GIS software to facilitate integration of assessment methods with spatial attribute databases and the computing environment.
- Establish more meaningful categories of vulnerability for assessment methods. Determine which processes are most important to incorporate into vulnerability assessments at different spatial scales.
- Obtain more information on the uncertainty associated with vulnerability assessments and develop ways to display this uncertainty. Methods are needed that can identify and differentiate among more sources of uncertainty.
- Develop methods for accounting for soil macropores and other preferential flow pathways that can affect vulnerability. These investigations should include evaluations of the uncertainty in methods and measurements as they affect the assessment.
- Develop method for incorporating process-based, statistical, and qualitative information into an integrated or hybrid assessment.
- Identify counterintuitive situations leading to greater true vulnerability than commonly perceived. For example, develop greater understanding of the circumstances in which low-permeability materials that overlay aquifers can transmit contaminants to groundwater.

From Committee on Techniques for Assessing Ground Water Vulnerability, *Ground Water Vulnerability Assessment — Predicting Contamination Potential Under Conditions of Uncertainty*, National Research Council, National Academy Press, Washington, D.C., 1993, pp. 1–11 and 42–86. With permission.

categorizations are often made without taking into account the types of sources and types of pollutants. The evaluation of vulnerability in relation to the subsurface behavior of inorganic and organic pollutants makes it necessary to classify pollutants according to their chemical characteristics and environmental behavior. Also, vulnerability assessments should be periodically updated on the basis of new knowledge and data.

Finally, it should be noted that the most appropriate approach for assessing agricultural sources of groundwater pollution should involve the combined consideration of pollutant characteristics, fertilizer application rates and practices, hydrogeological factors basic to vulnerability assessment, and the usage of the groundwater resource. In other words, vulnerability mapping, while very useful, does not provide a complete assessment of the overall risks related to fertilizer applications and their significance as sources of nitrates in groundwater resources.

SELECTED REFERENCES

Alexander, W.J. and Liddle, S.K., "Ground Water Vulnerability Assessment in Support of the First Stage of the National Pesticide Survey," *Proceedings of the Conference on Agricultural Impacts on Ground Water* August, 1986, National Water Well Association, Dublin, Ohio, pp. 77–87.

Aller, L., Bennett, T., Lehr, J.H., Petty, R.J., and Hackett, G., "DRASTIC: A Standardized System for Evaluating Ground Water Pollution Potential Using Hydrogeologic Settings," EPA/600/2-87/035, May, 1987, U.S. Environmental Protection Agency, Ada, Oklahoma, pp. 11–25, 68, and 101.

Bachmat, Y. and Collin, M., "Mapping to Assess Ground Water Susceptibility to Pollution," in *Vulnerability of Soil and Ground Water to Pollutants*, Van Duijvenbooden, W. and Van Waegeningh, H.G., Eds., Proceedings and Information No. 38, International Conference, Noodwijk aan zee, National Institute of Public Health and Environmental Hygiene, The Hague, The Netherlands, 1987, pp. 297–307.

Canter, L.W., "Vulnerability Mapping of Ground Water Resources," *Proceedings of a National Symposium on Agricultural Chemicals and Ground Water Pollution Control*, Kansas City, Missouri, March 25–27, 1987, pp. 191–231.

Canter, L.W. and Sabatini, D.A., "Use of Aquifer Vulnerability Mapping in Ground Water Management," *Proceedings of the International Conference on Groundwater Resources Management*, Bangkok, Thailand, November 5–7, 1990, pp. 227–251.

Carter, A.D., Palmer, R.C., and Monkhouse, R.A., "Mapping the Vulnerability of Ground Water to Pollution from Agricultural Practice, Particularly with Respect to Nitrate," in *Vulnerability of Soil and Ground Water to Pollutants*, Van Duijvenbooden, W. and Van Waegeningh, H.G., Eds., Proceedings and Information No. 38, International Conference, Noodwijk aan zee, National Institute of Public Health and Environmental Hygiene, The Hague, The Netherlands, 1987, pp. 333–342.

Committee on Techniques for Assessing Ground Water Vulnerability, *Ground Water Vulnerability Assessment – Predicting Contamination Potential Under Conditions of Uncertainty*, National Research Council, National Academy Press, Washington, D.C., 1993, pp. 1–11 and 42–86.

DeSmedt, P., et al., "Ground Water Vulnerability Maps," *Aqua*, Vol. 5, 1987, pp. 264–267.

Edmunds, W.M. and Kinniburgh, R.E., "The Susceptibility of United Kingdom Ground Waters to Acidic Deposition," *Journal of the Geological Society*, Vol. 143, 1986, pp. 707–720.

Fobe, B. and Goossens, M., "The Groundwater Vulnerability Map for the Flemish Region: Its Principles and Uses," *Engineering Geology*, Vol. 29, No. 4, 1990, pp. 355–363.

Foster, S.S.D., "Fundamental Concepts in Aquifer Vulnerability, Pollution Risk and Protection Strategy," in *Vulnerability of Soil and Ground Water to Pollutants*, Van Duijvenbooden, W. and Van Waegeningh, H.G., Eds., Proceedings and Information No. 38, International Conference, Noodwijk aan zee, National Institute of Public Health and Environmental Hygiene, The Hague, The Netherlands, 1987, pp. 69–86.

Holmberg, M., Johnston, J., and Maxe, L., "Assessing Aquifer Sensitivity to Acid Deposition," in *Vulnerability of Soil and Ground Water to Pollutants*, Van Duijvenbooden, W. and Van Waegeningh, H.G., Eds., Proceedings and Information No. 38, International Conference, Noodwijk aan zee, National Institute of Public Health and Environmental Hygiene, The Hague, The Netherlands, 1987, pp. 373–380.

Jaffe, M. and Dinovo, F., *Local Ground Water Protection*, American Planning Association, Washington, D.C., 1987, pp. 141–160, 225–250.

LeGrand, H.E., "A Standardized Method for Evaluating Waste Disposal Sites," 1983, National Water Well Association, Worthington, Ohio.

Pettyjohn, W.A., Savoca, M., and Self, D., "Regional Assessment of Aquifer Vulnerability and Sensitivity in the Conterminous United States," EPA/600/S 2-91/043, October, 1991, U.S. Environmental Protection Agency, Ada, Oklahoma.

Porcher, E., "Ground Water Contamination Susceptibility in Minnesota," February, 1988, Minnesota Pollution Control Agency, St. Paul, Minnesota.

Ramolino, L., "Development of a Nitrate Pollution Index for Ground Water," Master's Thesis, 1988, University of Oklahoma, Norman, Oklahoma.

Ramolino, L. and Canter, L.W., "Aquifer Pollution Vulnerability Mapping," August, 1990, Environmental and Ground Water Institute, University of Oklahoma, Norman, Oklahoma.

Roux, P., DeMartinis, J., and Dickson, G., "Sensitivity Analysis for Pesticide Application on a Regional Scale," *Proceedings of the Conference on Agricultural Impacts on Ground Water*, National Water Well Association, Dublin, Ohio, 1986, pp. 145–158.

Sauriol, J.J., "Terrain Suitability Rating for the Attenuation of Septic System Effluents," *Proceedings of the Sixth National Ground Water Quality Symposium*, National Water Well Association, Dublin, Ohio, 1982, pp. 270–275.

Swanson, G.J., "Sweden Finds Answers in United States," *Water Well Journal*, April, 1990, pp. 60–61.

5 DEVELOPMENT OF A NITRATE POLLUTION INDEX

INTRODUCTION

Several of the vulnerability assessment techniques described in Chapter 4 related to geographically based mapping of calculated numerical indices (Committee on Techniques for Assessing Groundwater Vulnerability, 1993). The purpose of this chapter is to describe the formulation and testing of a specific Nitrate Pollution Index for groundwater in agricultural areas in the U.S. The chapter begins with conceptual information related to the development of an index. The major focus of the chapter is on the developed index and its testing via the use of groundwater nitrate data for three geographical areas in the U.S. The chapter concludes with a summary of the advantages and limitations of the developed Nitrate Pollution Index.

CONCEPTUAL FRAMEWORK FOR DEVELOPMENT OF AN INDEX

Several generic steps are associated with the development of numerical indices or classifications of environmental quality or media vulnerability or the pollution potential of man's activities (Canter, 1996). Figure 5.1 depicts five steps: (1) factor identification, (2) assignment of importance weights, (3) establishment of scaling functions or other methods for factor evaluation, (4) determination of the aggregation approach, and (5) application and field verification.

Factor identification basically consists of delineating key factors that can be used as indicators of environmental quality, or susceptibility to pollution, or the pollution potential of the source type. Regarding a pollution index related to the nitrate concentration in groundwater, the central focus should be on influencing factors related to societal activities and the hydrogeological setting. Factor identification should be based on the collective professional judgment of knowledgeable individuals relative to the environmental media or pollution source category. Organized procedures such as the Delphi approach can be used to aid in the solicitation of this judgment and the aggregation of the results (Linstone and Turoff, 1978); another approach could involve the assemblage of case study

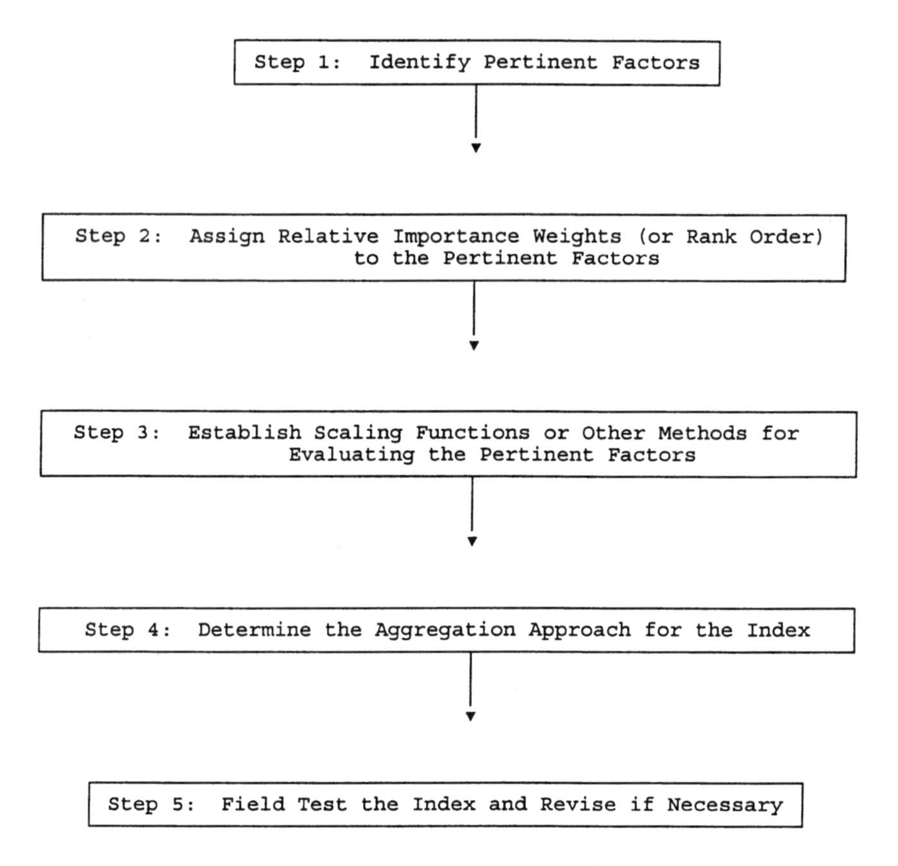

Figure 5.1 Generic steps for developing an environmental index.

information related to the issue being addressed. For the Nitrate Pollution Index described herein, a review was conducted of 22 case studies in terms of contributing factors to nitrate pollution of groundwater.

The second step in the development of an index is the assignment of relative importance weights to the environmental media and/or source-transport factors, or at least the arrangement of them in a rank ordering of importance. Some importance weighting or ranking techniques that could be used to achieve this step include: professional judgment, the Delphi approach, unranked pairwise comparisons, multi-attribute utility measures, ranking, rating, and the nominal group process (Canter, 1996). The techniques used for the index described herein were based on professional judgment derived from relative risk considerations.

Several approaches have been used to scale or evaluate the data associated with factors in index methodologies. Examples of techniques for scaling or evaluation for this purpose include the use of: (1) linear scaling or categorization based on the range of data, (2) letter or number assignments designating data categories, (3) functional curves, or (4) the unranked paired comparison technique. The development of scaling or evaluation approaches should be based on the collective professional judgment of knowledgeable individuals relative to the environmental media or pollution source category. The approaches can be based on published approaches used by others, or on the application of structured techniques such as the nominal group process or the Delphi approach. Data categorization and published approaches used by others were utilized in the Nitrate Pollution Index described herein.

Aggregation of the information on the weighted and scaled (or evaluated) factors into a final numerical index (or classification) is the important final step in the development of the index (or classification). The aggregation may include simple additions, multiplication, and/or the use of power functions. Details on the features of various aggregation approaches are given by Ott (1978). As a minimum, the collective professional judgment of knowledgeable substantive area individuals should be used. The aggregation approach used for the Nitrate Pollution Index involved multiplication of importance weights by factor ratings, and their summation into a final numerical score.

A final step in the development of an index (or classification) should include field verification of its applicability. This may involve simple to complex data collection and statistical testing. At a minimum, the useability of the index should be explored in terms of data needs and the availability of such data. Statistical testing based on field data was used for the Nitrate Pollution Index.

IDENTIFICATION OF PERTINENT FACTORS

A case study approach was used as the basis for the identification of pertinent factors for the Nitrate Pollution Index. Factors delineated as having an influence on nitrates in groundwater in agricultural areas were the focus of this approach. A total of 22 case studies were used for documenting nitrate behavior under different conditions of farming production such as climate, soil type, cropping pattern, and fertilizer type, amount, and timing of application. Table 5.1 summarizes the key features of the 22 case studies in terms of eight potential influencing factors and the resultant nitrate-nitrogen concentration in groundwater. In addition to land use, the eight factors in Table 5.1 include fertilization intensity, extent of irrigation, soil type, depth to water table, crop type, depth of wells, confined or unconfined aquifer, and aquifer environmental conditions (oxidized zone). The groundwater nitrate-nitrogen concentration classifications are expressed as very low (<3 mg/l), low (3 to 10 mg/l), high (10 to 20 mg/l), and very high (>20 mg/l). A tabular display of the eight factors and several other influencing factors mentioned in the 22 case studies is included in Table 5.2 (Ramolino, 1988).

A risk-based concept can be used for estimating the risk to groundwater of nitrate pollution resulting from agricultural activities. Risk is defined herein as related to the probability of nitrate-nitrogen in groundwater exceeding 3 mg/l. The 3-mg/l value is considered as the approximate concentration beyond which human activities could be contributing nitrogenous compounds to the groundwater (Madison and Brunett, 1984). However, it should be noted that concentrations of 3 mg/l or more have been found to occur naturally in some areas.

Table 5.1 Summary of 22 Case Studies

Case study	Land use	Factors[a]								Groundwater NO_3–N concentration[b]
		F_1	F_2	F_3	F_4	F_5	F_6	F_7	F_8	
Case A	Cropland	N fertilizer	Irrigation water applied	Sandy loam	Depth to water table, 36 m					High
	Cropland	None	Irrigation water applied	Sandy	Depth to water table, 35 m	Deep rooted crop (alfalfa)				Very low
	Animal industry	Cattle feed lot	Average annual precipitation, 0.45 in.	Sandy	Depth to water table, 48 m					Low
	Rangeland pasture	None	Average annual precipitation, 0.40 in.	Sandy loam	Depth to water table, 39 m					Very low
	Dairy herd	Pit storage	Large quantity of water for dairy operation	Sandy loam	Depth to water table, 34 m					Very high
Case B	Cropland	N fertilizer rate, 224–286 kg N/ha	Small amount of rainfall	Poorly drained soil	Shallow depth to water table		Shallow well			Low
	Cropland	N fertilizer rate, 224–286 kg N/ha	Large amount of rainfall	Poorly drained soil	Shallow depth to water table		Shallow well			High
	Cropland	N fertilizer rate, 58–186 kg N/ha	Large amount of rainfall	Poorly drained soil	Shallow depth to water table		Shallow well			Very low
	Cropland	N fertilizer rate, 58–186 kg N/ha	Large amount of rainfall	Moderately well drained	Shallow depth to water table		Shallow well			Low

Table 5.1 Summary of 22 Case Studies (Continued)

Case study	Land use	Factors[a]								Groundwater NO_3–N concentration[b]
		F_1	F_2	F_3	F_4	F_5	F_6	F_7	F_8	
	Cropland	N fertilizer rate, 58–186 kg N/ha	Small amount of rainfall	Moderately to well-drained soil	Shallow depth to water table		Shallow well			Very low
Case C	Cropland	N fertilizer rate, 322 kg N/ha	Irrigation water applied, 101 cm	Well-drained soil	Shallow depth to water table	Shortly rooted crop (potatoes)				Very high
	Cropland	N fertilizer rate, 223 kg N/ha	Irrigation water applied, 83 cm	Well-drained soil	Shallow depth to water table	Shortly rooted crop (potatoes)				Very high
Case D	Cropland	N fertilizer					Shallow well			Low
	Cropland	N fertilizer					Deep well			Very low
Case E	Cropland	N fertilizer		Well-drained soil	Shallow depth to water table		Shallow well	Unconfined aquifer		High
	Cropland	N fertilizer		Well-drained soil	Shallow depth to water table		Deep well	Unconfined aquifer		Low
	Cropland	N fertilizer		Well-drained soil			Deep well	Confined aquifer		Very low
	Septic tank Chicken house	Animal waste			Depth water table		Deep well	Unconfined aquifer		Very high

Table 5.1 Summary of 22 Case Studies (Continued)

Case study	Land use	Factors[a]								Groundwater NO_3–N concentration[b]
		F_1	F_2	F_3	F_4	F_5	F_6	F_7	F_8	
Case F	Cropland	N fertilizer		Well-drained soil			Shallow well	Unconfined aquifer		High
	Cropland	N fertilizer		Well-drained soil			Deep well	Unconfined aquifer		High
	Cropland	N fertilizer		Poorly drained soil				Confined aquifer		Very low
	Cropland	N fertilizer		Well-drained soil				Unconfined aquifer	Very large aquifer oxidized zone	High
	Woodland	None	Rainfall	Poorly drained				Confined aquifer	Very thin aquifer oxidized zone	Low
Case G	Woodland	None	Rainfall	Well-drained soil	Depth to water table, 3–12 m		Well depth less than 6 m	Unconfined aquifer		Very low
	Permanent pasture	None	Irrigation water applied	Poorly drained soil	Depth to water table, 3–12 m		Well depth less than 6 m	Unconfined aquifer		Very low
	Cropland	N fertilizer rate, 168–247 kg/ha	Large amount of water applied	Well-drained soil	Depth to water table, 3–12 m	Shortly rooted crop (potatoes)	Well depth less than 6 m	Unconfined aquifer		Very high

Table 5.1 Summary of 22 Case Studies (Continued)

Case study	Land use	Factors[a]								Groundwater NO_3–N concentration[b]
		F_1	F_2	F_3	F_4	F_5	F_6	F_7	F_8	
	Cropland	N fertilizer rate, 168–247 kg/ha	Large amount of water applied	Well-drained soil	Depth to water table, 3–12 m	Shortly rooted crop (potatoes)	Well depth more than 6 m	Unconfined aquifer		Low
Case H	Forest land	None	Large amount of water applied	Well-drained soil	Shallow depth to water table					Very low
	Cropland	Over-fertilized	Large amount of water applied	Well-drained soil	Shallow depth to water table					Very high
	Undisturbed land	None	Large amount of water applied	Well-drained soil	Shallow depth to water table					Very low
Case I	Cereal field	N fertilizer rate, 34–54 kg/ha	Rainfall, 51 cm	Well-drained soil						Low
	Cereal field	N fertilizer rate, 34–109 kg/ha	Rainfall, 56 cm	Well-drained soil						High
	Cereal field	N fertilizer rate, 107–113 kg/ha	Rainfall, 65 cm	Well-drained soil						Very high
	Cereal field	N fertilizer rate, 107–113 kg/ha	Rainfall, 66 cm	Poorly drained soil						Low

Table 5.1 Summary of 22 Case Studies (Continued)

Case study	Land use	Factors[a]								Groundwater NO_3–N concentration[b]
		F_1	F_2	F_3	F_4	F_5	F_6	F_7	F_8	
	Cereal field	N fertilizer rate, 107–113 kg/ha	Rainfall, 66 cm	Poorly drained soil with crack planes						Very high
Case J	Forest land	None	Rainfall	Well-drained soil						Very low
	Cropland	N fertilizer	Irrigation water applied	Well-drained soil						High
	Cropland	N fertilizer	Irrigation water applied	Poorly drained soil						Very low
Case K	Forest land	None	Rainfall	Well-drained soil						High
	Cropland	N fertilizer	Irrigation water applied	Well-drained soil						High
	Cropland	N fertilizer	Irrigation water applied	Poorly drained soil						Very low
Case L	Cropland	N fertilizer	Irrigation water applied	Well-drained soil						Very high

Table 5.1 Summary of 22 Case Studies (Continued)

Case study	Land use	Factors[a]								Groundwater NO_3–N concentration[b]
		F_1	F_2	F_3	F_4	F_5	F_6	F_7	F_8	
	Cropland and habitation	N fertilizer	Dry season	Well-drained soil						High
	Cropland	N fertilizer	Heavy rainfall flushing N amount out of soil profile	Well-drained soil						Very high
Case M	Cropland	N fertilizer	Annual average precipitation, 80 cm	Well-drained soil	Shallow depth to water table, 40–200 cm					Very high
	Cropland	Slurry and manure	Annual average precipitation, 80 cm	Well-drained soil	Shallow depth to water table, 40–200 cm					Very high
Case N	Cropland	N fertilizer		Well-drained soil			Depth of well less than 25 m			High
	Cropland	N fertilizer		Well-drained soil			Depth of well more than 35 m			Low
Case O	Cornfield	N fertilizer rate, 179 kg/ha; single application at planting period	Irrigation water applied	Well-drained soil	Depth to water table, 4.5 m					High

Table 5.1 Summary of 22 Case Studies (Continued)

Case study	Land use	Factors[a]								Groundwater NO_3–N concentration[b]
		F_1	F_2	F_3	F_4	F_5	F_6	F_7	F_8	
	Cornfield	N fertilizer rate, 179 kg/ha; splitting application at growing period	Irrigation water applied	Well-drained soil	Depth to water table, 4.5 m					Low
Case P	Cropland	Over-fertilized	Irrigation water applied	Well-drained soil	Depth to water table, 3 m	Alfalfa				Very high
	Cropland	None	Irrigated with wastewater	Well-drained soil	Depth to water table, 75 m					Very high
	Cropland	Over-fertilized	Irrigation water applied	Well-drained soil	Depth to water table, 75 m					Low
Case R	Cropland	Over-fertilized	Heavily irrigated	Well-drained to excessively drained soils	Depth to water table, less than 3 m		Depth of well less than 22 m			Very high
	Undisturbed land	None	Rainfall	Well- to excessively well-drained soil	Depth to water table, less than 3 m		Depth of well less than 22 m			Very low

Table 5.1 Summary of 22 Case Studies (Continued)

Case study	Land use	Factors[a]								Groundwater NO_3–N concentration[b]
		F_1	F_2	F_3	F_4	F_5	F_6	F_7	F_8	
	Cropland	Over-fertilized	Heavily irrigated	Poorly drained soil	Depth to water table less than 3 m		Depth of well less than 22 m			Low
Case S	Cropland	N fertilizer applied	Annual average precipitation, 110 cm	Well-drained soil	Depth to water table, 1.5–4.6 m		Depth of well less than 18 m			High
	Cropland	N fertilizer applied	Annual average precipitation, 110 cm	Well-drained soil	Depth to water table, 1.5–4.6 m		Depth of well more than 18 m			Low
	Cropland	N fertilizer applied	Annual average precipitation, 110 cm	Poorly drained soil	Depth to water table, 1.5–4.6 m		Depth of well more than 18 m			Very low
Case T	Cropland	N fertilizer	Rainfall	Poorly drained soil				Confined aquifer		Very low
	Cropland	N fertilizer	Rainfall	Well-drained soil				Unconfined aquifer		High
Case U	Cropland	None						Unconfined aquifer	Reducing vadose zone	Very low
	Cropland	N fertilizer						Unconfined aquifer	Oxidizing vadose zone	High

Table 5.1 Summary of 22 Case Studies (Continued)

Case study	Land use	Factors[a]								Groundwater NO_3–N concentration[b]
		F_1	F_2	F_3	F_4	F_5	F_6	F_7	F_8	
Case V	Cropland	N fertilizer						Confined aquifer	Reducing aquifer	Very low
	Cropland	N fertilizer						Unconfined aquifer	Oxidizing aquifer	High
Case W	Experimental field	N fertilizer and bromide injected into aquifer		Sandy soil	Depth to water table, 4 m			Unconfined aquifer	Decrease of nitrate/bromide ratio (denitrification process)	Low

[a] F_1 = fertilization; F_2 = irrigation; F_3 = soil type; F_4 = depth to water table; F_5 = crop type; F_6 = depth of wells; F_7 = confined or unconfined aquifer; F_8 = aquifer environmental conditions.

[b] Very low, NO_3–N concentration <3 mg/l; Low, NO_3–N concentration 3–10 mg/l; High, NO_3–N concentration 10–20 mg/l; Very high, NO_3–N concentration >20 mg/l.

From Ramolino, L., "Development of a Nitrate Pollution Index for Ground Water," Master's Thesis, 1988, University of Oklahoma, Norman, pp. 138–262. With permission.

Table 5.2 Factors Influencing Nitrate in Groundwater as Mentioned in 22 Case Studies

Factors	Case A	Case B	Case C	Case D	Case E	Case F	Case G	Case H	Case I	Case J	Case K	Case L	Case M	Case N	Case O	Case P	Case R	Case S	Case T	Case U	Case V	Case W
N fertilizer intensity and rate	X	X	X	X	X	X	X	X	X	X	X	X	X	X	X	X	X	X	X	X	X	X
Time of N application		X										X			X							
Type and source of N fertilizer	X	X	X	X	X	X	X	X	X	X	X	X	X	X	X	X	X	X	X	X		
N plant uptake	X	X	X				X		X	X	X	X	X		X	X	X	X				
Cropping pattern	X	X	X			X	X	X	X	X	X	X	X	X	X	X	X	X	X	X	X	X
Climate	X	X	X	X	X	X	X	X	X	X	X	X	X	X	X	X	X	X	X	X	X	X
Irrigation rate and quantity	X	X	X	X	X		X	X	X	X	X			X	X	X	X	X				
Soil texture	X	X	X	X	X	X	X	X	X	X	X	X	X	X	X	X	X	X	X	X	X	X
Soil organic matter												X	X				X	X				
Depth to water table	X	X	X	X	X	X	X	X				X	X		X	X	X					
Depth of well	X	X	X	X	X	X	X			X	X	X	X	X			X					
Type of aquifer	X	X			X	X	X			X	X	X	X	X		X	X	X	X		X	X
Aquifer lithology				X		X				X	X	X					X		X	X	X	X
Transport process	X		X	X	X	X	X	X		X	X	X	X						X		X	X
Biochemical process		X		X	X	X	X			X	X	X	X				X	X	X	X	X	X

From Ramolino, L., "Development of a Nitrate Pollution Index for Ground Water," Master's Thesis, 1988, University of Oklahoma, Norman, pp. 138–262. With permission.

Preliminary List of Factors

Based upon the information in Tables 5.1 and 5.2, and the collective information in Chapters 1 through 3, the following eight key factors (refined from the lists in Tables 5.1 and 5.2) were identified as most likely to exert an influence on nitrates in groundwater: (1) fertilizing intensity, (2) time of fertilizer application, (3) nitrogen fertilizer use efficiency, (4) rainfall patterns and irrigation practices, (5) soil texture, (6) depth to water table and type of aquifer, (7) well depth, and (8) chemical environment.

Fertilizing intensity is based on the fact that in many cases a relationship has been noted between the amount of nitrate leached and the amount of fertilizer applied. If the amount of fertilizer applied is in excess of that required by crops, nitrate available from oxidation may be leached from the root zone. This, in turn, could increase nitrate concentrations in groundwater.

Proper timing of fertilizer application can be used to reduce nitrate leaching toward groundwater. Nitrogen fertilizer should be applied as close as possible to the time of use by the crops in order to achieve maximum use by the crop and minimum accumulation and losses. If the nitrogen application does not exceed crop needs, there will be little nitrate available for leaching. However, when the same amount is applied too early in the season, leaching from the root zone may occur because of either an early rainy season or early irrigation practices. Therefore, the optimum timing for fertilizer application should be determined by considering the need to make nutrients available over the proper period for the crop.

Nitrogen fertilizer use efficiency (NUE) can be defined as the way crops recover nitrogen fertilizer that is applied to the soil (Barber, 1975). This agronomic NUE is also important from an environmental perspective, since nitrogen that is not utilized by crops can be lost from soil-plant systems and can adversely affect the environment. Nitrogen lost from soil-plant systems may contribute to groundwater nitrate contamination or eutrophication of surface water (Keeney, 1982). The NUE is sometimes limited by a crop's ability to recover nitrogen that remains in a chemically available form in the root zone. This may be due to either the distance between the nitrogen and the roots combined with the rate of nitrogen movement to the roots, or perhaps a lower than optimum rate of nitrogen absorption by the roots (Barber, 1975).

There are other factors that can strongly influence the NUE. For example, climatic factors that determine growth rate will also influence the rate of nitrogen uptake from the soil. Various soil properties affecting the retention of water and plant responses through proliferation of roots in zones of high water and nitrogen availability can influence the plant uptake. The greatest amount of nitrogen is typically required by crops in their early stages of development; therefore, the timing of fertilizer application can maximize the efficiency of crop uptake and minimize nutrient losses by leaching. Finally, any factor that affects crop growth influences the physiological efficiency since plant growth and nitrogen uptake are closely related.

Accurate estimates of crop water use are essential for *irrigation scheduling* in relation to typical *rainfall patterns*. When applied water (rainfall or irrigation) exceeds the crop evapotranspiration, most of it is lost from the root zone by deep percolation. The rate at which nitrate is lost from the soil as a result of downward movement depends upon the volume of water that moves downward and the concentration of nitrate in that water. Coarse-textured soils will not retain as much water in the root zone as poorly drained soils, and nitrogen is more susceptible to leaching below the root zone in sandy soils.

Variations of nitrate concentrations in groundwater have also been shown to be related to soil drainage characteristics. Permeability and water holding capacity of a soil affect the amount of water passing through the root zone and, indirectly, the movement of solutes (i.e., nitrate) carried in it. The water holding capacity of a soil is dependent on its texture, structure, organic matter content, and apparent bulk density. *Soil texture* refers to the relative percentage of sand, silt, and clay. Soils containing a large amount of sand are sandy; those with a high content of silt are silty; and soil that does not exhibit the properties of either sand, silt, or clay is called loam. The more common soil textural names listed in order of increasing fineness are: sand, loam sandy, sandy loam, silt loam, silt, sand clay, clay loam, silty clay loam, sandy clay, silt clay, and clay. Generally, the larger the percentage of fine particles in a soil, the higher the water holding capacity.

Areas with poorly drained soils may have a lower nitrate concentration, either because poorly drained soil may block its entrance into the aquifer or because poorly drained soil is associated with a reducing environment that promotes denitrification. In addition, roots in poorly drained soils can penetrate deeper and more plant nitrogen uptake can take place. In sandy soils, root depths rarely exceed 40 to 60 cm, and nitrogen below this is exposed to leaching because it is unavailable to the roots. Nevertheless, it should be noted that despite clear differences between nitrate concentrations under different soil types, nitrate may be present in all soil groups. The fundamental difference between soils is the rate at which nitrate leaching occurs and the time it takes to reach groundwater. The transit time of the leachate is related to the hydraulic conductivity of the soil profile, the degree of soil saturation, and the depth to the water table.

Groundwater contamination due to the use of nitrogen fertilizers in irrigated areas that have sandy soils and a shallow depth to the water table has been documented in many of the 22 case studies. *Depth to the water table* is an important factor because it determines the depth of subsurface material through which nitrates must travel before reaching the aquifer, and the amount of time during which nitrate contact with the surrounding media is maintained. The processes contributing to changes in the concentration of nitrates in the vadose zone include ion exchange, plant uptake, dispersion-convection, and nitrogen transformations which may consist of nitrification, denitrification, mineralization, or immobilization. In general, the relationship between nitrate concentration and depth to the water table can be considered as a function of time.

If the groundwater system is confined, nitrate may be lower because the entrance into the aquifer has been retarded or prevented by the confining layer, or it may have been diluted as it moves along the flow path. In areas where the aquifer is unconfined or where the confining layer is not continuous, recharge to the aquifer can occur readily, and the nitrate pollution potential is enhanced.

Another factor affecting the magnitude of nitrate concentrations in groundwater is the general inverse relationship between nitrate concentration and *well depth*. High nitrate concentrations at or near the water table and declining with depth have been observed in many instances and interpreted as an indication of denitrification. Even though the nitrate concentration in the saturated zone tends to decrease with increasing depth, in unconfined aquifers high nitrate concentrations can reach deeper zones. However, the presence of significant components of vertical groundwater flows, the absence of clay or silt clay confining layers, and nearby sources of nitrate favor elevated nitrate concentrations in deeper portions of saturated zones.

Nitrate concentrations tend to vary inversely with the *concentration of some chemical constituents* characteristic of reducing environments; examples include organic nitrogen, dissolved iron and dissolved silica. If the environment is reducing, nitrate concentrations in groundwater may be low because the nitrate has been denitrified. Biological denitrification is one of the processes responsible for the attenuation of nitrate in the vadose zone and for low nitrate levels in the groundwater of some hydrogeological environments.

The amount of organic matter rather than oxygen is considered to be the most important ecological factor that determines denitrification in natural systems. The higher the demand for electron acceptors, the greater are the chances for denitrification even under aerobic conditions (Ottow and Fabig, 1985). In the presence of large amounts of easily decomposable organic matter, the demands for external electron acceptors may greatly surpass the supply of oxygen and the microflora will switch to other electron acceptors such as nitrate, and manganese (Mn IV) and iron (Fe III) oxides.

Final Selection of Factors

As illustrated by the discussion of the eight key factors, the probability of nitrates entering groundwater is a function of a complex set of interactive hydrogeological, physical, and biochemical factors. Although all eight factors and even others are important for completely addressing the risk of groundwater pollution by nitrates, a smaller set of four key factors was identified; they include (1) nitrogen fertilizer application, (2) soil texture, (3) net recharge from rainfall or irrigation water, and (4) depth to the water table. The selection of the four final factors was based on the following considerations (Ramolino, 1988):

1. They play a primary role in controlling the magnitude of nitrate leaching. If nitrogen fertilizer is applied as a function of crop needs, there will be minimal nitrate leaching below the root zone. On the other hand, if nitrogen is applied in excess of that which the crop can utilize, the probability of nitrates entering the groundwater can be high; however, the nitrate concentration reaching the

groundwater is strongly controlled by the amount of water moving in the soil, the soil water holding capacity, and the distance to the water table.
2. They can be used as common factors to assess agricultural impacts on nitrates in groundwater at any site in any geographical area.
3. They can be used to compare different sites in the evaluation of alternative nitrogen management practices.
4. There is a minimal need for monitoring networks. Use of the index represents an economical and quick way to predict nitrate pollution potential.
5. Data on the four factors are available and relatively easy to procure and evaluate.
6. Data on the four factors can be communicated easily to individuals with diverse backgrounds and levels of expertise.
7. Preventive control measures can be based on the four factors; knowledge about these factors can help in planning management practices.
8. Data on the four factors can help determine and characterize zones of contamination and aquifer vulnerability.
9. The four factors are associated directly or indirectly with several other factors (crop factors, cropping pattern, and physical, biological, and chemical factors). Their selection does not completely exclude the relative importance of other factors.
10. The selection of the four factors (nitrogen fertilizer intensity, soil texture, net recharge, and depth to groundwater) represents an initial and simple way to predict or evaluate quickly, and with minimum data, the risk associated with nitrogen leaching. Other models that consider additional factors could be developed. Examples of such mathematical models are described in Chapter 6.

ASSIGNMENT OF RELATIVE IMPORTANCE WEIGHTS

The second step in the development of the Nitrate Pollution Index required the assignment of importance weights to the four selected risk factors. While the four factors (nitrogen fertilizer intensity, soil texture, net recharge, and depth to groundwater) are interconnected in the nitrate groundwater pollution process, they do not have the same relative importance. Therefore, a relative importance weight was assigned to each factor to reflect its importance with respect to the others in forecasting the magnitude of nitrate pollution of groundwater.

The assignment of importance weights was based on professional judgment derived from the results of the general literature review and the 22 case studies mentioned earlier. There are positive relationships between nitrate concentrations in groundwater and each of the four selected factors or combinations thereof. The relative importance of each factor is a function of its relative influence on the entrance of nitrates into groundwater. The potential nitrate pollution risk related to each of the four factors is depicted in Table 5.3 (Ramolino, 1988). A score was assigned to each level of risk as follows: high risk, 10; moderate to high risk, 9; moderate risk, 7; low to moderate, 5; low risk, 3; very low risk, 1. The resultant range of the total score (lowest to highest) for each factor reflects its relative importance in influencing the nitrate concentration reaching ground-

Table 5.3 Risk Analyses Matrix

	Nitrogen fertilizer		Net recharge		Soil texture		Depth to water table		
	Over-fertilized	Meets crop requirements	>FC[a]	FC	Well-drained (WD)	Poorly-drained (PD)	Shallow (S)	Deep (D)	Total score
Nitrogen fertilizer									
Worst (over)	—		10[b]		10		10		30
Best (meets)		—		1		9		7	17
Net recharge									
Worst (>FC)	10		—		10		10		30
Best (FC)		1		—		5		5	11
Soil texture									
Worst (WD)	9		10		—		10		29
Best (PD)		1		1		—		7	9
Depth to water table									
Worst (S)	9		10		9		—		28
Best (D)		1		1		5		—	7

[a] FC = field capacity of soil.

[b] High risk = 10; moderate to high risk = 9; moderate risk = 7; low to moderate risk = 5; low risk = 3; and very low risk = 1.

From Ramolino, L., "Development of a Nitrate Pollution Index for Ground Water," Master's Thesis, 1988, University of Oklahoma, Norman, pp. 138–262. With permission.

water. A final relative importance weight was then assigned to each of the four factors as shown in Table 5.4; the weight was based on 5.0 as a maximum and rounding to the nearest 0.5 based on proportionality (Ramolino, 1988).

Table 5.4 Final Assignment of Importance Weights to Four Factors

Factor	Range of total score	Importance weight[b]	Description
Nitrogen fertilization	17–30 (23.5)[a]	5.0	Nitrogen fertilizer application is the most important factor contributing to nitrate in groundwater under agricultural land use
Net recharge	11–30 (20.5)	4.5	Water (net recharge from irrigation and/or rainfall) contributes to the downward movement of nitrate in soils; without excess water moving through the soil profile, there will be minimal nitrate leaching below the root zone
Soil texture	9–29 (19.0)	4.0	The texture of a soil has a very important influence on the flow of soil water, the circulation of air, and the rate of chemical and biochemical transformations
Depth to water table	7–28 (17.5)	3.5	The depth to water table is the expression of thickness of the vadose zone and the time to reach groundwater; nitrate levels in groundwater will depend on the processes occurring in the vadose zone, such as transport and biochemical processes

[a] Numbers in parentheses reflect average of range.
[b] Based on 5.0 being assigned to 23.5, and the remainder proportioned to the nearest 0.5 based on their average of the range of total score.

From Ramolino, L., "Development of a Nitrate Pollution Index for Ground Water," Master's Thesis, 1988, University of Oklahoma, Norman, pp. 138–262. With permission.

ESTABLISH METHODS FOR EVALUATING PERTINENT FACTORS

Each of the four selected factors in the Nitrate Pollution Index must be rated based on site data. The potential data for each factor were divided into ranges that could be evaluated with respect to pollution potential and assigned a rating that varied between 1 and 10. The results of this approach are shown in Tables 5.5 through 5.8 (Ramolino, 1988). The ratings for net recharge and depth to groundwater were based on the DRASTIC methodology described in Chapter 4 (Aller, et al., 1987). The ranges and ratings for soil texture were based on the earlier discussion of this factor.

It should be noted that the nitrogen fertilizer ranges are not given in numerical values (Table 5.5) because crop requirements for nitrogen vary with the nature and the type of crops, the relative amounts and distribution of inorganic forms of nitrogen in the soil profile, climatic factors, and the type and number of management practices. In addition, while fertilizer recommendations serve as useful guides, they do not take into account the characteristics of individual farms or fields.

Table 5.5 Ranges and Ratings for Nitrogen Fertilization

Range	Rating (fi)
Over-fertilized	10
Fertilized to meet crop need	6
No fertilizer applied	1

From Ramolino, L., "Development of a Nitrate Pollution Index for Ground Water," Master's Thesis, 1988, University of Oklahoma, Norman, pp. 138–262. With permission.

Table 5.6 Ranges and Ratings of Net Recharge

Range (in.)	Rating (ri)
0–2	1
2–4	3
4–7	6
7–10	8
10+	10

Note: Annual net recharge ranges and ratings used were adapted from the DRASTIC methodology (Aller, et al., 1987).

From Ramolino, L., "Development of a Nitrate Pollution Index for Ground Water," Master's Thesis, 1988, University of Oklahoma, Norman, pp. 138–262. With permission.

Table 5.7 Ranges and Ratings of Soil Texture (Drainage Characteristics)

Range	Rating (si)
Well-drained soils (sand to loam sand)	10
Moderately to well-drained soils (loam to loam silt loam)	6
Poorly drained soil (silt clay to clay soils)	2

From Ramolino, L., "Development of a Nitrate Pollution Index for Ground Water," Master's Thesis, 1988, University of Oklahoma, Norman, pp. 138–262. With permission.

Table 5.8 Ranges and Ratings of Depth to Groundwater

Range (ft)	Rating (di)
0–5	10
5–15	9
15–30	7
30–50	5
50–75	4
75–100	3
100+	2

Note: Ranges and ratings used were adapted from the DRASTIC methodology (Aller, et al., 1987).

From Ramolino, L., "Development of a Nitrate Pollution Index for Ground Water," Master's Thesis, 1988, University of Oklahoma, Norman, pp. 138–262. With permission.

Specific nitrogen crop requirements can be calculated using realistic yield goals and information on available soil nitrogen. Realistic goals are based on the experience of the yield history of each field. Two approaches are often used (Johnson, 1987). One is to set the yield goal to the best yield of the last 5 years. The second is to set the yield goal equal to 50% above the long-term yield average for the field. Determination of available soil nitrogen involves obtaining reliable soil samples and submitting them to soil testing in the laboratory (Johnson, 1987). The difference between the nitrogen requirement for a given yield goal and available soil nitrogen indicates the fertilizer nitrogen requirements. For the selected yield goal, the amount of nitrogen requirements must also take into consideration the time of application, the method of application, and the type of nitrogen fertilizer to be used.

DETERMINE AGGREGATION APPROACH FOR THE INDEX

The Nitrate Pollution Index as described herein was based on a multiplicative and additive model as follows (Ramolino, 1988).

$$Ffi + Rri + Ssi + Ddi = \text{Nitrate Pollution Index}$$

where F, R, S, and D = importance weights for nitrogen fertilization (F), net recharge (R), soil texture (S), and depth to water table (D)

fi, ri, si, di = factor ratings for the four factors

The above model can be used to determine the nitrate pollution potential for various geographical locations. Based upon the model, the Nitrate Pollution Index can range numerically from 24 to 170. The higher the nitrate pollution potential, the greater is the concentration of nitrate expected to reach the groundwater. It should be noted that the Nitrate Pollution Index is an empirical methodology and, as such, does not yield absolute answers; it provides a relative estimation of whether nitrate from agricultural activities will reach groundwater in significant amounts. It also permits prioritizing areas with respect to groundwater vulnerability, and it can aid in planning management practices.

FIELD TESTING OF THE INDEX

The Nitrate Pollution Index was tested based on nitrate groundwater from three different areas of the U.S. They include the state of Oklahoma (69 wells, numbered as wells 1 through 69); the central Platte Region of Nebraska (100 wells, numbered as wells 70 through 169); and the Columbia aquifer, central Delmarva Peninsula, Maryland (92 wells, numbered as wells 170 through 261). Detailed information on each of the 261 wells was assembled from the following: (1) the state of Oklahoma (Oklahoma State Department of Agriculture, 1987; Oklahoma Water Resources Board, 1986); (2) central Platte Region of Nebraska

(Gormly and Spalding, 1978; U.S. Department of Agriculture Soil Conservation Service soil surveys for Buffalo, Hall, and Merrick Counties in Nebraska); and (3) Columbia aquifer of the central Delmarva Peninsula of Maryland (Bachman, 1984; U.S. Department of Agriculture Soil Conservation Service soil surveys for Caroline, Dorchester, Wicomico, and Worchester Counties in Maryland). A general source of information was the U.S. Geological Survey (1985). Table 5.9 depicts an example set of data assembled for the 261 wells; it applies to wells 1–7 from Oklahoma, wells 101–106 from Nebraska, and wells 257–261 from Maryland (Ramolino, 1988).

Oklahoma State Department of Agriculture (1987) conducted an exploratory study on the extent of groundwater contamination from agricultural use of selected pesticides in Oklahoma. In this study, nitrate and other chemical parameters were monitored; for this reason, Oklahoma was chosen as one of the three selected study areas. The selection of wells sampled was made on the basis that existing conditions around those wells would be most favorable to pesticides (or nitrates) reaching groundwater. The 69 wells chosen were located in or near alluvial and terrace aquifers because sandy soils normally associated with these aquifers allow more freedom of movement of water and possible contaminants. Recharge to aquifers in Oklahoma is predominantly from precipitation, which ranges from about 16 in./year in the western panhandle of the state to about 54 in./year in southeastern Oklahoma. Most of the precipitation is returned to the atmosphere by evapotranspiration, which ranges from 16 in. in the west to more than 36 in. in the east. Consequently, recharge from precipitation ranges from less than 0.25 in./year in the west to about 10 in./year in the east (U.S. Geological Survey, 1985).

Nitrogen is the fertilizer nutrient needed in the largest quantity for crop production in Oklahoma. It is used for the production of wheat, cotton, grain sorghum, corn, and Bermuda grass. Nitrogen usage in Oklahoma declined from a high of 273,520 metric tons in 1979/1980 to 249,260 metric tons for 1984/1985 (Westerman, 1987). Oklahoma agricultural producers currently add approximately 570 million pounds of inorganic nitrogen fertilizer to their fields each year. Calculations on the amount of nitrogen required by the crops in Oklahoma show that more nitrogen is required by the crop plant than is applied as inorganic nitrogen fertilizer. The difference is apparently accounted for by the recycled nitrogen contained in the nonharvested portion (crop residue) of the plant and soil organic matter (Johnson, 1987).

Information on the remaining two areas, the central Platte Region of Nebraska, and the Columbia aquifer in Maryland are summarized in case studies R and E, respectively, as shown in Table 5.1. Most recharge to the aquifer in central Nebraska comes from irrigation water. The soil drainage within investigated areas varies from poorly to excessively well-drained soils. Nitrogen fertilizer application rates are reported to vary from 75 lb/acre (84 kg/ha) to 300 lb/acre (336 kg/ha), and 75% of the study area has groundwater within 10 ft (3 m) of the surface.

In the central Delmarva Peninsula, Maryland, the annual precipitation is about 43 in. Recharge to the aquifer is from precipitation; and it varies from

Table 5.9 Example of Data Used in Nitrate Pollution Index Applied to Three Study Areas (Oklahoma; Central Platte Region, Nebraska; and Columbia Aquifer, Central Delmarva Peninsula, Maryland)

Well	Location	Nitrogen fertilizer	Net recharge (in.)	Soil texture (drainage characteristics)	Depth to water table (ft)	Index number	NO_3–N concentration (mg/l)
1	NE4,Sec 2, T3N,R26ECM	N crop rate requirement (30)[a]	0.28 (4.5)	Well-drained[b] (40)	20 (24.5)	99.0	2.26
2	N2,SE4,S27,T4N,R28	No N fertilizer applied (5)	0.28 (4.5)	Well-drained (40)	10 (31.5)	81.0	0.14
3	SE,NE,NE27,T8N,22W	N crop rate requirement (30)	0.44 (4.5)	Well-drained (40)	10 (31.5)	106.0	9.7
4	NE,NE,NE15,16N,12W	N crop rate requirement (30)	0.44 (4.5)	Well-drained (40)	30 (17.5)	92.0	0.6
5	SE,SE,SE23,18N, 13W	N crop rate requirement (30)	0.44 (4.5)	Moderately well-drained (24)	18 (17.5)	76.0	0.50
6	SW4,SE4,SE4, 36, T8S,R10E	No N fertilizer applied (5)	7.78 (36)	Well-drained (40)	11.0 (31.5)	112.5	2.8
7	SW,SW,SW33,11N, 11W	N crop rate requirement (30)	5.18 (27)	Well-drained (40)	6 (31.5)	128.0	5.7
101	15N,5W,34cb	N applied > N crop rate requirement (50)	10+ (45)	Well-drained (40)	10 (31.5)	166.5	29.7
102	14N,5W,3dd	N applied > N crop rate requirement (50)	10+ (45)	Moderately well-drained (24)	10 (31.5)	150.5	17.9
103	14N,5W,4dd	N applied > N crop rate requirement (50)	10+ (45)	Poorly drained (8)	10 (31.5)	134.5	5.5

Table 5.9 Example of Data Used in Nitrate Pollution Index Applied to Three Study Areas (Oklahoma; Central Platte Region, Nebraska; and Columbia Aquifer, Central Delmarva Peninsula, Maryland) (Continued)

Well	Location	Nitrogen fertilizer	Net recharge (in.)	Soil texture (drainage characteristics)	Depth to water table (ft)	Index number	NO_3–N concentration (mg/l)
104	13N,5W,5dd	N applied > N crop rate requirement	10+	Moderately well-drained	10		
		(50)	(45)	(24)	(31.5)	150.5	9.9
105	13N,5W,7ad	N applied > N crop rate requirement	5.37	Moderately well-drained	10		
		(50)	(27)	(24)	(31.5)	132.5	1.2
106	15N,6W,8da	N applied > N crop rate requirement	3.02	Well-drained	10		
		(50)	(13.5)	(40)	(31.5)	135.0	3.5
257	WO H35	N applied > N crop rate requirement	10	Moderately well-drained	88		
		(50)	(36)	(24)	(10.5)	120.5	<0.2
258	WO H48	N applied > N crop rate requirement	10	Well-drained	70		
		(50)	(36)	(40)	(14.0)	140.0	0.6
259	WO H52	N applied > N crop rate requirement	10	Poorly drained	25		
		(50)	(36)	(8)	(24.5)	118.5	0.5
260	WO H54	N applied > N crop rate requirement	10	Poorly drained	47		
		(50)	(36)	(8)	(17.5)	111.5	0.5
261	WO H57	N applied > N crop rate requirement	10	Poorly drained	98		
		(50)	(36)	(8)	(7.0)	101.0	0.2

[a] Numbers in parentheses are the importance weight multiplied by the rating for the factor.
[b] Well-drained: sandy loam to sand; moderately well-drained: loam-slit to loam; poorly drained: silt clay to clay.

From Ramolino, L., "Development of a Nitrate Pollution Index for Ground Water," Master's Thesis, 1988, University of Oklahoma, Norman, pp. 138–262. With permission.

about one fourth to one third of the annual precipitation. Soil textures associated with the study area vary from coarse to very fine. Nitrogen fertilizer is considered to be the only extensive source of nitrogen in the groundwater, but other sources exist in the Maryland study area, including manure applications and some point sources (i.e., chicken houses, livestock feeding operations, and septic tank systems).

Information for each of the four parameters (nitrogen fertilizer applied, net recharge from rainfall and irrigation water, soil texture, and depth to the groundwater) was procured for each of the 269 wells in the three study areas. This information was then converted into appropriate ratings based on Tables 5.5 through 5.8. The rating for each factor for each well was then multiplied by the importance weight for each factor and the products summed to yield the Nitrate Pollution Index for each of the 269 wells. Table 5.9 includes the Index number and the corresponding measured nitrate concentration in the example wells.

The Nitrate Pollution Index number and groundwater nitrate concentration data for each of the 269 wells were statistically analyzed to determine if there was a correlation between the Index number and the nitrate concentration. Two statistical methods were used: (1) the Pearson Product Moment Correlation Coefficient and (2) the Spearman Rank Correlation Coefficient.

The Pearson Product Moment Coefficient of Correlation (r) is a measure of strength of the linear relationship between two variables (x and y) (Mendenhall and Sincich, 1984). In this case, x is the Index number and y is the nitrate concentration in groundwater. It was computed (for a sample of n measurements on x and y) as follows:

$$r = \frac{SS_{xy}}{SS_{xx}SS_{yy}}$$

where

$$SS_{xy} = \sum x_i y_i - \frac{\left(\sum x_i\right)\left(\sum y_i\right)}{n}$$

$$SS_{xx} = \sum x_i^2 - \frac{\left(\sum x_i\right)^2}{n}$$

$$SS_{yy} = \sum y_i^2 - \frac{\left(\sum y_i\right)^2}{n}$$

n = sample size

The value of r is always between –1 and +1, no matter the units of x and y. A value of r near or equal to 0 implies little or no linear relationship between x and y; if y equals +1 or –1, all of the points fall exactly on a least squares line.

Positive values of r imply that y increases as x increases. When a high correlation is observed in the sample data, this indicates that a linear relationship may exist between x and y.

The Pearson Coefficient (r) allows an estimation of a population correlation coefficient, ρ, and consequently a delineation of a statistical hypothesis (H_o $\rho = 0$ against H_a $\rho \neq 0$). The hypothesis is that x contributes no information for the prediction of y using the straight line model against the alternative that the two variables are at least linearly related.

The null hypothesis is stated as the one the user hopes to reject. Two types of errors can be committed upon a decision related to the acceptance or rejection of the null hypothesis. A type I error is committed if the null hypothesis is rejected when, in fact, it is true. A type II error is committed if the null hypothesis is accepted when, in fact, it is false. The probability of committing a type I error is called "the level of significance" and referred to as "alpha." The probability of committing a type II error is called "beta" and requires the formulation of an alternative hypothesis. A test is said to be significant if the null hypothesis is rejected at alpha = 0.05, and considered highly significant if the null hypothesis is rejected at alpha = 0.01. Stated differently, the number that results from a mathematical relationship:

$$(1 - \alpha)\ 100$$

is equal to the confidence level, probability value, at which it can be said that a type I error has not been committed.

The Nitrate Pollution Index is an empirical index and, as developed, the results are based on numerical assignments to both qualitative and quantitative data. Due to the possibility of a nonlinear relationship in the total points assigned, a nonparametric statistical text can be used to evaluate the relationship between Index numbers and groundwater nitrate concentrations.

The nonparametric method used in the field testing was the Spearman Rank Correlation Coefficient; the method is basically a modified linear regression. In this test the data in each set are ranked with the lowest data point being assigned a 1, the next lowest being assigned a 2, and so on. Once the data are ranked, the Correlation Coefficient can be found according to the following relationship (Walpole and Myers, 1978):

$$r_s = \frac{\sum \left(r_i - \bar{R}\right)\left(S_i - \bar{S}\right)}{\left(\sum \left(R_i - \bar{R}\right)^2 \sum \left(S_i - \bar{S}\right)^2\right)^{1/2}}$$

where:

r_s = Spearman Rank Correlation Coefficient
R_i = rank of the "i"th x value
S_i = rank of the "i"th y value

$\overline{R}$ = mean of R ranks
$\overline{S}$ = mean of S ranks

Some advantages accrue from using the Spearman Rank Correlation Coefficient rather than linear regression coefficients. For example, there is no assumption that the underlying relationship between x and y is linear and, therefore, when the data possess a distinct curvilinear relationship, the Rank Correlation Coefficient compensates for the relationship, and subsequently will likely be more reliable than the conventional parametric measure. A second advantage is that no assumptions of normality are made concerning the distributions of x and y (Walpole and Myers, 1978).

Data from 261 wells were included in this field study. The numerical values acquired at each well site for the Nitrate Pollution Index were determined; Table 5.9 provides examples. The calculated Index values were then summarized using a Statistical Analysis System (SAS) on an IBM 3081 mainframe computer. The mean, standard deviation, median, minimum, and maximum for the Index and the field observations of nitrate concentrations are given in Table 5.10 (Ramolino, 1988).

As can be seen from Table 5.11 and according to both statistical analyses used, the Nitrate Pollution Index (PI) correlates with field data for nitrate concentrations in groundwater (C) (Ramolino, 1988). The Pearson and Spearman Coefficients range from 0.67 to 0.82, and the corresponding α values are equal to 0.0001. As previously mentioned, a high correlation coefficient and low α value indicate that a correlation exists between the parameters being compared. This rejects the null hypothesis of a zero linear correlation between the two variables. At the 0.0001 level, there is a 99.99% confidence that a type I error has not been committed. Therefore, a high positive correlation does exist between the concentration of nitrate (C) in groundwater and the Nitrate Pollution Index (PI).

Figure 5.2 is a scatter graph of the nitrate concentration in groundwater versus the Nitrate Pollution Index for all three study areas. The straight line through the data points is the best-fitting linear regression line. This line allows comparison of the predicted results (x axis) versus the detected results (y axis). By plotting the linear regression line, the general trend of nitrate concentration increases with Nitrate Pollution Index increases can be seen.

Although there is generally a good correlation between predicted PI values and detected nitrate concentrations in groundwater, it is easy to see from Figure 5.2 that the data are widely scattered. These varying results can be caused by several factors that were not considered in either the Nitrate Pollution Index methodology nor controlled in the field sampling program. Several important factors that may have affected the results are as follows:

1. Exact Well Location: The location of wells can be reported by either the latitude or by the legal descriptions (township and range). The more precisely the well location is described, the more accurate the determination of soil texture based on soil survey maps. The study areas in Oklahoma and Nebraska have well locations specified using legal descriptions. This system was not used for the

Table 5.10 Statistical Summary of the Nitrate Pollution Index and the Field Data for the Study Areas

Study area	Variable	Number of items of information	Mean	Standard deviation	Median	Minimum	Maximum
Oklahoma	NO_3–N concentration in groundwater (mg/l)	69	4.87	4.31	3.95	0.01	15.31
	Nitrate Pollution Index	69	102.60	19.60	99.00	60.00	137.50
Central Platte Region of Nebraska	NO_3–N concentration in groundwater (mg/l)	100	13.32	8.13	13.60	0.20	32.5
	Nitrate Pollution Index	100	141.52	18.24	148.50	81.0	166.5
Columbia Aquifer, Maryland	NO_3–N concentration in groundwater (mg/l)	92	4.66	4.46	3.55	0.20	19.00
	Nitrate Pollution Index	92	129.0	13.71	127.50	98.50	150.50
Combined data	NO_3–N concentration in groundwater (mg/l)	261	8.10	7.41	6.70	0.01	32.50
	Nitrate Pollution Index	261	127.0	23.15	127.5	60.0	166.5

From Ramolino, L., "Development of a Nitrate Pollution Index for Ground Water," Master's Thesis, 1988, University of Oklahoma, Norman, pp. 138–262. With permission.

Table 5.11 Pearson and Spearman Correlation Coefficients for Nitrate Pollution Index (PI) and Concentration of Nitrate in Groundwater in the Three Study Areas

Study area	Pearson and Spearman correlation coefficients		
Oklahoma	Pearson correlation coefficients/Prob > IRI under HO:RHO = 0; N = 69		
		PI	**C**
	PI	1.000	0.668
		0.0000	0.0001
	C	0.668	1.000
		0.0001	0.0000
	Spearman correlation coefficients/Prob > IRI under HO:RHO = 0; N = 69		
		PI	**C**
	PI	1.000	0.677
		0.0000	0.0001
	C	0.677	1.000
		0.0001	0.0000
Central Platte Region of Nebraska	Pearson correlation coefficients/Prob > IRI under HO:RHO = 0; N = 100		
		PI	**C**
	PI	1.000	0.818
		0.0000	0.0001
	C	0.818	1.000
		0.0001	0.0000
	Spearman correlation coefficients/Prob > IRI under HO:RHO = 0; N = 100		
		PI	**C**
	PI	1.000	0.813
		0.0000	0.0001
	C	0.813	1.000
		0.0001	0.0000
Columbia Aquifer, Central Delmarva Peninsula, Maryland	Pearson correlation coefficients/Prob > IRI under HO:RHO = 0; N = 92		
		PI	**C**
	PI	1.000	0.719
		0.0000	0.0001
	C	0.719	1.000
		0.0001	0.0000
	Spearman correlation coefficients/Prob > IRI under HO:RHO = 0; N = 92		
		PI	**C**
	PI	1.000	0.745
		0.0000	0.0001
	C	0.745	1.000
		0.0001	0.0000
Combined data of the three areas	Pearson correlation coefficients/Prob > IRI under HO:RHO = 0; N = 261		
		PI	**C**
	PI	1.000	0.729
		0.0000	0.0001
	C	0.729	1.000
		0.0001	0.0000
	Spearman correlation coefficients/Prob > IRI under HO:RHO = 0; N = 261		
		PI	**C**
	PI	1.000	0.757
		0.0000	0.0001
	C	0.757	1.000
		0.0001	0.0000

From Ramolino, L., "Development of a Nitrate Pollution Index for Ground Water," Master's Thesis, 1988, University of Oklahoma, Norman, pp. 138–262. With permission.

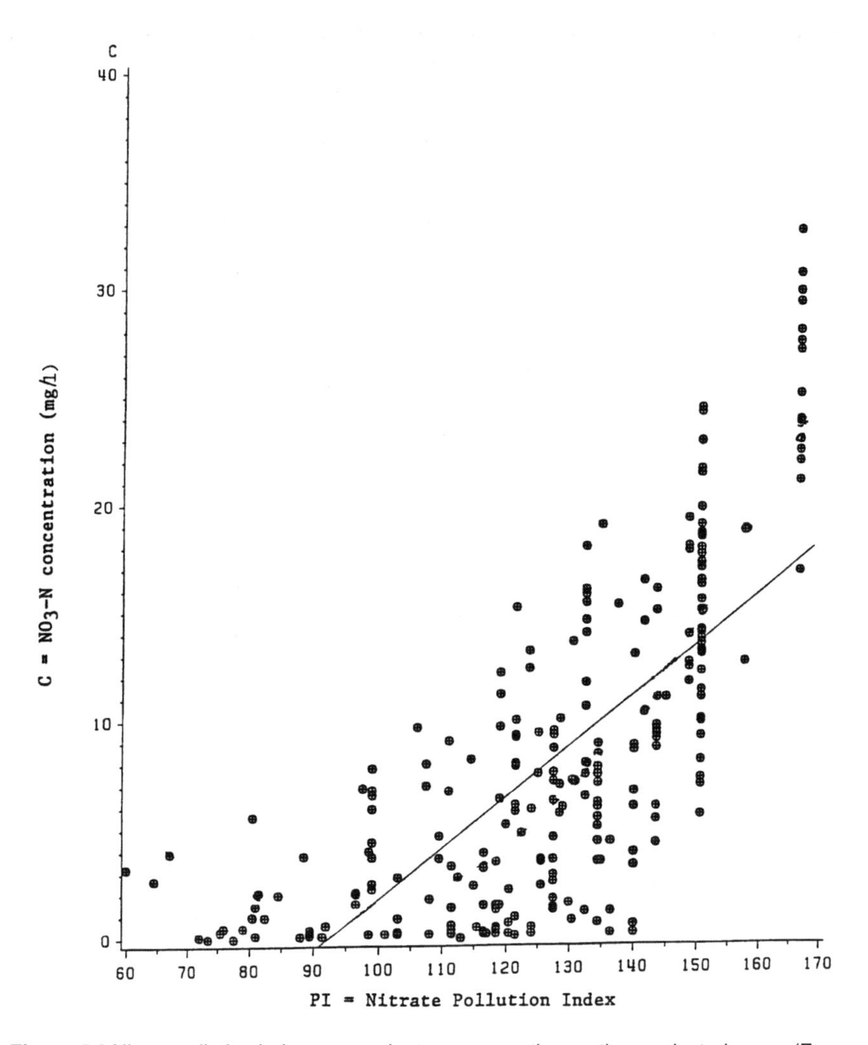

Figure 5.2 Nitrate pollution index versus nitrate concentrations— three selected areas. (From Ramolino, L., "Development of a Nitrate Pollution Index for Ground Water, "Master's Thesis, 1988, University of Oklahoma, Norman, pp. 138–262. With permission.)

Columbia aquifer (Table 5.8); thus, the degree of precision in locating wells on soil survey maps was not very high.

2. Potential Point Sources of Nitrate Pollution: This information was not reported for each well site. Wells contaminated by nitrate point sources may be characterized by excessively high concentrations of nitrates. A section may have been rated as having low pollution potential, but because the well is located near a point source, its nitrate level could be high.
3. Type of Well Construction: This information was only reported for Oklahoma. Knowing the type of casing can be useful in identifying wells that serve as conduits for the entry of contamination into the groundwater. Wells that have

not been properly maintained and/or constructed may have high nitrate levels because they allow runoff rich in nitrogen to pass directly to the aquifer.

4. Rate and Extent of Nitrogen Fertilizer Application: The nitrogen fertilizer application rate generally varies from field to field. In this study, the amount of fertilizer applied at or near each well site was not specifically recorded. The rate considered in calculating the Nitrate Pollution Index was a regional rate instead of field specific. For some crops such as legumes, nitrogen fertilization is not required. In Oklahoma, for example, alfalfa, soybeans, and peanuts are grown. But the exact crop associated with each well site was not reported; therefore, it had to be assumed that nitrogen fertilizer was applied to every well area, and it was the sole source of nitrate in groundwater.
5. Net Recharge: Where low values of recharge are assumed, lower amounts of nitrogen are introduced to the aquifer because the degree of movement through the unsaturated zone is largely dependent upon the volume of the recharge. In Oklahoma only the net recharge from precipitation was considered; net recharge from irrigation was not reported.
6. Depth to Groundwater Table: Depth to the groundwater table was not reported for each well in the central Platte Region of Nebraska. It was assumed that all groundwater in this area was within 10 ft (3 m) of the surface.
7. Nitrate Data: Concentrations may be reported as nitrate or as nitrate-nitrogen. Some methods of analysis measure nitrate alone, and others measure nitrate plus nitrite. Nitrate concentrations for the Columbia aquifer in the central Delmarva Peninsula in Maryland were reported as nitrate plus nitrite. While this may be of some relevance, it was not considered to be a significant limitation to the results of this study.
8. Type of Aquifer: The Columbia aquifer, in Maryland, is mostly unconfined, but it is likely to be partially confined to the southeastern parts of the study area. Some wells situated in the southeastern part of the Columbia aquifer had very low nitrate concentrations and high pollution potential values. Thus, the pollution potential did not reflect the nitrate field data. This is due to the presence of confining layers that prevented the entrance of nitrate to groundwater; in addition, as confined aquifers are generally associated with reducing environments, nitrate could have been denitrified. Therefore, it might be desirable to incorporate the variable, type of aquifer, in the Nitrate Pollution Index.

SUMMARY

The objective of the study described herein was to develop an index to provide information for predicting the potential risk to groundwater from nitrates applied as fertilizers in agricultural areas. Twenty-two case studies were reviewed to document nitrate behavior under different types of farming activities and different hydrogeological conditions. Eight basic factors that could affect the magnitude of nitrate concentrations reaching groundwater were identified. Among these, four significant factors were then selected and used in the Nitrate Pollution Index. These four factors were: (1) nitrogen fertilization (intensity), (2) soil texture (drainage characteristics), (3) net recharge from irrigation and/or rainfall, and (4) depth to the water table. These four factors were assessed and rated for each of 261 well sites from three different areas of the U.S., that is, the state of Oklahoma, the central Platte Region of Nebraska, and the Columbia aquifer of the central Delmarva Peninsula in Maryland. The resultant ratings were multiplied by importance weights defined for each factor and then summed for each well to yield its Nitrate Pollution Index. The Index values for each well site were then correlated statistically with nitrate measurements from groundwater from the 261 wells.

Two statistical methods were used to correlate the Nitrate Pollution Index and the field data: (1) the Pearson Product Moment Correlation Coefficient and (2) the Spearman Rank Correlation Coefficient. According to both statistical analyses used, the Nitrate Pollution Index correlates with field data for nitrate concentrations in groundwater.

Although the Nitrate Pollution Index methodology does not produce a one-to-one relationship between predicted Index values and detected nitrate concentrations, it does reflect with reasonable statistical accuracy the nitrate pollution potential. The significance of this statistical evidence indicates that the Nitrate Pollution Index can be applied to any agricultural site and the general trend in groundwater pollution potential can be predicted. This empirical methodology can be rapidly applied, and it is more economically feasible than groundwater quality monitoring. It can be used as a beginning step for identifying management recommendations to prevent further nitrate pollution or to correct existing problems.

SELECTED REFERENCES

Aller, L., Bennett, T., Lehr, J.H., Petty, R.J., and Hackett, G., "DRASTIC: A Standardized System for Evaluating Ground Water Pollution Potential Using Hydrogeologic Settings," EPA 600/2-87/035, May, 1987, U.S. Environmental Protection Agency, Ada, Oklahoma.

Bachman, J.L., "Nitrate in the Columbia Aquifer, Central Delmarva Peninsula, Maryland," Water Resources Investigations No. 84-4322, 1984, U.S. Geological Survey, Baltimore, Maryland.

Barber, S.A., "Efficient Fertilizer Use," in *Agronomic Research for Food*, ASA Special Publication No. 26, F.L. Patterson, et al., Eds., American Society of Agronomy, Madison, Wisconsin, August, 1975, pp. 13–29.

Canter, L.W., *Environmental Impact Assessment*, McGraw-Hill Book Company, Inc., New York, New York, 1996, pp. 140–143.

Committee on Techniques for Assessing Ground Water Vulnerability, *Ground Water Vulnerability Assessment – Predicting Contamination Potential Under Conditions of Uncertainty*, National Research Council, National Academy Press, Washington, D.C., 1993, pp. 1–11.

Gormly, J.R. and Spalding, R.F., "Nitrate Source Identification," 1978, Cooperative Project with Old West Regional Commission, Central Platte Natural Resources District, and Conservation and Survey Division, University of Nebraska, Lincoln, Nebraska.

Johnson, G.V., "Soil Testing As a Guide to Prudent Use of Nitrogen Fertilizer in Oklahoma Agriculture," in *Ground Water Quality and Agricultural Practices*, D.M. Fairchild, Ed., Lewis Publishers, Inc., Boca Raton, Florida, 1987, pp. 127–135.

Keeney, D.R., "Nitrogen Management for Maximum Efficiency and Minimum Pollution," in *Nitrogen in Agricultural Soils*, F.J. Stevenson, Ed., No. 22, American Society of Agronomy, Inc., Madison, Wisconsin, 1982, pp. 605–650.

Linstone, H.A. and Turoff, M., *The Delphi Method – Techniques and Applications*, Addison-Wesley Publishing Company, Reading, Massachusetts, 1978.

Madison, R.J. and Brunett, J., "Overview of the Occurrence of Nitrate in Ground Water of the U.S.," in *National Water Summary*, Water Supply Paper 2275, 1984, U.S. Geological Survey, Washington, D.C., pp. 93–104.

Mendenhall, W. and Sincich, T., *Statistics for the Engineering and Computer Sciences*, 1st ed., Dellen Publishing Company, San Francisco, California, 1984, pp. 395–442.

Oklahoma State Department of Agriculture, "Exploratory Study on the Extent of Ground Water Contamination from Agricultural Use of Selected Pesticides in Oklahoma," 1987, Oklahoma City, Oklahoma.

Oklahoma Water Resources Board, *Oklahoma's Water Atlas*, Publication No. 76, 1986, Oklahoma City, Oklahoma.

Ott, W.R., *Environmental Indices: Theory and Practice*, Ann Arbor Science Publishers, Inc., Ann Arbor, Michigan, 1978, pp. 2, 5, 135–169, and 202–213.

Ottow, J.C.G. and Fabig, W., "Influence of Oxygen Aeration on Denitrification and Redox Level in Different Bacterial Batch Cultures," in *Planetary Ecology*, D.E. Cardwell, J.A. Brierley, and C.L. Brierley, Eds., Van Nostrand Reinhold Company, Inc., New York, New York, 1985, pp. 427–441.

Ramolino, L., "Development of a Nitrate Pollution Index for Ground Water," Master of Environmental Science Thesis, 1988, University of Oklahoma, Norman, Oklahoma, pp. 138–262.

U.S. Geological Survey, "Hydrologic Events and Surface-Water Resources," *National Water Summary*, Water Supply Paper 2300, 1985, Washington, D.C.

Walpole, R.E. and Myers, R.H., "Nonparametric States," in *Probability and Statistics for Engineers and Scientists*, 2nd ed., Macmillan Publishing Company, New York, New York, 1978, pp. 492–495.

Westerman, R.L., "Efficient Nitrogen Fertilization in Agricultural Production Systems," in *Ground Water Quality and Agricultural Practices*, D.M. Fairchild, Ed., Lewis Publishers, Inc., Boca Raton, Florida, 1987, pp. 137–151.

6 MODELS FOR NITRATES IN THE SUBSURFACE ENVIRONMENT

INTRODUCTION

This chapter highlights a number of models and submodels for describing the transport and fate of nitrogen and nitrates in the subsurface environment. Sections are included on source characterization models, nitrate transport models, a plume model for a nitrogen point source, and management models useful for prevention and control of nitrate pollution of groundwater. Fundamental information on the basic aspects of flow and solute transport modeling is available in several hydrogeological or environmental engineering books. The primary purpose of this chapter is to demonstrate the range in the types of models potentially useful for addressing nitrates in the subsurface environment. The range encompasses statistical models as well as process submodels and coupled models for unsaturated and saturated zones (Committee on Techniques for Assessing Groundwater Vulnerability, 1993). A summary section concludes this chapter.

SOURCE CHARACTERIZATION MODELS

Table 6.1 contains summary comments on 15 references related to characterizing sources of nitrogen/nitrates in the subsurface environment. These 15 references can be subdivided into four categories dealing with soil processes, recharge, leaching, and input/output approaches.

Soil Processes

Soil processes can include infiltration, evapotranspiration, nitrogen transformations, nitrogen losses due to ion exchange, and nitrogen uptake by crops. Models for soil processes can address one to all of these processes. For example, nitrogen transformations are addressed by Shaffer, Dutt, and Moore (1969), and by Walter, et al. (1974). Shaffer, Dutt, and Moore (1969) developed a digital computer program to model soil-water systems with respect to nitrogen transformations, including hydrolysis of urea, immobilization-mineralization of ammonia

Table 6.1 References Involving Source Characterization Models

Author(s) (year)	Comments
Soil Processes	
Dutt, Shaffer, and Moore (1972)	Computer model incorporating infiltration and redistribution of soil water, evapotranspiration, nitrogen transformations, solute changes in soil water due to ion exchange, and nitrogen uptake by crops
Shaffer, Dutt, and Moore (1969)	Computer model for nitrogen transformations in soil-water systems
Walter et al. (1974)	Computer model for nitrogen transformations and nitrate leaching from heavy spring applications of liquid dairy manure on coarse soils
Watts and Martin (1981)	Mechanistic model of the soil-plant-water-nitrogen system associated with the irrigation of corn grown in sandy areas
Recharge	
Porter and Shoemaker (1978)	Simulation model for recharge and nitrogen in the recharge for Long Island, New York
Leaching	
Barry et al. (1985)	Solute transport model for partitioning total measured fertilizer loss into that from each fertilizer application
Khanif, Cleemput, and Baert (1984)	Simple model for predicting the distribution of nonadsorbed solutes, like nitrates, subject to leaching or upward movement
Mathew, Ho, and Newman (1984)	Simple model for determining the effects of infiltration rate and periods of flooding and drying on the removal of nitrogen from wastewater applied to soil
Shaffer, Halvorson, and Pierce (1991)	Description of a computerized nitrate leaching model (NLEAP) that can be used as a screening tool to identify significant potential nitrate-nitrogen leaching under agricultural crops
Stemmler (1982)	Use of CREAMS model to quantify the relation between fertilization intensity and nitrate concentrations in water percolated below the root zone
Tanner and Gardner (1973)	Simple model for nitrate leaching that incorporates climatological information
White (1987)	Development and testing of transfer function model for nitrate leaching under various rainfall scenarios
Input/Output	
Bashkin and Kudeyarov (1983)	Correlation of nitrate concentration in groundwater with fertilizer application rates and cattle populations
Belamie et al. (1984)	Input/output budgets for nitrogen leaching from cultivated lands and forested areas
Farajalla, Deyle, Vieux, and Canter (1993)	Statistical correlations of nitrate levels in Oklahoma groundwater with agricultural land usage and hydrogeological characteristics

and organic nitrogen, and immobilization of nitrate-nitrogen. As a part of the effort, comparisons were made of predicted and observed data for several soils having different textures and various moisture contents, temperatures, and fertilizer applications. This procedure yielded simple correlation coefficients of 0.99, 0.97, and 0.97 for the urea, organic, and ammonia nitrogenous types, respectively.

Walter et al. (1974) developed a model for predicting nitrate movement from the application of manure on experimental plots on a plainfield sand and in a plano silt loam soil. The model was designed to estimate, for specified intervals of soil depth and time after manure application, the delay to initial nitrification of manurial ammonium, and the ammonium nitrification rate, soil organic nitrogen mineralization rate, nitrate dispersion, nitrate and ammonium content, and soil moisture.

Additional soil processes were included in a model, by Watts and Martin (1981), of the soil-plant-water-nitrogen system for irrigated corn on sands. It simulates the biological transformation of various nitrogen forms in the soil, nitrogen uptake by the crop, and the loss of water and nitrate from the root zone. Corn, having a 105- to 107-day relative maturity range, was simulated. The soil for which the simulations were performed was a Valentine very fine sand, with an available water holding capacity of about 10% by volume. Two nitrogen amounts, 168 and 253 kg/ha, were used in the studies. The nominal irrigation frequency in the simulations was 4 days. A maximum system capacity of 0.94 cm/day was assumed. Four important factors that were identified as affecting nitrate leaching loss and that were controllable included irrigation quantity, nitrogen source, nitrogen amount, and timing of nitrogen application.

Finally, the model by Dutt, Shaffer, and Moore (1972) simulated nonsteady-state chemical, physical and biological changes in the unsaturated soil matrix and percolating water; the processes considered were infiltration and redistribution of soil water, evapotranspiration, nitrogen transformation, changes in solute concentration of soil water due to ion exchange, and nitrogen uptake by crops. The verified model predicts the time distribution and concentration of a number of inorganics, including chloride, organic and ammonia nitrogen, and nitrate.

Recharge

One factor that obviously affects nitrate movement in the subsurface environment is the quantity of recharge water. In order to quantify both the amount of recharge water and the nitrogen concentration in the recharge, a simulation model was developed for Long Island, New York, by Porter and Shoemaker (1978). The model can be used to calculate a mass-balance of water and nitrogen on 762 cells, each of which is 1.5 square miles . The calculations, which can be computed daily or monthly, are based on land use, soil type, temperature, precipitation, and sewerage in each grid cell. Detailed soil moisture data were collected at several sites. Data from the early part of the year were used to calibrate the model. Validation of the model was achieved by comparison with independent data collected at a later time.

Leaching

An initiating mechanism for nitrogen or nitrate movement toward the ground-water system involves leaching. Several models have been developed to address various features of the leaching process. For example, Tanner and Gardner (1973) developed a simple model, applicable to both humid and arid regions, of nitrate leaching that incorporated an analysis of weather records.

The influences of fertilizer application periods and alternate wetting and drying cycles were addressed by Barry, et al. (1985), and Mathew, Ho, and Newman (1984), respectively. Barry, et al. (1985) demonstrated that a solute transport model could be used to partition the total measured fertilizer loss into that from each fertilizer application. The model allows estimation of the mean depth of solute penetration from any sequence of infiltration and evapotranspiration events and of the apparent dispersion in relation to the mean depth. The model by Mathew, Ho, and Newman (1984) addressed the removal of nitrogen from sewage effluent during soil percolation by alternate flooding and drying. The effects of infiltration rate and periods of flooding and drying on removal efficiency can be determined by usage of this model.

More detailed models for leaching and subsequent subsurface transport have been developed. For example, Khanif, Cleemput, and Baert (1984) noted that nitrate can be translocated by leaching during the wet season and by upward movement during the dry season. A simple model to predict the distribution of nonadsorbed solutes subject to leaching and upward movement was tested to predict the redistribution of nitrate-nitrogen during the winter in soil profiles of 1-m depth. The measured and model-based calculated nitrate concentrations in 100 cm wet sandy soil profiles were compared, and the results showed reasonably good agreement between the two levels. Better agreement was obtained when only samples from above the groundwater table were considered.

A field-scale model for chemicals, runoff, and erosion from agricultural management systems, CREAMS, has been developed by the U.S. Department of Agriculture to model field losses of sediment, nutrients, and pesticides. The CREAMS model is a physically based, daily simulation model that can be used to estimate runoff, erosion/sediment transport, plant nutrient, and pesticide yield from field-sized areas. The plant nutrient submodel of CREAMS has a nitrogen component that considers mineralization, nitrification, and denitrification processes. Plant uptake can be estimated, and nitrate leached by percolation out of the root zone can be calculated. Stemmler (1982) tested the CREAMS model in the Mussum water reserve area of northwestern Germany. Through the use of the model, it was possible to quantify the relation between the fertilization intensity (up to 480 kg N per hectare) of arable land and nitrate concentrations in the water percolated below the root zone. However, the results were applicable only to the thin and uniformly distributed aquifers in the study area.

The nitrate leaching and economic analysis package (NLEAP) computer model is user-friendly and designed to develop site-specific estimates of nitrate-nitrogen leaching potential under agricultural crops along with potential aquifer impacts of such leaching (Shaffer, Halvorson, and Pierce, 1991). The NLEAP model uses a phased approach beginning with an annual screening analysis that provides initial estimates of potential nitrate-nitrogen leaching, while monthly and event-by-event water and nitrogen budgets are used for more detailed analyses. The initial screening analysis provides a rapid means of identifying potential leaching problems that should lead to additional detailed analysis.

Detailed analyses in the NLEAP model can be based on monthly or event-by-event approaches throughout the year to compute water and nitrogen budgets (Shaffer, Halvorson, and Pierce, 1991). The model is based on mass balances around two soil horizons (the upper 1 ft and the remaining portion, if any, from 1 ft down to the bottom of the root zone or a root-restricting layer) that account for the following sources, influencing factors, or sinks: (1) nitrate-nitrogen available for leaching, (2) ammonium-nitrogen nitrification, (3) soil temperature, (4) soil water, (5) soil organic matter mineralization, (6) crop residue and other organic matter mineralization, (7) crop nitrogen uptake, (8) soil nitrogen uptake by legumes, (9) nitrogen loss to ammonia volatilization, (10) nitrogen loss to denitrification, (11) water available for leaching, and (12) potential evapotranspiration. Table 6.2 summarizes the mathematical expressions used in the NLEAP model for each of these 12 items (Shaffer, Halvorson, and Pierce, 1991).

The resultant nitrate-nitrogen leached (NL expressed in lb/acre) during a time step is computed using the following exponential relationship (Shaffer, Halvorson, and Pierce, 1991):

$$NL1 = (NAL1)\{1 - EXP[(-K)(WAL1)/POR1]\}$$

$$NAL = NAL2 + NL1$$

$$NL = (NAL)\{1 - EXP[(-K)(WAL)/POR2]\}$$

where NL1 is nitrate-nitrogen leached from the upper 1 ft (lb/acre); K is the leaching coefficient (unitless); POR1 is the porosity of the upper 1 ft (in.); NAL is nitrate-nitrogen available for leaching from the root zone (lb/acre); NL is nitrate-nitrogen leached from the bottom of the root zone (lb/acre); and POR2 is the porosity of the lower horizon (in). The total nitrate-nitrogen leached for any month or year is computed by summing the leaching obtained from each time step during the period of interest (Shaffer, Halvorson, and Pierce, 1991).

The NLEAP model then addresses the potential impact of the leached nitrate-nitrogen (NL) on underlying groundwater via an aquifer risk index (ARI) defined as follows (Shaffer, Halvorson, and Pierce, 1991):

$$ARI = 0.369[N_o + (NL)(A) + N_{s1} - N_1]/AMV$$

where AMV is the aquifer mixing volume (acre-ft); N_o is the initial nitrate-nitrogen content of the AMV (lb); NL is the soil nitrate-nitrogen leached to the aquifer (lb/[acre time step]); A is the area of the field or farm (acre); N_{s1} is nitrate-nitrogen entering the AMV from sources outside the farm or field of interest (lb/time step); N_1 is nitrate-nitrogen leaving the AMV in pumped wells, tile drains, and other flows (lb/time step); and 0.369 converts lb/acre-ft to parts per million (ppm) or mg/l. It is assumed that the upper portion (usually a few feet in depth) of a shallow aquifer (called the AMV) can be defined where an approximate complete mix is occurring with respect to the sources and sinks of nitrate-nitrogen; this can be calculated using:

$$N_o = 2.71(N_c)(AA)(W)$$

where N_c is the initial nitrate-nitrogen concentration in the AMV (mg/l); AA is the surface area of the aquifer (acre); W is thickness of the AMV (ft) multiplied by its porosity; and 2.71 converts mg/l·acre-ft to lb/acre-ft. N_{s1} is calculated by

Table 6.2 Mathematical Relationships in NLEAP Model

Nitrogen source, influencing factor, or sink	Mathematical expression
Nitrate–nitrogen available for leaching (NAL in lb/acre)	$NAL = N_f + N_p + N_{rsd} + N_n - N_{plt} - N_{det} - N_{oth}$ where N_f is NO_3–N added to the soil from fertilizers (lb/[acre time step]), N_p is NO_3–N added from precipitation and irrigation water (lb/[acre time step]), N_{rsd} is residual NO_3–N in the soil profile (lb/acre), N_n is NO_3–N produced from nitrification of ammonium-N (NH_4–N)(lb/[acre time step]), N_{plt} is NO_3–N uptake by the crop (lb/[acre time step]), N_{det} is NO_3–N lost to denitrification (lb/[acre time step]), and N_{oth} is NO_3–N lost to runoff and erosion (lb/[acre time step]).
Ammonium–nitrogen nitrification (N_n)	$N_n = K_n(TFAC)(WFAC)(ITIME)$ subject to the constraint $N_n \leq NAF$, where K_n is the zero-order rate coefficient for nitrification (lb/acre/d), TFAC is the temperature stress factor (0-1), WFAC is the soil water stress factor (0-1), ITIME is the length of the time step (d), and NAF is the NH_4–N content of the top foot (lb/acre) at the end of the time step. The use of nitrification inhibitors is simulated by reducing the magnitude of the rate coefficient, K_n. NAF is calculated using the equation: $NAF = NAF_f + NAF_p + NAF_{rsd} + NOMR + NRESR + NMANR - NPLTA - N_{NH_3} - NAF_{oth}$ where NAF_f is NH_4–N added from fertilizers (lb/[acre time step]), NAF_p is NH_4–N added from precipitation and irrigation (lb/[acre time step]), NAF_{rsd} is residual soil NH_4–N from the previous step (lb/acre), NOMR is NH_4–N mineralized from soil organic matter (lb/[acre time step]), NRESR is net mineralization of NH_4–N from crop residues (lb/[acre time step]), NMANR is net mineralization from manure plus other organic wastes (lb/[acre time step]), NPLTA is plant uptake of NH_4–N (lb/[acre time step]), N_{NH_3} is NH_3–N volatilization (lb/[acre time step]), and NAF_{oth} is NH_4–N lost to runoff and erosion (lb/[acre time step]). Here the assumption is made that NH_4–N does not leach or move with the water.
Soil temperature stress factor (TFAC)	$TFAC = 1.68E9 (\exp\{-13.0/[(1.99E-3)(TMOD + 273)]\})$ where TMOD equals (T – 32)/1.8 when T ≤ 86°F, and TMOD equals 60 – (T – 32)/1.8 when T > 86°F, and T is soil temperature (°F). The TFAC has a range of 0.0 to 1.0.
Soil water stress factor (WFAC)	For aerobic processes such as mineralization and nitrification, $WFAC = 0.0075(WFP)$ where WFP is percent water-filled pore space for WFP ≤20, $WFAC = -0.253 + 0.0203(WFP)$ for WFP ≥20 and <59, and $WFAC = 41.1\{\exp[(-0.0625(WFP)]\}$ for WFP ≥59, and for anaerobic processes such as denitrification, $WFAC = 0.000304\{\exp[(0.0815(WFP)]\}$
Soil organic matter mineralization (NOMR)	$NOMR = k_{omr}(OMR)(TFAC)(WFAC)(ITIME)$ where NOMR is the NH_4–N mineralized (lb/[acre time step]), k_{omr} is the rate coefficient, and OMR is soil organic matter (lb/acre).

Table 6.2 Mathematical Relationships in NLEAP Model (Continued)

Nitrogen source, influencing factor, or sink	Mathematical expression
Crop residue and other organic matter mineralization	$CRES = P_c(RES)$ where RES represents residues (lb/acre), P_c is the residue fraction that is carbon, and CRES is the C content of the residues (lb/acre), and $CRESR = k_{resr}(CRES)(TFAC)(WFAC)(ITIME)$ where CRESR is the residue C metabolized (lb/[acre time step]), and k_{resr} is the first order rate coefficient (1/d); the residue C is updated after each time step using $CRES = CRES - CRESR$ constrained by $CRESR \leq CRES$, and net mineralization-immobilization is determined using $NRESR = (CRESR)(1/CN - 0.042)$ constrained by $-NRESR \leq NAF + N1T1$, when $NRESR < 0.0$, where NRESR is the net residue-N mineralized (lb/[acre time step]), CN is the current C:N ratio of the residues, and N1T1 is the NO_3–N content of the top foot; the N content of the decaying residues is updated after each time step using $NRES = NRES - NRESR$ constrained by $NRESR \leq NRES$ and a new value for CN is computed for the next time step $CN = CRES/NRES$ where NRES is the N content of the crop residues, manure, or other organic wastes (lb/[acre time step]).
Crop nitrogen uptake (N_{plt})	$N_{dmd} = (YG)(TNU)(fNU)(ITIME)$ where N_{dmd} is N uptake demand (lb/[acre time step]), YG is yield goal or maximum yield in appropriate units, TNU is total N uptake (lb/harvest unit), and fNU is fractional N uptake demand at the midpoint of the time step. The N uptake demand is proportioned between the upper and lower soil horizons according to the relative water uptake. Nitrogen available for uptake in each horizon is computed as follows: $Navail_1 = NAF + N1T1$ for the upper horizon, and $Navail_2 = N1T2$ for the lower horizon, where N1T2 is the NO_3–N content in the lower horizon (lb/acre). In each case, the uptake demand for each layer is constrained by the N availability. Therefore, N_{plt} is set equal to the smaller of N_{dmd} or ($Navail_1$ + $Navail_2$). Plant uptake of NH_4–N (NPLTA) is calculated from total N uptake in the upper foot according to the fraction of NO_3–N plus NH_4–N that is NH_4–N.
Soil nitrogen uptake by legumes	Soil N uptake by legumes is taken as either the N demand by the crop or the sum of $Navail_1$ and $Navail_2$, whichever is smaller. If the N demand is greater than the N available in the soil, the plant is assumed to obtain the difference from N_2 fixation.
Nitrogen loss to ammonia volatilization (N_{NH_3})	$N_{NH_3} = k_{af}(NAF_s)(TFAC)(ITIME)$ subject to the constraint, $N_{NH_3} \leq NAF_s$, where N_{NH_3} is NH_3–N volatilized (lb/[acre time step]), k_{af} is the rate constant for NH_3 volatilization, and NAF_s is the NH_4–N content of the surface (lb/acre). The particular value for k_{af} is a function of fertilizer application method, occurrence of precipitation, cation exchange capacity of surface soil, and percent residue cover. In the case of manure, k_{af} is a function of the type of manure and application method.

Table 6.2 Mathematical Relationships in NLEAP Model (Continued)

Nitrogen source, influencing factor, or sink	Mathematical expression
Nitrogen loss to denitrification (N_{det})	$N_{det} = k_{det}(N1T1)(TFAC)[NWET + WFAC(ITIME - NWET)]$ subject to the constraint, $N_{det} \leq N1T1$, where N_{det} is NO_3–N denitrified (lb/[acre time step]), k_{det} is the rate constant for denitrification, N1T1 is the NO_3–N content of the top foot (lb/acre), and NWET is the number of days with precipitation or irrigation during the time step. The value assigned to k_{det} is a function of percent soil organic matter, soil drainage class, type of tillage, presence of manure, tile drainage, type of climate, and occurrence of pans.
Water available for leaching (WAL)	Water available for leaching (WAL) is calculated after each precipitation and irrigation event using the two-layer soil model and the following relationships, $WAL1 = P_e - ET1 - (AWHC1 - S_{t1})$ constrained by $WAL1 \geq 0.0$, and $WAL = WAL1 - ET2 - (AWHC2 - S_{t2})$ constrained by $WAL \geq 0.0$, where WAL1 is water available for leaching from the top foot (in.), ET1 is potential evapotranspiration associated with the top foot (in./time step), AWHC1 is the available water-holding capacity of the top foot (in.), WAL is water available for leaching from the bottom of the soil profile (in.), P_e is effective precipitation (in.), ET2 is potential evapotranspiration from the lower horizon (in.), S_{t1} is available water in the top foot at the end of the previous time step (in.), AWHC2 is the available water-holding capacity of the lower horizon (in.), and S_{t2} is available water in the lower horizon at the end of the previous time step.
Potential evapotranspiration (ET_p)	$ET_p = [(EV_p)(k_{pan})(k_{crop}) + (EV_p)(k_{pan})(1 - k_{crop})](ITIME)$ where ET_p is potential evapotranspiration (in./time step), EV_p is average daily pan evaporation during the time step (in./day), k_{pan} is the pan coefficient, and k_{crop} is the crop coefficient. ET_p is proportioned between potential evaporation at the soil surface, ET_{ps}, and potential transpiration, ET_{pt}, using normalized curves for each crop to compute k_{crop}. ET_{pt} is then proportioned between the upper and lower soil horizons according to the relative root distributions.

After Shaffer, M. J., Halvorson, A. D., and Pierce, F. J., in *Managing Nitrogen for Groundwater Quality and Farm Profitability*, Follett, R. F., Keeney, D. R., and Cruse, R. M., Eds., Soil Science Society of America, Madison, WI, 1991, chap. 13, pp. 285–322.

multiplying associated flows (acre-ft/time step) times their concentration of nitrate-nitrogen (mg/l) times 2.71. N_l is computed in a similar fashion by multiplying N_c times the corresponding discharge volumes (acre-ft/time step) times 2.71. For steady-state conditions, aquifer discharge volume equals input volume. In general, for shallow aquifers used for drinking water, any values for ARI > 10 would indicate a need for increased monitoring and study (Shaffer, Halvorson, and Pierce, 1991).

Input/Output

Statistical models based on relating certain input variables to output concentrations of nitrates have also been developed. For example, Belamie, et al. (1984)

developed long-term input/output budgets to display the effects of agriculture on water quality in an experimental watershed in France. Bashkin and Kudeyarov (1983) established a significant, but not linear, correlation of the nitrate content of groundwaters with nitrogen fertilizer application rates and cattle populations in four agricultural areas in the USSR.

Farajalla, et al. (1993) assessed correlations of agricultural practices, climate, and geologic conditions with nitrate contamination of groundwater in Oklahoma. Data on nitrate concentrations were analyzed for 184 wells in 55 of Oklahoma's 77 counties for the period 1983 to 1988. Nitrate concentrations consistently exceeded the drinking water standard of 10 mg/l for the entire period in 7.0% of the wells tested. The Oklahoma data also revealed a trend of increasing contamination; for example, nitrate concentrations increased in 23 of the 55 counties during the 1983 to 1986 period.

This study employed bivariate correlation analysis (Pearson's r) and multiple regression to measure correlations between groundwater nitrate concentrations and the following variables: (1) fertilizer use, (2) crops grown, (3) soil characteristics, (4) permeability of the overlying geologic substrate, (5) aquifer type, (6) depth to groundwater, and (7) net water balance. The geographic unit of analysis was the county because this is the level at which records were available for agricultural practices (Farajalla et al., 1993).

Direct measures of application rates of nitrogenous fertilizers were not available. The annual tonnage of nitrogen fertilizer shipped into each county was therefore employed as a proxy. Two sets of variables were tested to assess the importance of crop type: (1) total annual acreage planted or harvested per county and (2) total irrigated acreage per county. Values for total acreage were included for the six most widely grown crops — cotton, oats, peanuts, sorghum, soybeans, and wheat. Values for total irrigated acreage were included for cotton, peanuts, sorghum, and wheat.

Soil association maps from county soil surveys were used to estimate soil texture for the approximate recharge area around the 184 well sites. Seven soil textures were coded as indicator variables: (1) clay, (2) clay loam, (3) loam, (4) sandy loam, (5) silty loam, (6) loamy sand, and (7) sand. Characteristics of regional geologic formations overlying the aquifer at each well site were used to define an ordinal measure of geologic permeability as follows: 1 = least permeable; 2 = moderately permeable; 3 = relatively high permeability. Data on depth to groundwater and aquifer type were obtained for the 184 wells. Aquifer type was represented by two indicator variables: (1) alluvial and (2) bedrock. Finally, a water balance variable was defined as the difference between annual precipitation and annual evapotranspiration (ET). Data for precipitation and evaporation were obtained from the Oklahoma Climatological Survey's weather station network (Farajalla et al., 1993).

The results of the bivariate analysis were generally consistent with other models developed to predict nitrate leaching and groundwater vulnerability based on soil and geologic characteristics. Nitrate concentrations were negatively correlated with depth to groundwater, bedrock aquifer material, and the heavier textured soils, while they were positively correlated with geologic permeability, alluvial aquifer material, and the more permeable soil types. Negative coefficients for the water balance and precipitation variables were observed, and these were counter-intuitive. They probably reflect the limitations of using annual climatological data (Farajalla et al., 1993).

As expected, nitrate concentrations were positively correlated with the annual tonnage of nitrogenous fertilizer shipped to the county within which a well was located. Concentrations were also positively correlated with total acres of cotton, oats, and wheat, and total irrigated acres of cotton and wheat. The negative correlation between nitrate concentrations and soybean acreage may reflect the nitrogen-fixing capacity of this legume. The variables with the greatest explanatory power included geological permeability, total acreage of cotton and oats, irrigated acreage of cotton, and the presence of sandy soils (Farajalla et al., 1993).

Based on the correlation coefficients of the bivariate and multivariate analyses, the parameters that most influenced the nitrates concentration in Oklahoma, ranked from high to low, were (Farajalla, et al., 1993): (1) geologic permeability of the overlying strata, (2) aquifer type, (3) depth to groundwater, (4) soil texture of the zone above the aquifer, and (5) amount of fertilizer shipped into the county.

NITRATE TRANSPORT MODELS

Table 6.3 contains summary comments on 26 references dealing with transport models for nitrates in the subsurface environment. These references (models) have been grouped into those related to the unsaturated zone, water management and drainage, the saturated zone, and combined consideration of both the unsaturated and saturated zones.

Unsaturated Zone Models

Five models related to nitrogen or nitrate movement and/or removal through the unsaturated portion of the subsurface environment are identified in Table 6.3. Ho, Mathew, and Newman (1985) described a model for the removal of ammonium in the vadose zone when secondarily treated wastewater is applied to spreading basins with alternate periods of flooding and drying. The processes involved in the removal of ammonium-nitrogen are adsorption by the soil, oxidation of the adsorbed ammonium to nitrate by nitrifying bacteria in the subsequent drying period (thus freeing adsorption sites in the soil for further ammonium adsorption), and reduction of nitrate to nitrogen gas by denitrifying bacteria during flooding. The extent of modeled nitrogen removal was governed by the rate of infiltration, the lengths of the flooding and drying periods, the adsorption capacity of the soil for ammonium, and the rates of nitrification and denitrification.

Jury (1978) proposed two models for calculating the time required for dissolved chemicals to move from the soil surface to either underlying groundwater in the case of free drainage, or to tile drain outlets in the case of artificial drainage. The models were based on the assumption that dissolved substances are trans-

Table 6.3 References Involving Nitrate Transport Models

Author(s) (year)	Comments
Unsaturated zone	
Barker and Foster (1981)	Mathematical model for a diffusion exchange mechanism for solute transport in fissured porous media
Ho, Mathew, and Newman (1985)	Model for ammonium removal in the vadose zone when secondary wastewater is applied to spreading basins
Jury (1978)	Model for calculating the time required for dissolved chemicals to move from the soil surface through the unsaturated zone
Kanwar, Johnson, and Baker (1983)	Hydrologic and nitrate-transport simulation model for an agricultural watershed during the crop growth period
Wellings and Bell (1980)	Piston displacement model for the movement of water and nitrate in the unsaturated zone of the Chalk aquifer in England
Water management/drainage	
Skaggs and Gilliam (1981)	Water management computer model to simulate soil water conditions and nitrate outflow from artificially drained soil
Saturated zone	
Kirkham and Affleck (1977)	Piston flow calculations for solute travel times to wells
Rice and Raats (1980)	Model for nitrate movement in a deep aquifer subjected to newly infiltrated wastewater
Schmid (1981)	Stream function approach for modeling seepage flow in extremely thin aquifers
Nitrate transport in unsaturated and saturated zones	
Carey and Lloyd (1985)	Numerical distributed transport model for simulating nitrate concentrations in groundwater
DeCoursey (1991)	Description of root zone water quality model (RZWQM) for simulating the movement of water and solutes from layer to layer in the soil profile
Frimpter, Donohue, and Rapacz (1990b)	Mass-balance accounting model for predicting nitrates in groundwater from septic tank systems and fertilizer applications; can be used in land development planning
Frimpter, Donohue, and Rapacz (1990a)	Detailed information on a mass-balance model for predicting nitrates in groundwater from a variety of urban and agricultural land uses
Hantzsche and Finnemore (1992)	Use of mass-balance approach for determining the potential cumulative effect of septic tank systems on nitrates in groundwater; graphical solutions provided as an aid in analysis
Hendry, Gillham, and Cherry (1983)	Numerical simulation of groundwater flow and solute transport, and an analytical solution of solute transport, to document role of denitrification
Keeney (1986)	Advantages and limitations of mass-balance models
Lerner and Papatolios (1993)	Analytical model for predicting nitrate concentrations in pumped groundwater in agricultural areas
Mehran, Noorishad, and Tanji (1983–1984)	Numerical models for vadose zone and aquifer system to predict nitrate pollution potential in groundwater
Mercado (1976)	Single-cell model for regional chloride and nitrate pollution patterns in the coastal aquifer of Israel
Oakes (1982)	Combined vertical transport model and catchment model for predicting changes in groundwater nitrates
Oakes, Young, and Foster (1981)	Simulation model for the vertical distribution of nitrate derived from agricultural land in British aquifers
Phillips and Gelhar (1978)	Convective transport model for conservative contaminant transport to deep wells

Table 6.3 References Involving Nitrate Transport Models (Continued)

Author(s) (year)	Comments
Spalding (1984)	Solute transport model for temporal variations and vertical stratification of nitrate in groundwater
Tinker (1991)	Application of three nitrogen mass-balance models for evaluating sources of nitrate-nitrogen in groundwater beneath unsewered subdivisions
Young (1981)	Simulation model for the distribution and movement of nitrate from agricultural land in the principal aquifers of the U.K.
Young, Oakes, and Wilkinson (1976)	Models for predicting the rate of nitrate movement through the unsaturated and saturated Chalk aquifer in England

ported primarily by moving the soil solution, which displaces soil water initially present in the wetted pore space (piston flow approximation). For free drainage, this results in a single travel time equation that is a function of soil water content and drainage volume. For tile drainage, the travel time also depends on the surface entry point.

Barker and Foster (1981) developed a model that accounts for a diffusion exchange mechanism for solute transport in fissured porous media. The model can be used to describe solute movement in an idealized, fissured, porous medium, involving diffusion exchange between mobile fissure water and immobile pore-water, for a range of values of various input parameters. The results of the model, when applied to such carbonate aquifers as the Chalk in the U.K., indicate that once solutes (pollutants) enter the Chalk matrix by some mechanism, it would take an extended time period for them to be completely eluted.

In a specific model tied to agricultural practices, Kanwar, Johnson, and Baker (1983) developed and used a hydrologic and nitrate-transport simulation model for the major water and nitrogen-transport processes occurring in a typical agricultural watershed during the crop growth period. Finally, Wellings and Bell (1980) utilized a piston displacement model for nitrate fluxes in the Chalk aquifer in the U.K. The average annual rate of downward movement of nitrate was found to be about 0.9 m per year, with a small lag of solute with respect to water. In the summer, upward water fluxes of 5 to 6 m are capable of moving solutes back up the profile by 10 cm or more.

Water Management/Drainage Model

Skaggs and Gilliam (1981) simulated soil water conditions and nitrate outflow from artificially drained soils using a water management computer model. Simulations were conducted for various drainage system designs and operational criteria for a poorly drained soil. The simulation model can be used to predict water table position, water content distribution, surface runoff, subsurface drainage volume, and seepage volumes at the end of selected time increments. If the nutrient concentration of each of the outflow sources is known or can be predicted, total mass outflows can be calculated on a day-to-day basis for a given drainage system design and mode of operation.

Saturated Zone Models

Three models were identified that deal with solute, or nitrate, movement in the saturated zone of aquifer systems. For example, theoretical travel times were given by Kirkham and Affleck (1977) for a solute to reach a well from an injection point for wells that fully penetrate confined aquifers of constant thickness and generally constant conductivity and porosity. The solute is assumed to move along streamlines with the water in piston flow under steady flow conditions. Further, it was noted that travel times can affect the extent of chemical changes of the solute (contaminant) during flow to the wells.

Schmid (1981) used a groundwater flow model that considered only a vertical cross-section of a thin aquifer in order to simplify an investigation of the source of nitrate concentrations of 130 mg/l at a water-works in West Germany. The stream function approach to describing seepage flow was used, since the velocity field and flow path were considered to be of primary importance. The approach resulted in a one-dimensional solution that neglected vertical flow components.

Rice and Raats (1980) analyzed the underground movement of renovated wastewater and presented equations to predict the minimum underground detention time for groundwater recharge systems. Analysis of the flow in a deep aquifer system showed that a nitrate peak, associated with the newly infiltrated wastewater, would be considerably flattened by the time the renovated water reached a collection well.

Models for Nitrate Transport in Unsaturated and Saturated Zones

The most comprehensive (complete) models for nitrates in the subsurface environment are those that address transport in both the unsaturated and saturated zones. The models often combine submodels for the separate zones; examples of both submodels and combined models are described herein.

The role of denitrification in a thin (3–7 m) phreatic sand aquifer beneath agricultural land was examined by Hendry, Gillham, and Cherry (1983). The study consisted of the application of physical hydrogeological methods of investigations, geochemical studies, environmental isotope studies, and numerical simulations of groundwater flow and solute transport. It was shown that denitrification was the principal cause of the nitrate distribution in the study area. The nonconservative nature of nitrates was also noted by Spalding (1984); he suggested that this nonconservative nature invalidates the predictive usage of classical solute transport models.

In order to model the observed nitrate contamination of wells in Long Island, New York, analytical and numerical techniques were used for addressing the convective transport of a conservative contaminant to a deep, partially screened pumping well overlain by a zone of contaminated water (Phillips and Gelhar, 1978). The three-dimensional flow to the well was treated as a point sink in an anisotropic medium. The effects of a regional downward flow, a phreatic surface above the well, and an impervious lower boundary were evaluated. It was shown that the local convective transport produced due to pumping greatly accelerates the process of contamination and decontamination of the wells.

Mercado (1976) used a single-cell model to study the regional chloride and nitrate pollution patterns in part of a coastal aquifer in Israel. The model integrated pollution sources on the land surface, hydrologic parameters of the aquifer and the unsaturated zone, and variations of chloride and nitrate concentration distributions in pumping wells. Through usage of the developed model, Mercado demonstrated that linear relationships exist between nitrogen quantities released on the land surface and the resultant quantities reaching the water table.

Mehran, Noorishad, and Tanji (1983–1984) used two numerical models, one for the vadose zone and the other for the aquifer zone, to predict the migration of nitrates in both zones as influenced by agricultural and groundwater management parameters. The model can be used to predict long-term impacts of agricultural activities on groundwater nitrate concentrations and to evaluate the effects of potential groundwater management alternatives. Figure 6.1 displays a schematic diagram for the overall computational steps used in the coupled models developed by Mehran, Noorishad, and Tanji (1983–1984). The processes contributing to changes in concentration of nitrates in the vadose zone included: dispersion-convection, plant uptake, ion exchange, and nitrogen transformations. These transformations incorporated nitrification, denitrification, mineralization, and immobilization processes.

Figure 6.2 shows the nitrogen transformation pathways used in the model developed by Mehran, Noorishad, and Tanji (1983–1984). The concentration of nitrate in the aquifer was assumed to be affected only by dispersion-convection phenomena. The transport equations, applied to the vadose zone and to the aquifer, were solved by finite difference and finite element methods, respectively. Although the primary reason for this approach lies in the chronological development of the computer codes, the advantages of the approach are numerous. The finite difference method, used for the vadose zone, where biophysicochemical processes are numerous and complex but the geometry is one-dimensional and simple, has the advantage of being computationally efficient. In contrast, the finite element method is suited for the aquifer zone in which the primary mechanism of transport is dispersion-convection, but the transport is two-dimensional and the geometry is more complex.

Mehran, Noorishad, and Tanji (1983–1984) concluded by noting that for field-scale problems, one-dimensional transport in the vadose zone with two-dimensional flow in the aquifer allows a realistic description of transport processes in the overall flow region. This type of modeling approach provides an efficient and practical tool for long-term predictions of the impact of agricultural activities on aquifer systems and evaluation of potential nitrate management alternatives.

Several comprehensive models have been developed for predicting the impacts of agricultural practices on nitrate concentrations in the Chalk aquifer system in the U.K. An early model was developed to predict the rate of movement of nitrate through the unsaturated and saturated zones at a farm site near Winchester, England (Young, Oakes, and Wilkinson, 1976). After this, Oakes, Young, and Foster (1981) described several mathematical models developed to simulate

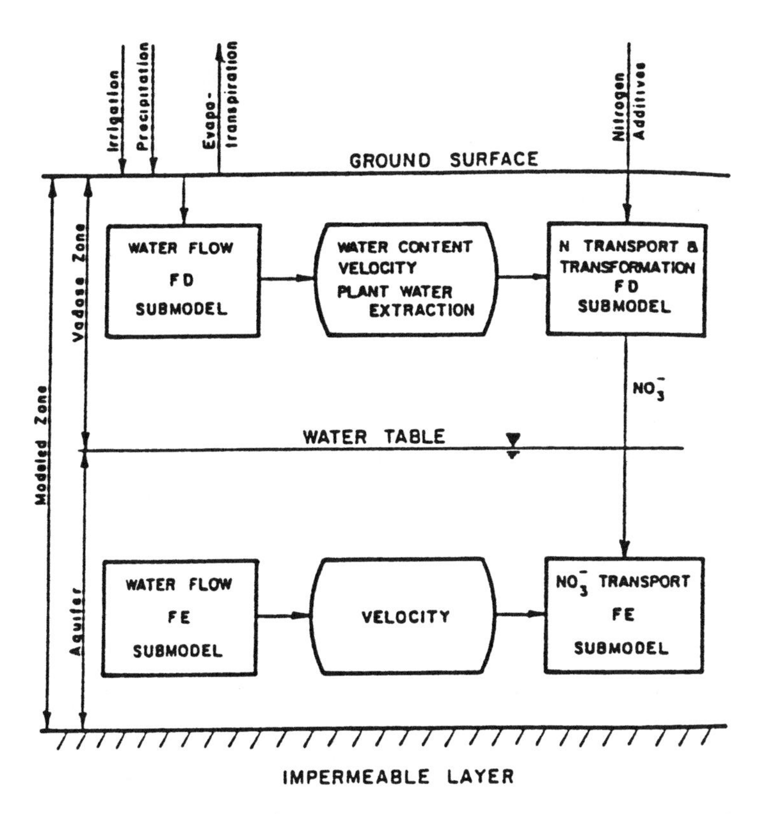

Figure 6.1 Schematic diagram of the overall computational steps used in coupled model. (From Mehran, M., Noorishad, J., and Tanji, K. K., *Environmental Geology*, Vol. 5, No. 4, 1983/4, pp. 213–218. With permission.)

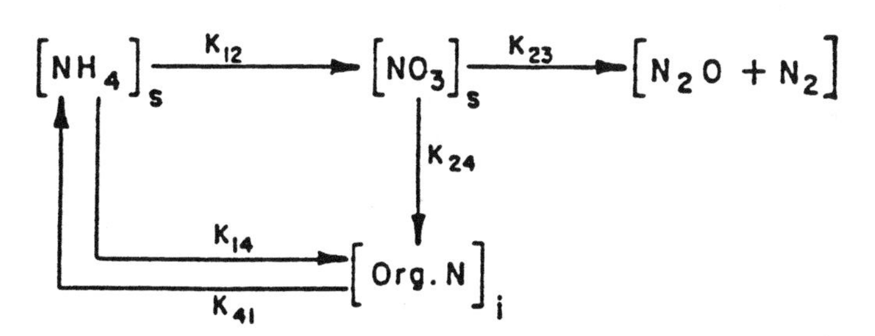

Figure 6.2 Nitrogen transformations pathways included in coupled model. (From Mehran, M., Noorishad, J., and Tanji, K. K., *Environmental Geology*, Vol. 5, No. 4, 1983/4, pp. 213–218. With permission.)

the vertical distribution of mobile solutes, notably nitrate and tritium, at a number of sites overlying the Chalk aquifer system.

Two mathematical models to simulate the movement of solutes derived from diffuse sources through aquifers were addressed by Young (1981). The first was a one-dimensional model for simulating solute distributions in the unsaturated zone and requiring, as inputs, estimates of annual infiltration, land use history and fertilizer application rates, and variations in pore-water content and hydraulic properties with depth. Control rules governing the release of nitrate from arable crops and grassland were promulgated based on published data and farming experiments; these rules indicated that a quantity of nitrate equivalent to approximately 50% of the fertilizer applied to arable crops became available for leaching, and that the nitrogen removed from the plowing of grassland varied with the age of the grass, but was in the range of 120 to 380 kg N/ha. The relatively accurately known rainfall tritium inputs have been used to obtain a mass-balance of inputs against quantities measured in Chalk profiles. The balances indicate that between 5 and 15% of infiltration bypasses the fine pore structure of the Chalk, via the fissure systems, suggesting comparable frequencies of intense infiltration events. The model produced satisfactory simulations of measured nitrate and tritium profiles at 15 sites in the Chalk, one of which is shown in Figure 6.3 (Young, 1981). In these simulations, nitrate is treated as a conservative solute. Although the presence of bacterial populations, potentially capable of nitrogen transformations, have been reported from the unsaturated Chalk to depths of 50 m, evidence of nitrate removal by such processes was not substantiated.

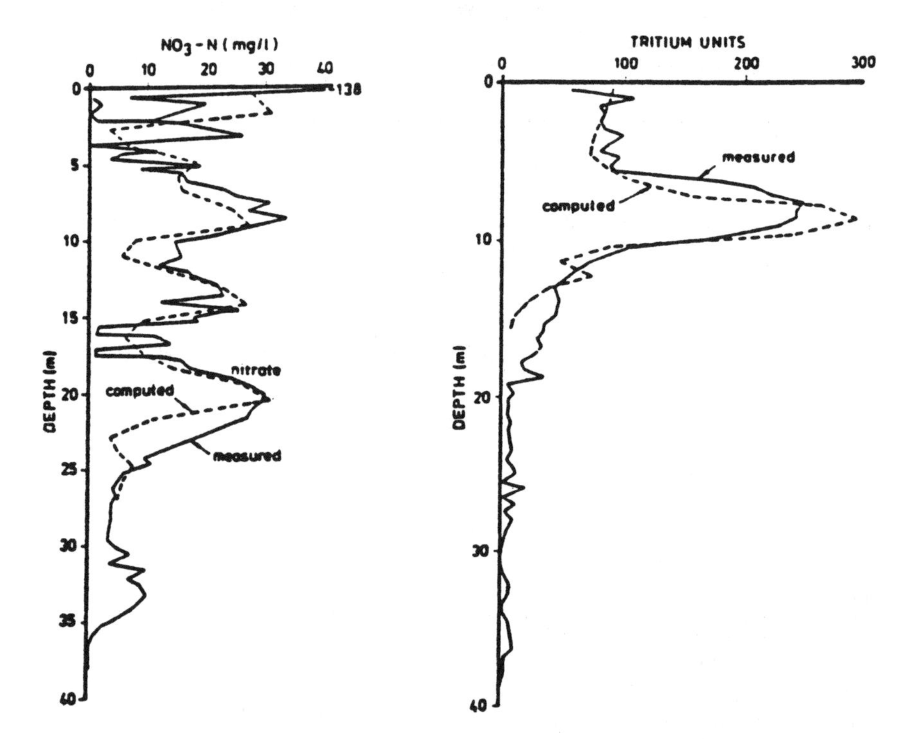

Figure 6.3 Vertical flow model simulations of nitrate and tritium profiles, arable site on Chalk, near Winchester, Hampshire, U.K. (From Young, C. P., *Water Science and Technology,* Vol. 13, No. 4, 1981, pp. 1137–1152. With permission of Elsevier Science.)

The second model described by Young (1981) allowed simulations of nitrate fluxes within the groundwater system; the model consists of a series of the one-dimensional vertical flow model described above and fed into a two-dimensional fully mixing aquifer system flow model. Calibration of the vertical flow routine against measured profiles and verification of the combined model by comparison with historical groundwater quality data was completed for three study areas.

Oakes (1982) expanded on the modeling work by Young (1981) in terms of additional site work and details relative to modeling components. Again, a vertical transport model for the unsaturated zone was coupled with a model from the groundwater system (catchment area). The assumption basic to the vertical flow model was that solutes originating at the soil surface were leached downward at a rate dependent on infiltration and the pore-water content of the rock. It was assumed that a small fraction, typically 5 to 15% of the infiltrating water and solutes, moves quickly to the water table via larger fissures. The remaining water and solutes fill the pore spaces at the top of the unsaturated zone, thus displacing downward water and solutes already in the profile. Some attenuation of peak concentrations was assumed to occur and was modeled by a dispersion mechanism. The model can be used for up to yearly time increments based on the following inputs (Oakes, 1982):

1. Annual infiltration rates for the period of simulation
2. Land-use history and fertilizer application rates for the period of simulation
3. Pore-water content, assumed to be constant in time and with depth

The mass of nitrogen released each year in the soil layers for uptake by infiltrating water was assumed to depend on present and antecedent field use and fertilizer application. About 50% of the fertilizer applied to root crops and cereals was not taken up by the crop; thus, it was assumed that this quantity was leached downward. However, not all of this material is available in the year of application; for example, Table 6.4 provides an estimate of the nitrate available to infiltration as a fraction of the nitrogen application rate (Oakes, 1982).

Table 6.4 Model Control Rules for Roots and Cereals

	N (kg/ha) available as a fraction of the application rate of N (kg/ha)		
Crop	Year of application	Following year	Next following year
Roots	0.35	0.10	0.05
Cereals	0.25	0.15	0.10

From Oakes, D. B., *Proceedings of an IIASA Task Force Meeting on Nonpoint Nitrate Pollution of Municipal Water Supply Sources: Issues of Analysis and Control*, Series No. CP-82-54, 1982, International Institute of Applied Systems Analysis, Laxenburg, Austria, pp. 207–230. With permission.

Further, it was determined that the result of varying the time distribution of the mineralization rate in the vertical transport model was small; the major contribution to nitrate leaching came from the plowing of grassland. By matching the model results with observed nitrogen profiles in a number of boreholes, it was possible to estimate the amount of nitrogen released by plowing grasslands of various ages. Table 6.5 summarizes the corresponding model control rules (Oakes, 1982). For each year of simulation, the following sequence of computations was undertaken (Oakes, 1982):

1. Evaluate the mass of nitrate per unit surface area available for leaching from the soil zone using the land-use history and model-control rules.
2. Divide the mass of nitrate leached by the infiltration to obtain the mean annual concentration.
3. Route a fixed fraction of the leachate directly to the water table, conceptually through the larger fissures.
4. With the remaining leachate, fill the available pore space at the top of the unsaturated zone, displacing downward water and solutes already in the profile.
5. Aggregate water and solutes so displaced from the base of the unsaturated zone with those moving rapidly downward via the larger fissures to obtain the next fluxes at the water table.
6. Apply the dispersion equation to the solute profile.

Table 6.5 Model Control Rules for the Plowing of Grass

Years in grass prior to plowing	N(kg/ha) released by plowing			
	Total	Year of plowing	Following year	Next following year
1	100	60	30	10
2	190	114	57	19
3	240	144	72	24
4 or more	280	168	84	28

From Oakes, D. B., *Proceedings of an IIASA Task Force Meeting on Non-point Nitrate Pollution of Municipal Water Supply Sources: Issues of Analysis and Control,* Series No. CP-82-54, 1982, International Institute of Applied Systems Analysis, Laxenburg, Austria, pp. 207–230. With permission.

The model of nitrate movement in the saturated zone had as inputs the nitrate fluxes across the water table generated by the vertical flow model (Oakes, 1982). The catchment area was divided into 500 × 500 m squares, and nitrates and water routed through the system according to the prevailing hydraulic gradients. Nitrate reaching the water table during a yearly increment was assumed to completely mix with nitrate already in the saturated zone. This type of model is generally called a "fully mixed cell model"; it can be used to address diffuse source inputs.

The generation of nitrate fluxes across the water table using the vertical flow model is not a necessary prerequisite to groundwater modeling. The principal effect of the unsaturated zone is to delay solutes in their movement to the water table. In the case of the Chalk aquifer, the delay is about 2 years per meter of unsaturated thickness (Oakes, 1982). For groundwater modeling, the nitrate fluxes across the water table may be generated directly from the time series of nitrate leached from the soil zone by applying lags dependent on unsaturated zone thickness and infiltration. The advantages of using the vertical flow model to simulate measured unsaturated zone profiles is the check on the hydrological data and the application of the model control rules. The groundwater model can also be used for yearly time increments when the following inputs are provided (Oakes, 1982):

1. Annual infiltration rates for the period of simulation
2. Land-use history and fertilizer application rates for the period of simulation for distinct land units within the catchment
3. Distribution of groundwater levels and pumped abstractions
4. Depths to water table
5. Pore-water contents of the unsaturated and saturated zones
6. Effective depth of flow in the saturated zone

For each year of simulation, the following sequence of computations should be undertaken (Oakes, 1982):

1. Evaluate groundwater flows from the water level distribution, infiltration rates, and groundwater abstractions starting at the highest water level in the catchment and working through the nodes in order of decreasing water level.
2. For each node, evaluate the nitrate flux at the water table either using the vertical flow model or directly from the historical land use data by applying a lag time dependent on the unsaturated zone water content and thickness, and the infiltration.
3. Starting at the node with the highest water level and working through the nodes in order of decreasing level, evaluate the nitrate concentrations assuming that nitrate entering each node, either from above or with subsurface flow, mixes fully with nitrate already stored in the unsaturated zone.

The next comprehensive model to be described also represents an expansion of previously described work. Carey and Lloyd (1985) updated the work of Oakes (1982) and Young (1981) in developing a distributed transport model to simulate nitrate concentrations in groundwater for an area of approximately 600 km^2 of the Chalk aquifer. The method involves the use of a number of simplified equations that allow the potential nitrate load — that is, the quantity of nitrate available for leaching from the soil — to be calculated for different land-use types. The components of the methodology include the following (Carey and Lloyd, 1985):

1. Arable Land: Nitrate losses from arable soils were assumed to be equivalent to a percentage of the total input of nitrogen to the soil from inorganic and organic fertilizers, from atmospheric precipitation, and from the quantity of nitrate made available by mineralization of soil organic nitrogen; the assumptions are depicted in the following equation:

$$NA = (FERT(m) + ATN) \cdot (100 - CU(m))/100$$

where NA = nitrate available for leaching from arable crops (kg/ha·yr N); m = crop type; FERT = fertilizer (inorganic and organic) applied to crop (kg/ha·yr N); ATN = nitrate available from precipitation and mineralization of organic matter (kg/ha·yr N); and CU = percentage of nitrate removed at harvest.

2. Grassland and Woodland: The nitrate concentrations of leachate waters from beneath unfertilized grass, heathland, and woodland are typically less than 2.0 mg/l NO_3–N. Higher nitrate losses are observed for fertilized grassland (4–10 mg/l NO_3–N), and it is only when fertilizer nitrogen applications are excessive, that is, greater than 400 kg/ha·yr N, that nitrate losses become very high, that is, greater than 50 mg/l NO_3–N. Average fertilizer applications in the modeled area are less than 200 kg/ha·yr N, so that high nitrate losses from grassland are unlikely.
   The quantity of nitrate available for leaching was calculated based on the following equation:

$$NG = 0.01 \cdot AI \cdot (GU + GR + GW)$$

   where NG = nitrate input from grass (kg/ha·yr N); AI = potential recharge (mm/yr); GU = nitrate concentration of drainage from unfertilized grass (mg/l NO_3–N); GR = nitrate concentration of drainage from fertilized grass (mg/l NO_3–N); and GW = nitrate concentration of drainage from woodland (mg/l NO_3–N).
   In this approach, the nitrate concentration of recharge was assumed to be constant and the quantity of nitrate leached dependent on the amount of recharge. For fertilized grass, a simple rule was applied where for every 50 kg/ha·yr N applied, 2.0 mg/l NO_3–N was leached.
3. Plowing of Grassland: The plowing of grassland has been shown to release large quantities of nitrate. The amount of nitrate released depends predominantly on the age of the grass and soil type. Because agricultural data typically differentiate between grass older and younger than 4 years, a number of assumptions were necessary to calculate the input from the plowing of grass. The amount of nitrate released by the plowing of temporary grass and permanent grass was taken as 110 and 260 kg/ha N, respectively. The area of permanent grass plowed each year was taken as the difference in area in successive years; 35% of temporary grass was assumed to become permanent grass and 50% of the remainder was plowed each year. The calculation was based on:

$$NP = (PG(I + 1) - PG(I) + 0.35 \cdot TG(I + 1) - TG(I)) \cdot PP$$
$$+ 0.50 \cdot (1.0 - 0.35) \cdot TG(I) \cdot TP$$

   where NP = nitrate released by plowing of grassland in year I (kg/ha·yr N); PG = area of permanent grass in year I (ha); TG = area of temporary grass in year I (ha); PP = nitrate released by plowing of permanent grass (kg/ha·yr N); and TP = nitrate released by plowing of temporary grass (kg/ha·yr N).
4. Potential Nitrate Load: The potential nitrate load (NPL) was defined as follows:

$$NPL = NA + NG + NP$$

5. Actual Nitrate Load: The actual nitrate load to the aquifer (ANL) was assumed to be dependent on the influence of soils and superficial deposits on recharge, and on denitrification losses as described by the following expression:

$$ANL = ((PNL \cdot AR)/AI) \cdot DF$$

where ANL = actual nitrate load (kg/ha·yr N); PNL = potential nitrate load (kg/ha·yr N); AR = actual recharge (mm/yr); AI = potential recharge, precipitation minus evapotranspiration and direct surface runoff (mm/yr); and DF = empirical factor to account for denitrification.
The weighting factors to account for denitrification were determined by calibrating the transport model; the resultant values are given in Table 6.6 (Carey and Lloyd, 1985). Soil and superficial deposits were classified according to thickness and clay content.

Table 6.6 Empirical Weighting Factors (DF) to Account for Denitrification Occurring in Different Soil and Drift Types

Drift or soil type	Thickness (m)	Clay content (%)	Weighting factor, DF
Soils derived from Chalk or sand and gravels	0–2.0	0–10	0.99
	0–1.0	10–20	0.90
	0–1.0	20–25	0.85
	0–1.0	25–30	0.80
	0–2.0	30–35	0.75
Boulder clay derived soils	0–5.0	35–45	0.45
	0–5.0	>45	0.10
	>5.0	>45	0.05

Note: Clay content <2 μm.

From Carey, M. A. and Lloyd, J. W., *Journal of Hydrology,* Vol. 78, No. 1/2, 1985, pp. 83–106. With permission of Elsevier Science, Amsterdam.

Following the determination of the actual nitrate load, the next issue was related to mechanisms controlling the movement of nitrate in the unsaturated and saturated zones of the Chalk aquifer (Carey and Lloyd, 1985). There has been considerable debate as to whether fissure flow or intergranular flow is the predominant mechanism of water movement in the unsaturated zone of this aquifer. The debate has occurred because research was conducted at different sites in England, and depending on climate, soil type, weathering, and the local nature of the Chalk, the controlling processes could differ from site to site. To illustrate the findings for a site in Hampshire, water movement was by intergranular flow and therefore the downward migration of nitrate in the unsaturated zone was by a piston displacement mechanism; thus, the rate of transport could be calculated as follows (Carey and Lloyd, 1985):

$$V = (AI)/\theta$$

where

V = downward rate of migration of nitrate (mm)
AI = actual recharge (mm)
θ = porosity

In forecasting nitrate concentrations in groundwater, the worst-case situation was considered and a piston-flow mechanism assumed, allowing for a component of bypass (10%). The average downward rate of movement of nitrate in the model area was calculated as 0.45 m/yr (Carey and Lloyd, 1985), and the calculated delay times for nitrate to pass through the Chalk unsaturated zone to the water table are generally greater than 50 years.

The input to the saturated zone is from the flushing of nitrate in pore-water in the Chalk unsaturated zone by the groundwater flow in the zone of water level fluctuations and from the bypass component. The data available on the distribution of nitrate in the Chalk saturated zone indicated that there is a vertical stratification of nitrate with higher concentrations in the top 20 to 40 m of the saturated zone, decreasing to lower concentrations (often <1.0 mg/1 NO_3–N) at depth, which may reflect limited groundwater flow or denitrification (Carey and Lloyd, 1985). A feature of long-term historical nitrate concentrations measured in the Chalk groundwater is the lack of seasonal variations; this is in contrast to the seasonal fluctuations of groundwater levels and the seasonal nature of the nitrate input to the saturated zone, thus indicating that the system is buffered by a storage larger than that provided by fissure water (Carey and Lloyd, 1985).

Carey and Lloyd (1985) then modeled nitrate concentrations in the saturated zone of the Chalk aquifer by using a distributed groundwater flow model that was modified to include a mixing cell calculation to examine the dilution of the nitrate input entering the saturated zone. For each cell, a nitrate load from the soil zone was calculated and a delay calculation was included to account for the transit time for nitrate to pass through the unsaturated zone. The input to the saturated layer was calculated as the sum of the component of nitrate transmitted rapidly through the unsaturated zone via fissures (10%) and the component of nitrate moving slowly downwards by piston flow (90%) (Carey and Lloyd, 1985).

A mass-balance calculation was performed for each modeled cell as follows:

$$d(S_{ij}C_{ij})/dt = \Sigma Q_{Xij}C_{ij} + \Sigma Q_{Yij}C_{ij} + R_{ij}C_{Rij} - A_{ij}C_{Aij}$$

where

ΣQ_{Xij}, Q_{Yij} = groundwater flows into and out of cell, respectively
S_{ij} = storage of cell (mixing depth multiplied by effective porosity)
R_{ij} = recharge to cell
A_{ij} = abstraction (pumping) from cell
C_{ij} = nitrate concentration of cell
C_{Rij} = nitrate concentration of recharge to cell
C_{Aij} = nitrate concentration of abstraction
i,j = coordinates of cell
t = time

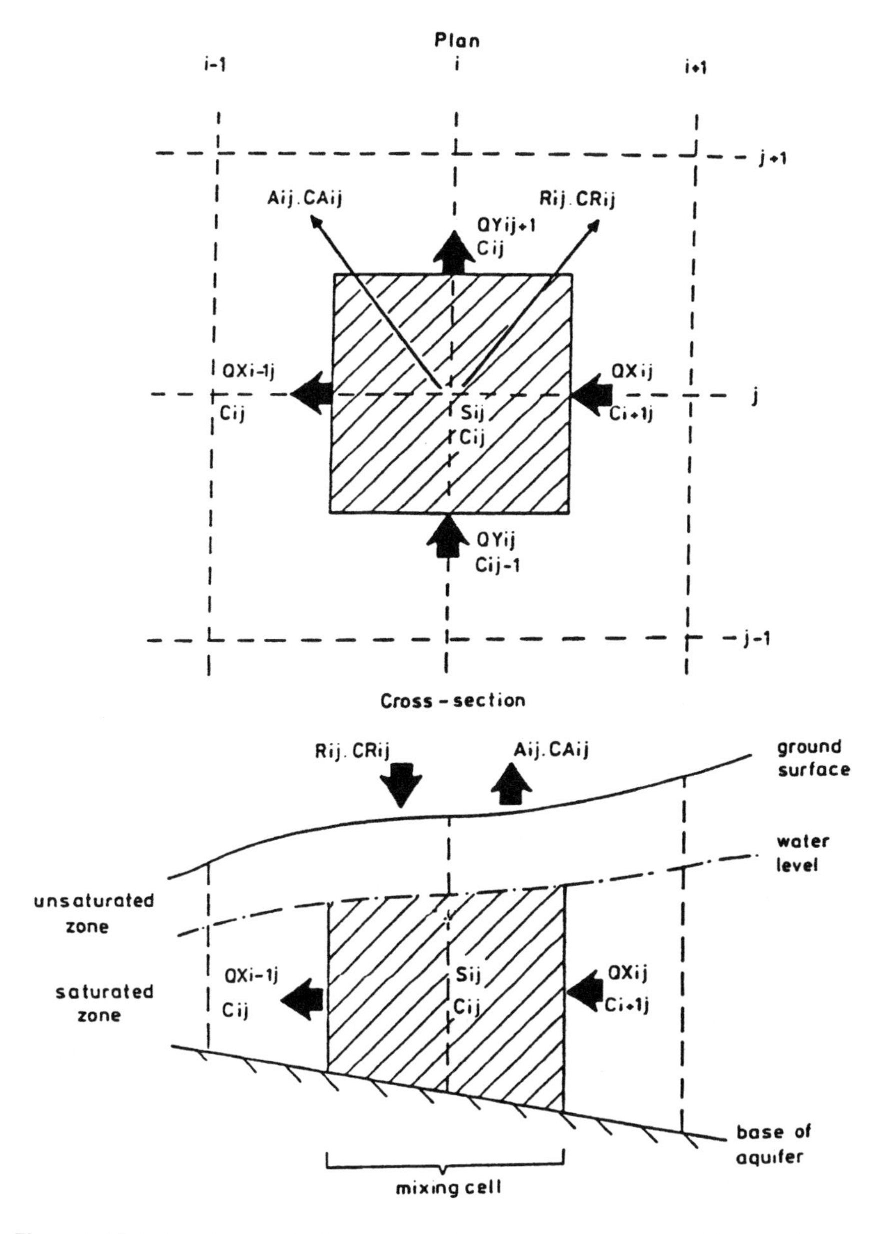

Figure 6.4 Details of mixing cell showing mass-balance components for cell ij. (From Carey, M. A. and Lloyd, J. W., *Journal of Hydrology*, Vol. 78, No. 1/2, 1985, pp. 83–106. With permission of Elsevier Science, Amsterdam.)

Figure 6.4 shows the details of the mixing cell and the various components used in the mass-balance calculation (Carey and Lloyd, 1985). Complete mixing of nitrate throughout the cell was assumed. Such a model is ideally suited to this study because of the diffuse nature of the nitrate input to the aquifer. Groundwater flow was assumed to occur to a maximum depth of 40 m, which was taken as the depth of nitrate mixing. This depth corresponds to the observed vertical distribution of nitrate and to the average depth of penetration of abstraction boreholes in the saturated zone. The nitrate concentration in fissure and pore-water was assumed to be in equilibrium. The effective porosity of the Chalk matrix was taken as 35%; therefore, there is a large storage available to buffer variations in the nitrate input.

The model by Carey and Lloyd (1985) was calibrated and applied to several locations on the Chalk aquifer. The results of the model application shown in Figure 6.5 indicate that the long-term trend of nitrate concentrations over the last 10 to 20 years can be simulated for the majority of sources (Carey and Lloyd, 1985). An exact simulation of the year-to-year variations of nitrate concentrations in groundwater was not obtained. This was probably due to the fact that each model cell represents the average condition in terms of the nitrate input, groundwater flow, and nitrate movement for the aquifer volume represented by that cell. As the purpose of the investigation was to simulate long-term trends of nitrate concentrations in groundwater, the model results were considered satisfactory (Carey and Lloyd, 1985).

The model was used to forecast the anticipated nitrate concentrations in groundwater to the year 2020. A number of options of future agricultural management was examined to determine the effects on nitrate input to the aquifer. Predicted nitrate concentrations in abstracted (pumped) groundwater are shown in Figure 6.6; these were calculated assuming no change in the present nitrate input from the soil zone, no change in agricultural cropping, but a 20% increase in fertilizer applications (Carey and Lloyd, 1985). The effect of differing agricultural practices on predicted nitrate concentrations for the abstraction sources modeled did not become apparent for 30 to 40 years, mainly because of the significant delay for nitrate to pass through the unsaturated zone.

In summary, relative to the modeling results of Carey and Lloyd (1985), forecasts for the groundwater supply sources in the modeled area showed that the present trend of nitrate concentrations, 0.05 to 0.1 mg/l per year nitrate-nitrogen, will increase gradually to 0.1 to 0.2 mg/l per year nitrate-nitrogen over the next 40 years. Higher rates of increase can be expected in areas with thinner unsaturated zones. Further, it should be noted that the worst-case model based on piston displacement was assumed for the downward movement of nitrate through the Chalk unsaturated zone. Calculation of the present nitrate concentration of leachate from arable land indicates that nitrate concentrations in groundwater will eventually exceed 22.6 mg/l nitrate-nitrogen. However, because of the considerable delay for nitrate to pass through the unsaturated zone, this concentration will not be reached until the middle part of the next century over a greater part of the study area.

The Agricultural Research Service of the U.S. Department of Agriculture developed the root zone water quality model (RZWQM) in 1990 (DeCoursey, 1991). The model simulates the movement of water and solutes (pesticides and nitrates) from layer to layer in the soil profile. It includes flow through such large pores as worm-holes and cracks. Nitrogen cycling is simulated using soil moisture

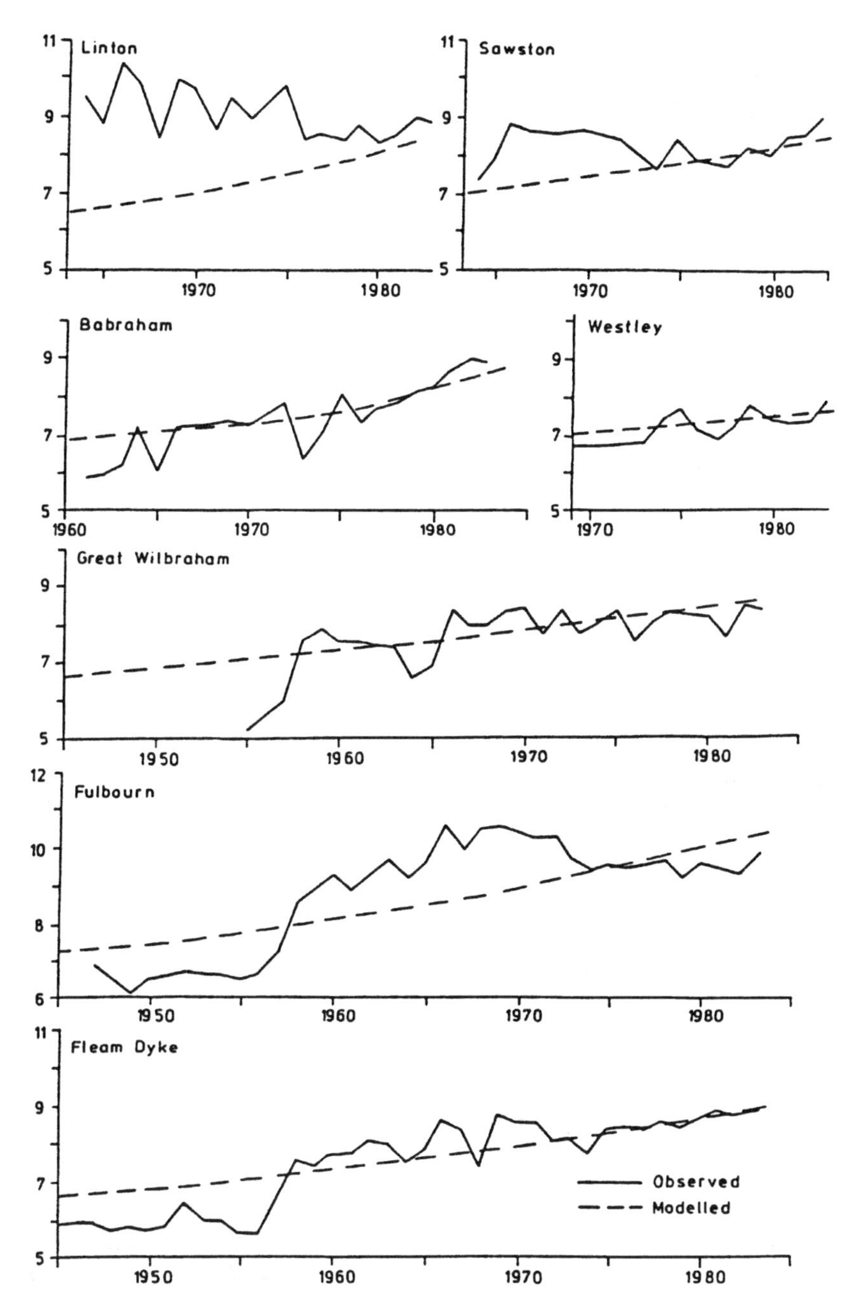

Figure 6.5 Comparison of simulated against observed nitrate concentrations of groundwater at public water supply sources. (From Carey, M. A. and Lloyd, J. W., *Journal of Hydrology*, Vol. 78, No. 1/2, 1985, pp. 83–106. With permission of Elsevier Science, Amsterdam.)

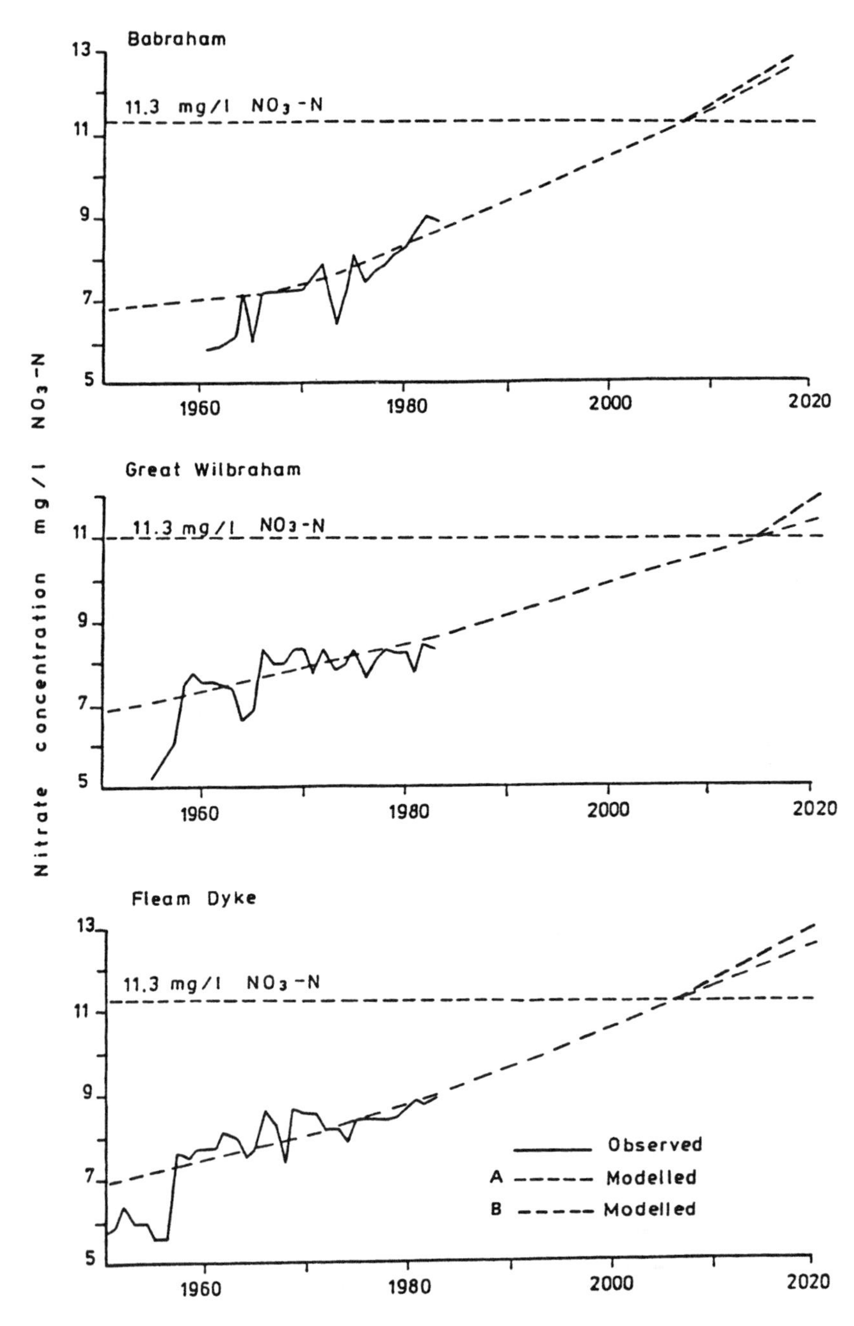

Figure 6.6 Forecast nitrate concentrations in groundwater for: Option A, no change in present nitrate input from the soil zone; and Option B, no change in agricultural cropping but a 20% increase in fertilizer applications. (From Carey, M. A. and Lloyd, J. W., *Journal of Hydrology*, Vol. 78, No. 1/2, 1985, pp. 83–106. With permission of Elsevier Science, Amsterdam.)

levels, temperature, bacterial populations, degree of acidity, amount and quality of organic matter, etc. The model includes a plant growth submodel that simulates root and surface vegetative growth and grain production. Plant stresses caused by temperature, nutrient, and soil water extremes are simulated. Also included are those physical properties of the soil that change as a result both of tillage and of the reconsolidating effects of rainfall.

Frimpter, Donohue, and Rapacz (1990a) developed a mass-balance accounting model that can be used to guide the management of septic tank systems and fertilizers to control the degradation of groundwater quality in zones of an aquifer that contribute water to public supply wells. A computerized spreadsheet for the model is available. The fundamental model for a defined zone of contribution area around the well is as follows (Frimpter, Donohue, and Rapacz, 1990a):

$$C_w = \frac{C_r\left(V_w - 0.9\left(V_1 + V_2 + \ldots + V_n\right)\right) + \left(L_1 + L_2 + \ldots + L_n\right)}{V_w}$$

where

C_w = nitrate concentration of groundwater at the well, in mg/l
V_w = volume of withdrawal from well, in liters
C_r = nitrate concentration in recharge from precipitation, in mg/l
$L_1+L_2+\ldots+L_n$ = nitrate load, in milligrams, from individual sources where $L = C \times V$, when load is calculated from the volume and nitrate concentration of effluent from the source
$C_1+C_2+\ldots+C_n$ = nitrate concentration from individual sources, in mg/l
$V_1+V_2+\ldots+V_n$ = volume of water used by each source before discharge to septic tank system, in liters

To illustrate the usage of the mass-balance accounting model, consider that 15 nitrogen sources in a zone of contribution area have been defined as shown in Tables 6.7 and 6.8 (Frimpter, Donohue, and Rapacz, 1990a). It is assumed that the well is pumped at 1.0 million gallons per day (3,785,000 liters per day); and that the recharge from precipitation has a nitrate concentration of 0.05 mg/l. The concentration of nitrate at the pumped well can be calculated as follows (Frimpter, Donohue, and Rapacz, 1990a):

$$(V_1 + V_2 + \ldots + V_{13}) = 426{,}860 \text{ liters}$$

$$\begin{aligned}(L_1 + L_2 + \ldots + V_{15}) &= 2{,}017{,}580 + 16{,}497{,}275 \\ &= 18{,}514{,}855\end{aligned}$$

$$C_w = \frac{0.05\left(3{,}785{,}000 - 0.9\,(426{,}860)\right) + 18{,}514{,}855}{3{,}785{,}000}$$

$$C_w = \frac{18{,}684{,}896}{3{,}785{,}000} = 4.94 \text{ mg/l}$$

Table 6.7 Summary of Nitrate Loads from Septic Tank Systems for 1-d Average Period in 1.0-mgd Well

Source	Flow (gal/d)	Units (variable)	Volume (l/d)	Nitrate as nitrogen conc. (mg/l)	Load (mg/d)
1. $^1/_2$-Acre housing	65/person	400 people	98,410	40	3,936,400
2. High school	20/student	1000 students	75,700	40	3,028,000
3. Fast food restaurant (counter seat)	150/seat	70 seats	39,740	40	1,589,700
4. Fast food restaurant (table seat)	350/seat	10 seats	13,250	35	463,750
5. One acre housing	65/person	200 people	49,210	40	1,968,400
6. Condominium	65/person	120 people	29,520	40	1,180,800
7. Shopping center	60/employee	50 employees	11,360	40	454,400
8. Office building	15/employee	25 employees	1,420	40	56,800
9. Gas station	500/island	2 islands	3,785	40	151,400
10. Church	3/seat	200 seats	2,270	40	90,800
11. Motel A	75/person	40 people	11,355	35	397,425
12. Motel B	75/person	160 people	45,420	35	1,589,700
13. Hospital	200/bed	60 beds	45,420	35	1,589,700
Totals		$(V_1 + V_2 + ... + V_{13}) =$	426,860	$(L_1 + L_2 + ... + L_{13}) =$	16,497,275

Note: gal/d, gallons per day; l/d, liters per day; mg/l, milligrams per liter; mg/d, milligrams per day.

From Frimpter, M. H., Donohue, J. J., and Rapacz, M. V., “A Mass-Balance Nitrate Model of Predicting the Effects of Land Use on Ground Water Quality,” Report 88-493, 1990a, U.S. Geological Survey, Boston, MA.

Table 6.8 Summary of Solid Nitrate Loads

		Nitrate as nitrogen		
Source	**Units**	**(lb/d)**	**(mg/lb)**	**Load (mg/d)**
14. Lawns (5000 ft²)	100 lawns	0.025[a]	454,000	1,135,000
15. Horses @ 1200 lb each	6 horses	0.027/100 lb of animal	454,000	882,580
Total			$(L_{14} + L_{15})$ =	2,017,580

Note: ft², square feet; lb/d, pounds per day; mg/d, milligrams per day; mg/lb, milligrams per pound.

[a] Based on 9 lb/year of nitrate leaching into the groundwater system from 5000 ft² of lawn.

From Frimpter, M. H., Donohue, J. J., and Rapacz, M. V., "A Mass-Balance Nitrate Model of Predicting the Effects of Land Use on Ground Water Quality," Report 88-493, 1990a, U.S. Geological Survey, Boston, MA.

The effects of alterations in pumping (V_w), the addition of other nitrate sources, and/or control measures on existing sources can be examined by the above mass-balance accounting model. Also, the model can be modified to account for induced infiltration from streams or groundwater discharge to streams. The report by Frimpter, Donohue, and Rapacz (1990a) contains detailed information on nitrate loads from restaurants, schools, parks/campgrounds, hospitals, recreation areas, commercial establishments, dwellings, animal feedlots, wastewater treatment facilities, lawn fertilizers, and golf courses.

The mass-balance accounting model for urban areas is based on the following simplifying assumptions (Frimpter, Donohue, and Rapacz, 1990a):

1. The approach assumes that, under steady-state withdrawal conditions, all of the water and nitrate withdrawn from the well are derived from the zone of contribution for the well, and that only some of the water withdrawn is returned to the zone of contribution as return flow. In those situations where a well derives some of its yield from induced infiltration from streams or other surface-water bodies, the quantity and quality of induced infiltration need to be entered in the accounting.
2. The model is useful for predicting the nitrate concentration at the well under steady-state conditions where all of the water from the zone of contribution is mixed. Individual plumes with elevated concentrations of contaminants would be expected to emanate from septic tank systems and other sources within the zone of contribution. Therefore, the prediction is not appropriate for determining contaminant concentration at other points within the aquifer, or for determining the concentration in any smaller (private-domestic supply) wells within the zone of contribution.
3. After entering the saturated zone, the contaminant (nitrate) is considered to be conservative. It is not precipitated or adsorbed by aquifer materials. Attenuation in the saturated zone is assumed to occur only through the process of dilution. Some diminishment of nitrate through other processes such as denitrification is known to occur, but the quantities affected are not large enough to be considered in these gross calculations. Any changes in water quality owing to renovation in the unsaturated zone need to be accounted for before load values are input to the mass-balance model. Reduction of source loads

in the unsaturated zone will be dependent on soil type, the thickness of the unsaturated zone, and the interaction of the source's variable components, which are specific to each zone of contribution. No renovation was assumed in the above model because the unsaturated zone in the study area was thin (10 to 30 ft) and composed of permeable coarse sand.

4. The zone of contribution to the well is assumed to remain constant in size and shape for application of the nitrate accounting approach. Actually, the size of the zone is expected to become smaller as more return flow from septic tank systems recharges the zone of contribution, but additional recalculations of the zone of contribution would most likely be expensive and have an unacceptably high cost-to-benefit ratio. Therefore, this assumption results in protection of a zone slightly larger than may actually contribute water to the well and is therefore considered conservative if sources are uniformly distributed. Recharge to the aquifer is assumed to be uniform over the zone of contribution. Where variations of aquifer properties or surface-drainage characteristics cause irregular distribution of recharge, both the delineation of the zone of contribution and the calculation of contaminant concentration would have to take those variations into account.

Tinker (1991) conducted a study of the sources of nitrate-nitrogen in five unsewered subdivisions in Eau Claire County and La Crosse County, Wisconsin. Three nitrogen mass-balance models were used: (1) the Wehrmann model, (2) the BURBS model, and (3) a combination of the Wehrmann and BURBS models. The Wehrmann model, which does not address input of nitrogen from lawn fertilizers, is as follows (Tinker, 1991):

$$V_b C_b + V_i C_i + V_s C_s - V_p C_p = (V_b + V_i + V_s - V_p) C_o$$

where

V_b = volume of groundwater entering the subdivision from upgradient area

C_b = concentration of nitrate-nitrogen contained in the groundwater entering the subdivision

V_i = volume of precipitation infiltrating beneath the subdivision

C_i = concentration of nitrate-nitrogen contained in the infiltrating precipitation

V_s = volume of septic tank system effluent introduced beneath the subdivision

C_s = concentration of nitrate-nitrogen contained in the septic tank system effluent

V_p = volume of groundwater pumped by wells beneath the subdivision

C_p = concentration of nitrate-nitrogen contained in the pumped groundwater

C_o = diluted concentration of nitrate-nitrogen leaving the subdivision

The BURBS model considers nitrogen input from turf, natural land, and impervious land; thus, lawn fertilizer is an input parameter to the model. The BURBS model predicts the nitrogen content recharging to the water table. The user must then decide an appropriate mixing zone between the recharge and the groundwater. The model combining the Wehrmann model with the BURBS model was as follows (Tinker, 1991):

$$V_b C_b + (V_t + V_i + V_n + V_s - V_p) C_{BURBS} = (V_b + V_t + V_i + V_n + V_s - V_p) C_o$$

where

V_t = volume of water recharged from turf
V_i = volume of water recharged from impervious land
V_n = volume of water recharged from natural land
C_{BURBS} = nitrate-nitrogen concentration in recharge to groundwater

Based upon the results of the use of the combined model in a spreadsheet format, Tinker (1991) concluded that nitrogen from lawn fertilizers and septic tank systems caused nitrate-nitrogen values to increase in the groundwater beneath and on the down-gradient side of the five subdivisions. Further, lot size, location, and depth of wells, location of septic tank systems, amount of lawn fertilizer, and the hydrogeology beneath the subdivision are some of the parameters that influence nitrate-nitrogen levels in water-supply wells in unsewered subdivisions.

In summary, mass-balance models are sometimes criticized due to their lack of inclusion of subsurface system dynamics and transport processes, and the difficulties in establishing system boundaries (Keeney, 1986). However, they offer the key advantage of simplicity in concepts and the use of generally available data.

PLUME MODEL FOR NITROGEN POINT SOURCE

A point source release of inorganic fertilizer occurred in a sand and gravel aquifer in western Illinois in 1978 (Naymik and Barcelona, 1981). The source was inorganic fertilizer stored in an uncovered bin located on a property adjacent to a liquid ammonia fertilizer distribution terminal. Within the resultant contaminant plume, initial ammonia studies indicated that ammonia concentrations were 285 to 2100 mg/l; and nitrate concentrations were 570 to 1885 mg/l. Groundwater monitoring in a subsequent study revealed that dissolved ammonium and nitrate levels exceeded 2000 and 13,000 mg/l, respectively (Barcelona and Naymik, 1984). A numerical solute transport model was applied to the system; the results indicated that approximately 420 days would be necessary, after source removal, to permit recovery of the groundwater to near background levels within the study site. Subsequent monitoring results generally supported the modeling results and demonstrated the usefulness of the model for addressing transport and transformations of inorganic forms of nitrogen.

MANAGEMENT MODELS FOR PREVENTION AND CONTROL

Table 6.9 contains summary comments on eight references describing management models that can be used to plan and implement prevention and/or control strategies for the problem of groundwater nitrates resulting from agricultural practices. Three of the eight models are related specifically to agricultural practices (Martin and Watts, 1980; Reeves, 1977; DeCoursey, 1991); four are associated with other sources of nitrogen input to the subsurface environment (Moosburner and Wood, 1980; Rajagopal, et al., 1975; Shoemaker, Pacenka, and Porter, 1980; Vasconcelos, 1976); and one addresses a risk assessment approach for management of nitrate in groundwater (Dahab and Bogardi, 1990).

Table 6.9 References Involving Planning and Management Models

Author(s) (year)	Comments
Dahab and Bogardi (1990)	Framework for assessing the health risk of carcinogenic nitrosamines as a result of nitrates in groundwater; approaches for risk management are also addressed
DeCoursey (1991)[a]	Description of GLEAMS (groundwater leaching effects of agricultural management systems) model and its usage in groundwater planning and management decisions
Martin and Watts (1980)	Field-calibrated computer model for studying irrigation and nitrogen management for corn production and groundwater nitrate pollution potential
Moosburner and Wood (1980)	Simulation/optimization model for land-use zoning to minimize nitrate contamination of groundwater
Rajagopal et al. (1975)	Nonlinear statistical models for evaluating effects of land use on nitrates in groundwater
Reeves (1977)	Simulation model for nitrates in groundwater based on mass-balance calculations
Shoemaker, Pacenka, and Porter (1980)	Simulation and optimization models for controlling nonpoint sources of nitrates in groundwater
Vasconcelos (1976)	Quality model for total dissolved solids and nitrates in groundwater

[a] Reference also listed for different example in Table 6.3.

Martin and Watts (1980) used a field-calibrated computer model to evaluate the effect of irrigation and nitrogen management upon corn production and upon the potential nitrate pollution or purification of groundwater resources. The study was conducted for conditions of the Central Platte Valley of Nebraska where the soils are sandy, the groundwater has 10 to 40 mg/l nitrate-nitrogen, and the water table is close to the soil surface. Pumped groundwater is the source for all irrigation water. Results indicated that both irrigation and nitrogen management have a strong influence upon corn production and upon the increase or decrease of the groundwater nitrate content.

Reeves (1977) developed a simulation model to address physical mechanisms occurring in a generalized Chalk aquifer area in the U.K. The mechanisms incorporated in the model included those controlling nitrate additions and losses and those controlling nitrate movement rates. The modeling effort was primarily based upon mass-balance calculations. Reeves (1977) noted that areas requiring additional research include rates of vertical movement and dispersion of nitrate in soils and unsaturated strata, variations in rainfall nitrogen content, nitrogen immobilized by unharvested crops, annual losses by soil denitrification, denitrification in aquifers, and nitrate losses in streams and rivers.

Land usage patterns were the focus of the models by Moosburner and Wood (1980) and Rajagopal, et al. (1975). Moosburner and Wood (1980) used a simulation/optimization approach for management modeling to establish land use zoning patterns to minimize the impact on water quality in the rapidly developing Jackson Township in the Pine Barrens area in New Jersey. Land use is predominantly suburban and rural residential with little agricultural or industrial development. Nitrates from septic tank systems are the primary source of pollution. The model consisted of a simulation submodel for nitrate transport from septic tank systems, and a multiobjective goal programming optimization submodel to determine population density restrictions.

Rajagopal, et al. (1975) used an analytical model to evaluate wastewater disposal and water supply alternatives on the basis of both economic and groundwater quality criteria for rural areas in transition to urbanized areas surrounding Grand Traverse Bay, Michigan. The assessment was based on the current state of groundwater quality, statistical correlations of water quality as related to land use, soil, well, and water table characteristics, population growth and distribution, and cost analyses of alternative supply and disposal systems. Residential density, cherry orchards acreage, and well and soil characteristics exhibited a statistical relationship with the groundwater concentrations of nitrate-nitrogen. The results indicated that for a set of existing groundwater flow and soil conditions, an optimal community size may be defined to meet water quality criteria at specific aquifer depths.

Wastewater management, along with other issues, was addressed by Shoemaker, Pacenka, and Porter (1980), and Vasconcelos (1976). Shoemaker, Pacenka, and Porter (1980) developed simulation and optimization models to evaluate alternative procedures for controlling nitrate pollution of groundwater in eastern Suffolk County on Long Island in New York. The management alternatives considered were restrictions on land development and augmented sewerage systems. The optimization model was developed to determine the most economical location of wastewater treatment plants and sewer force mains. The simulation model was used to estimate changes in nitrogen loading likely to occur as a result of future changes in land use.

Vasconcelos (1976) reported on a mathematical model to optimize the design and minimize total costs for a regional water supply-wastewater treatment management system for the Fresno groundwater basin in the San Joaquin Valley in California. Declining groundwater levels and increasing concentrations of total dissolved solids and nitrates in groundwater were the major water resource problems in the area. An iterative algorithm that included a linear programming optimization model, a groundwater quantity and quality model, and a data processing-costing program, was used to analyze four management alternatives.

Another example of a planning and management model is GLEAMS (Groundwater Leaching Effects of Agricultural Management Systems). GLEAMS is an outgrowth of the CREAMS model, incorporating many of its original process descriptions by adding numerous improvements, especially in the pesticide and nutrient subsystems (DeCoursey, 1991). GLEAMS was developed by the Agricultural Research Service of the U.S. Department of Agriculture for field-size areas to evaluate the effects of agricultural management systems on the movement of agricultural chemicals within and through the plant root zone. A hydrologic component simulates infiltration and soil water and solute movement through soil layers. A soil erosion component estimates soil loss by particle size and includes pesticides adsorbed to clay and organic matter particles.

Finally, Dahab and Bogardi (1990) developed a risk assessment framework for addressing carcinogenic nitrosamines in groundwater. The main steps in the framework are shown in Figure 6.7 (Dahab and Bogardi, 1990). Components are included in the framework for uncertainty analysis, risk calculations based on exposure assessment, and risk regulation. This approach can be useful in groundwater nitrate problem assessment, and the identification and evaluation of management strategies for risk reduction.

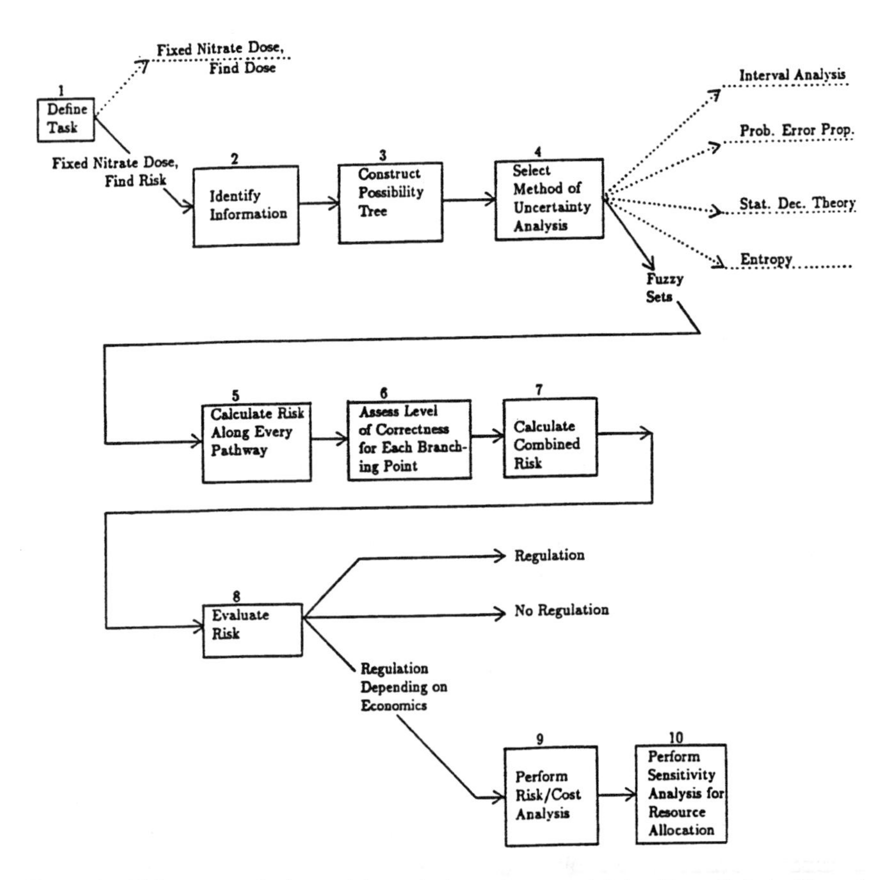

Figure 6.7 Main steps of nitrate risk analysis under uncertainty. (From Dahab, M. and Bogardi, I., "Risk Management for Nitrate-Contaminated Groundwater Supplies," 1990, U.S. Geological Survey, Reston, VA, pp. 21–54, 73–75.)

SUMMARY

This chapter has provided a qualitative description of a number of models and submodels for the transport and fate of nitrogen and nitrates in the subsurface environment. Brief information is presented on source characterization models, a plume model for a nitrogen point source, and management models useful for developing nitrate pollution prevention and control programs. More detailed information is included on nitrate transport models, particularly those that couple mass-balance considerations and models for the unsaturated and saturated zones.

SELECTED REFERENCES

Barcelona, M.J. and Naymik, T.G., "Dynamics of a Fertilizer Contaminant Plume in Groundwater," *Environmental Science and Technology*, Vol. 18, No. 4, April, 1984, pp. 257–261.

Barker, J.A. and Foster, S.S., "A Diffusion Exchange Model for Solute Movement in Fissured Porous Rock," *Journal of Engineering Geology*, Vol. 14, No. 1, 1981, pp. 17–24.

Barry, D.A. et al., "Interpretation of Leaching Under Multiple Fertilizer Applications," *Journal of Soil Science*, Vol. 36, No. 1, March, 1985, pp. 9–20.

Bashkin, V.N. and Kudeyarov, V.N., "Nitrate Content of Ground Waters in Agricultural Areas of the Oka River Basin," *Soviet Soil Science*, Vol. 15, No. 1, 1983, pp. 41–47.

Belamie, R. et al., "Methodological Investigations on the Transfer of Nitrogen to Groundwater in a Rural Setting," *Computes Rendus des Seances de l'Academie d'Agriculture de France*, Vol. 70, No. 12, 1984, pp. 1565–1576.

Carey, M.A. and Lloyd, J.W., "Modelling Non-point Sources of Nitrate Pollution of Groundwater in the Great Ouse Chalk, U.K.," *Journal of Hydrology*, Vol. 78, No. 1/2, May, 1985, pp. 83–106.

Committee on Techniques for Assessing Ground Water Vulnerability, *Ground Water Vulnerability Assessment – Predicting Contamination Potential Under Conditions of Uncertainty*, National Research Council, National Academy Press, Washington, D.C., 1993, p.6.

Dahab, M. and Bogardi, I., "Risk Management for Nitrate-Contaminated Groundwater Supplies," November, 1990, U.S. Geological Survey, Reston, Virginia, pp. 21–54, and 73–75.

DeCoursey, D.G., "Computer Models for Pesticide and Fertilizer Use," in *Agriculture and the Environment – the 1991 Yearbook of Agriculture*, U.S. Government Printing Office, Washington, D.C., 1991, pp. 160–172.

Dutt, G.R., Shaffer, M.J., and Moore, W.J., "Computer Simulation Model of Dynamic Bio-physico-chemical Processes in Soil," Technical Bulletin 196, October, 1972, Agricultural Experiment Station, University of Arizona, Tucson, Arizona.

Farajalla, N.S., Deyle, R.E., Vieux, B.E., and Canter, L.W., "Correlating Nitrate Levels in Ground Water with Agricultural Land Use in Oklahoma," *Proceedings of 1993 Joint CSCE-ASCE National Conference on Environmental Engineering*, July, 1993, American Society of Civil Engineers, New York, New York, pp. 469–476.

Frimpter, M.H., Donohue, J.J., and Rapacz, M.V., "A Mass-Balance Nitrate Model for Predicting the Effects of Land Use on Ground Water Quality," Report 88–493, 1990a, U.S. Geological Survey, Boston, Massachusetts.

Frimpter, M.H., Donohue, J.J., and Rapacz, M.V., "A Mass-Balance Model for Predicting Nitrate in Ground Water," *Journal of New England Water Works Association*, Vol. CIV, No. 4, December, 1990b, pp. 219–232.

Hantzsche, N.N. and Finnemore, E.J., "Predicting Groundwater Nitrate-Nitrogen Impacts," *Ground Water*, Vol. 30, No. 4, July–August, 1992, pp. 490–499.

Hendry, M.J., Gillham, R.W., and Cherry, J.A., "An Integrated Approach to Hydrogeologic Investigations — A Case History," *Journal of Hydrology*, Vol. 63, No. 3–4, June, 1983, pp. 211–232.

Ho, G.E., Mathew, K., and Newman, P.W., "Modelling of Nitrogen Removal in the Vadose Zone," *Proceedings of Second International Conference on Ground Water Quality Research*, 1985, Oklahoma State University, Stillwater, Oklahoma, pp. 112–114.

Jury, W.A., "Estimating the Influence of Soil Residence Time on Effluent Water Quality," *National Conference on Management of Nitrogen in Irrigated Agriculture*, 1978, University of California at Riverside, Riverside, California, pp. 265–290.

Kanwar, R.S., Johnson, H.P., and Baker, J.L., "Comparison of Simulated and Measured Nitrate Losses in Tile Effluent," *Transactions of the American Society of Agricultural Engineers*, Vol. 26, No. 5, September–October, 1983, pp. 1451–1457.

Keeney, D.R., "Sources of Nitrate to Ground Water," *CRC Critical Reviews in Environmental Control*, Vol. 16, No. 3, 1986, pp. 257–304.

Khanif, Y.M., Cleemput, O.V., and Baert, L., "Evaluation of the Burns' Model for Nitrate Movement in Wet Sandy Soils," *Journal of Soil Science*, Vol. 35, No. 4, December, 1984, pp. 511–518.

Kirkham, D. and Affleck, S.B., "Solute Travel Times to Wells," *Ground Water*, Vol. 15, No. 3, May–June, 1977, pp. 231–242.

Lerner, D.N. and Papatolios, K.T., "A Simple Analytical Approach for Predicting Nitrate Concentrations in Pumped Ground Water," *Ground Water*, Vol. 31, No. 3, May–June, 1993, pp. 370–375.

Martin, D.L. and Watts, D.G., "Potential Nitrogen Purification of Groundwater Through Irrigation Management," Paper 80–2027, 1980, American Society of Agricultural Engineers, St. Joseph, Michigan.

Mathew, K., Ho, G.E., and Newman, P.W., "Removal of Nitrogen from Wastewater by Soil Percolation with Alternate Flooding and Drying," *Asian Environment*, Vol. 6, No. 4, 1984, pp. 9–16.

Mehran, M., Noorishad, J., and Tanji, K.K., "Numerical Technique for Simulation of the Effect of Soil Nitrogen Transport and Transformations on Groundwater Contamination," *Environmental Geology*, Vol. 5, No. 4, 1983–1984, pp. 213–218.

Mercado, A., "Nitrate and Chloride Pollution of Aquifers: A Regional Study with the Aid of a Single-cell Model," *Water Resources Research*, Vol. 12, No. 4, August, 1976, pp. 731–747.

Moosburner, G.J. and Wood, E.F., "Management Model for Controlling Nitrate Contamination in the New Jersey Pine Barrens Aquifer," *Water Resources Bulletin*, Vol. 16, No. 6, December, 1980, pp. 971–978.

Naymik, T.G. and Barcelona, M.J., "Characterization of a Contaminant Plume in Ground Water, Meredosia, Illinois," *Ground Water*, Vol. 19, No. 5, September–October, 1981, pp. 517–526.

Oakes, D.B., Young, C.P., and Foster, S.S., "Effects of Farming Practices on Groundwater Quality in the United Kingdom," *Science of the Total Environment*, Vol. 21, November, 1981, pp. 17–30.

Oakes, D.B., "Nitrate Pollution of Groundwater Resources: Mechanisms and Modelling," *Proceedings of an IIASA Task Force Meeting on Nonpoint Nitrate Pollution of Municipal Water Supply Sources: Issues of Analysis and Control*, Series No. CP-82-S4, February, 1982, International Institute of Applied Systems Analysis, Laxenburg, Austria, pp. 207–230.

Phillips, K.J. and Gelhar, L.W., "Contaminant Transport to Deep Wells," *Journal of the Hydraulics Division, American Society of Civil Engineers*, Vol. 104, No. HY6, June, 1978, pp. 807–819.

Porter, K.S. and Shoemaker, C.A., "Recharge and Nitrogen Transport Models for Nassau and Suffolk Counties, N.Y.," OWRT-A-075-NY(1), January, 1978, Center for Environmental Research, Cornell University, Ithaca, New York.

Rajagopal, R. et al., "Water Quality and Economic Criteria for Rural Wastewater and Water Supply Systems," *Journal Water Pollution Control Federation*, Vol. 47, No. 7, 1975, pp. 1834–1837.

Reeves, M.J., "A Procedure for the Prediction of Nitrate Levels in Water Supplies in the United Kingdom," *Progress in Water Technology*, Vol. 8, No. 4–5, 1977, pp. 161–177.

Rice, R.C. and Raats, P., "Underground Travel of Renovated Wastewater," *Journal of the Environmental Engineering Division, American Society of Civil Engineers*, Vol. 106, No. 6, December, 1980, pp. 1079–1098.

Schmid, G., "Seepage Flow in Extremely Thin Aquifers," *Advances in Water Resources*, Vol. 4, No. 3, September, 1981, pp. 134–136.

Shaffer, M.J., Dutt, G.R., and Moore, W.J., "Predicting Changes in Nitrogenous Compounds in Soil-water Systems," *Collected Papers Regarding Nitrates in Agricultural Wastewaters*, Series 13030ELY, December, 1969, Federal Water Quality Administration, Washington, D.C., pp. 15–28.

Shaffer, M.J., Halvorson, A.D., and Pierce, F.J., "Nitrate Leaching and Economic Analysis Package (NLEAP): Model Description and Application," in *Managing Nitrogen for Groundwater Quality and Farm Profitability*, Follett, R.F., Keeney, D.R., and Cruse, R.M., Eds., Soil Science Society of America, Inc., Madison, Wisconsin, 1991, chap. 13, pp. 285–322.

Shoemaker, C.A., Pacenka, S., and Porter, K.S., "Mathematical Models for the Analysis of Management of Non-point Sources of Nitrogen Pollution," May, 1980, Department of Environmental Engineering, Cornell University, Ithaca, New York.

Skaggs, R.W. and Gilliam, J.W., "Effect of Drainage System Design and Operation on Nitrate Transport," *Transactions of the American Society of Agricultural Engineers*, Vol. 24, No. 4, July–August, 1981, pp. 929–934.

Spalding, M.E., "Implications of Temporal Variations and Vertical Stratification of Groundwater Nitrate-Nitrogen in the Hall County Special Use Area," September, 1984, Water Resources Center, University of Nebraska, Lincoln, Nebraska.

Stemmler, S., "Environmental Effects of Nitrogen Fertilization Exemplified by Groundwater Pollution as Simulated by CREAMS," *European and U.S. Case Studies in Application of the CREAMS Model*, Proceedings Series CP-82-S11, 1982, International Institute for Applied Systems Analysis, Laxenburg, Austria, pp. 49–62.

Tanner, C.B. and Gardner, W.R., "Relation of Climate to Leaching of Solutes and Pollutants Through Soil," June, 1973, Department of Soil Science, University of Wisconsin, Madison, Wisconsin.

Tinker, J.R., Jr., "An Analysis of Nitrate-Nitrogen in Ground Water Beneath Unsewered Subdivisions," *Ground Water Monitoring Review*, Winter, 1991, pp. 141–150.

Vasconcelos, J.J., "Optimization of a Regional Water Resource Quality Management System," Ph.D. Dissertation, 1976, University of California, Berkeley, California.

Walter, M.F. et al., "Evaluation of a Soil Nitrate Transport Model," Winter Meeting of American Society of Agricultural Engineers, 1974, Chicago, Illinois.

Watts, D.G. and Martin, D.L., "Effects of Water and Nitrogen Management on Nitrate Leaching Loss from Sands," *Transactions of the American Society of Agricultural Engineers*, Vol. 24, No. 4, July/August, 1981, pp. 911–916.

Wellings, S.R. and Bell, J.P., "Movement of Water and Nitrate in the Unsaturated Zone of Upper Chalk Near Winchester, Hants, England," *Journal of Hydrology*, Vol. 48, No. 1/2, August, 1980, pp. 119–136.

White, R.E., "Transfer Function Model for the Prediction of Nitrate Leaching Under Field Conditions," *Journal of Hydrology*, Vol. 92, No. 3–4, July, 1987, pp. 207–222.

Young, C.P., Oakes, D.B., and Wilkinson, W.B., "Prediction of Future Nitrate Concentrations in Ground Water," *Ground Water*, Vol. 14, No. 6, November–December, 1976, pp. 426–438.

Young, C.P., "The Distribution and Movement of Solutes Derived from Agricultural Land in the Principal Aquifers of the United Kingdom, with Particular Reference to Nitrate," *Water Science and Technology*, Vol. 13, No. 4/5, 1981, pp. 1137–1152.

7 MANAGEMENT MEASURES FOR NITRATE POLLUTION PREVENTION

INTRODUCTION

Several management measures are available for preventing, or at least minimizing, nitrate pollution of groundwater resulting from agricultural or waste disposal related practices. The purpose of this chapter is to provide a review of such measures, with the results of case studies presented in selected instances. General information as contained in four references related to this subject is summarized in Table 7.1. The management measures described herein will be addressed in two categories: (1) fertilizer management measures, including Best Management Practices; and (2) other management measures, including groundwater protection zones. A summary section concludes this chapter.

Table 7.1 References Addressing General Information on Management Measures

Author(s) (year)	Comments
Allee and Abdalla (1989)	Description of role of groundwater policy education provided by extension services in managing the risk of groundwater contamination
Batie, Kramer, and Cox (1989)	Discussion of legal bases and economic implications of state and/or federal strategies for managing agricultural pollution of groundwater
Conrad (1986)	General information on policies and regulations that can be used to minimize nitrates in ground- and drinking water
Lahl, Zeschmar, Gabel, Kozicki, and Podbielski (1983)	Nitrate reduction in groundwater via the avoidance of nitrate input in water production areas, mixing of water with differing nitrate contents, and/or supra-regional connection of water supplies

FERTILIZER MANAGEMENT MEASURES

Table 7.2 contains summary comments on 27 references on fertilizer management measures that can be used to prevent, or minimize, agriculturally related nitrate pollution of groundwater. Several types of measures are represented by the 27 references; thus, Table 7.3 contains a categorical listing of the ten identified

Table 7.2 References Involving Fertilizer Management Measures

Author(s) (year)	Comments
Andersson et al. (1984)	Alterations of agricultural practices in well recharge areas
Bock and Hergert (1991)	Discussion of importance of nitrogen application rate and timing in relation to crop need, improving nitrogen use efficiency, and the profitability of nitrogen management practices
Bouldin and Selleck (1977)	Scheduling fertilizer applications for potatoes so as to minimize nitrates in groundwater
Bourg (1984)	Steps associated with matching a crop yield goal with the crop requirement for nitrogen in relation to the application of nitrogen fertilizer
de Haen (1982)	Cost-benefit analysis for evaluating fertilizer applications and resultant nitrates in groundwater
Embleton et al. (1979)	Development of citrus fertilizer management programs that reduce the potential for nitrate pollution of groundwater
Fleming (undated)	Review of Best Management Practices for preventing groundwater pollution by nitrates; and public policies to reduce agrichemical usage
Fricker (1983)	Fertilizer restrictions in groundwater conservation areas
Hall (1992)	Field study of the effects of reduced manure and fertilizer applications on groundwater quality at an agricultural field site in Pennsylvania
Hills, Broadbent, and Fried (1978)	Timing and rate of fertilizer applications for sugarbeets to maximize nitrogen uptake and minimize nitrogen pollution potential
Johnson, Perry, and Adams (1989)	Study of the economic effects on an irrigated Columbia Basin farm in Oregon of adopting alternative strategies that reduce agricultural related nitrate pollution of groundwater
Kaap (1986)	Fertilizer applications as a function of crop need and season
Keeney (1986)	Use of agroecosystem approach in planning fertilizer applications
Kepler, Carlson, and Pitts (1978)	General information on irrigation and fertilization practices
Logan (1990)	Review of the effectiveness of agricultural Best Management Practices in minimizing groundwater pollution
McWilliams (1984)	Discussion of federal, state, and local action for implementing practices that prevent groundwater pollution
Munson and Russell (1990)	Discussion of the Arizona Regulated Agricultural Activities Program for reducing nitrates in the environment
National Governors Association (1991)	Review of four state programs for controlling agricultural sources of nitrate contamination
Office of Technology Assessment (1990)	Comprehensive discussion of best management practices via a range of fertilizer and other management practices to minimize nitrate pollution of groundwater
Pallares (1978)	Field experiment to evaluate fertilizer management of a lemon orchard to minimize nitrate pollution of groundwater
Peterson and Frye (1989)	Comparative information on fertilizer nitrogen carriers, application methods for fertilizers, timing of applications, and nitrogen stabilizers
Schneider and Day (1976)	Evaluation of policies for controlling nitrate pollution of groundwater resulting from field-applied manure and fertilizers
Shirmohammadi, Magette, and Shoemaker (1991)	Use of CREAMS model to simulate the long-term effects of seven BMPs on nitrate loadings to a shallow, unconfined groundwater system
Singh and Sekhon (1979)	Review of factors influencing nitrate pollution of groundwater from farm usage of fertilizers
Swoboda (1977)	Discussion of methods of reducing nitrate leaching in soils

Table 7.2 References Involving Fertilizer Management Measures (Continued)

Author(s) (year)	Comments
Timmons and Dylla (1979)	Evaluation of irrigation and fertilization of corn to produce maximum yields and minimize nitrate leaching
Zaporozec (1983)	Contribution of fertilizers as a source of nitrate pollution of groundwater, and matching application to crop needs

Table 7.3 Examples of Fertilizer Management Measures

Measure	References describing measure
Match timing of fertilizer application to meet crop needs	Bock and Hergert (1991) Bouldin and Selleck (1977) Embleton et al. (1979) Fricker (1983) Hills, Broadbent, and Fried (1978) Kaap (1986) Pallares (1978) Swoboda (1977)
Match fertilizer application rate to crop needs	Bock and Hergert (1991) Bourg (1984) Hills, Broadbent, and Fried (1978) Kaap (1986) Pallares (1978) Singh and Sekhon (1979) Zaporozec (1983)
Choose fertilizer application technique to minimize nitrogen losses	Embleton et al. (1979) Pallares (1978)
Limit fertilizer application in well recharge areas	Andersson et al. (1984) Fricker (1983) Hall (1992)
Regulate crop types to those with minimal fertilizer requirements	de Haen (1982)
Implement management policies intended to minimize nitrogen losses	Schneider and Day (1976)
Adopt legislation related to minimizing fertilizer usage	McWilliams (1984)
Implement taxation plan focused on minimizing fertilizer usage	de Haen (1982)
Implement Best Management Practices (BMPs)	Fleming (undated) Johnson, Perry, and Adams (1989) Keeney (1986) Kepler, Carlson, and Pitts (1978) Logan (1990) McWilliams (1984) Munson and Russell (1990) National Governors Association (1991) Office of Technology Assessment (1990) Peterson and Frye (1989) Shirmohammadi, Magette, and Shoemaker (1991)
Control the usage of irrigation	Kepler, Carlson, and Pitts (1978) Timmons and Dylla (1979)

types. The most frequently used measures are associated with matching the timing and rate of fertilizer application to the needs of the crops being fertilized.

Hills, Broadbent, and Fried (1978) studied the timing and rate of fertilizer nitrogen application to sugar beets in order to maximize nitrogen uptake and minimize the resultant surface and groundwater pollution potential. Efficient nitrogen fertilization is particularly important to sugar beet crops, as a deficiency of available nitrogen often limits yield, while an excess can result in unnecessary economic expenditures. Further, efficient fertilization is important to minimize energy input in crop production and to minimize additions of nitrogen that might contribute to pollution of groundwater. This study focused on two field experiments at Davis, California; the key findings were (Hills, Broadbent, and Fried, 1978):

1. There were no significant differences in root, top, or sugar yield when fertilizer nitrogen (135 kg N/ha) was applied at planting, at thinning, split equally between thinning and layby, or split equally between planting, thinning, and layby.
2. Fertilizer nitrogen recovery was 47% when 112 kg N/ha were applied to achieve maximum sugar yield. Roots removed as much nitrogen as that applied, and tops contained an additional 105 kg N/ha. When applied, nitrogen was 2.5 times the amount required for maximum sugar yield; tops and roots contained almost as much nitrogen as applied. Therefore, at fertilizer rates giving maximum sucrose yield, sugar beet storage roots usually took up as much nitrogen as was applied and, therefore, efficient fertilization contributed little to reducing the nitrogen pollution potential.
3. There is considerable nitrogen in sugar beet tops that should be considered in fertilizing the next crop in order to avoid pollution from excessive fertilization. Part of the nitrogen fertilizer requirement of the next crop can be met by leaving the tops in the field.

The advantages and limitations of various conventional nitrogen fertilizers (such as anhydrous ammonia, urea, ammonium nitrate, urea-ammonium nitrate solution, ammonium sulfate, monoammonium phosphate, and diammonium phosphate) have been discussed by Peterson and Frye (1989). Nitrogen fertilizers may be applied in many different ways and at various times. Table 7.4 summarizes commonly used methods of application and placement of nitrogen fertilizers for various crop stages (Peterson and Frye, 1989). These methods of application have implications for nitrate pollution of groundwater.

Kaap (1986) summarized the overuse of nitrogen fertilizer in a 1984 study conducted in northeast Iowa. The study inventoried land use, crops and their rotation, fertilization practices, livestock numbers, and manure management practices conducted by 209 farmers. The survey area included 32,900 acres of corn; the area was also a highly productive alfalfa and livestock area. Corn yields averaged 130 bu/ac; and the fertilizer nitrogen application rate averaged 150 lb N/ac. The study findings indicated that available nitrogen may exceed the nitrogen needs of corn by as much as 80 lb N/ac; this excess occurred because most farmers do not take manure and alfalfa nitrogen into consideration. These organic

Table 7.4 Methods of Applying Nitrogen Fertilizers at Various Crop Stages

Crop stage	Method of application and placement
Preplant	Broadcast on surface (Note 1)
	Broadcast, mixed into soil (Note 1)
	Injection, subsurface (Note 2)
At-plant or pre-emergence	In-row with seed (Note 3)
	Banded beside seed (Note 4)
	Injection subsurface away from seed
	Broadcast on surface
Post-emergence	Side-dress, surface banded (Note 5)
	Side-dress, injected subsurface
	Top-dress (broadcast), surface (Note 6)
	Foliar, sprayed on leaves of crop (Note 7)

Note:

1. A broadcast application is a relatively even distribution of the fertilizer material over the entire surface of a field. The fertilizer may be left on the surface or mixed into the soil by tillage.
2. Injection involves subsurface placement in any of several ways. Most commonly, anhydrous ammonia or liquid nitrogen is knifed-in. Solid fertilizer may be dropped in slots or channels opened by shanks or chisels. These injection methods necessarily band the fertilizer.
3. In-row fertilizer application places the fertilizer directly in contact with the seed during the planting operation.
4. Solid nitrogen fertilizer may be banded on the surface in narrow strips between rows at planting or post-emergence, or placed below the soil surface during the planting operation, usually a few centimeters to one or both sides of the seed and a few centimeters below the seed.
5. Side-dressing is often used to apply fertilizer after a row crop is established. The fertilizer is applied in a band or strip beside the row, either on the surface or injected into the soil. A surface-banded side-dressing may be subsequently mixed into the soil by cultivation.
6. Top-dressing is a broadcast application of solid or liquid fertilizer over the top of a growing crop. It is similar to foliar application except, in a top-dress application, most of the fertilizer is applied on the soil.
7. Foliar fertilization involves spraying a fertilizer solution directly onto plant foliage or applying it top-dress through sprinkler irrigation.

From Peterson, G. A. and Frey, W. W., in *Nitrogen Management and Ground Water Protection*, Follett, R. F., Ed., Elsevier, Amsterdam, 1989, chap. 7, pp. 183–219. With permission.

sources can supply up to 70 and 20 lb N/ac, respectively, to corn. Therefore, it was determined that farmers could cut their crop nitrogen, phosphorus, and potassium fertilizer needs by up to 66% while maintaining yields by utilizing alfalfa nitrogen and manure nitrogen, phosphorus, and potassium. Only 40% of the survey area farmers considered manure nitrogen credits, while 60% considered manure phosphorus and potassium credits in their nutrient additions.

The process of arriving at nitrogen recommendations on a field basis by matching the nitrogen requirement with a yield goal can be summarized in the following steps (Bourg, 1984):

1. Select yield goal.
2. Determine the total nitrogen required to attain yield goal selected in Step 1.
3. Determine the amount of carryover nitrate-nitrogen already available in the soil.
4. Determine the amount of nitrate-nitrogen available from the irrigation water to be applied during the season.
5. Subtract amounts determined in Steps 3 and 4 from 2. This gives the net amount of nitrogen to be applied in fertilizer to attain the selected yield goal.
6. Adjust the net nitrogen requirement (Step 5) for time and method of application. The result is the amount of supplemental nitrogen to be applied from all sources — preplant, starter, sidedress, in irrigation water, etc.

A procedure that can be used for determining the nitrogen fertilizer requirement at a location is shown in Table 7.5 (Bock and Hergert, 1991). The table also includes an example. The necessary parameters in the procedure include:

1. Unit nitrogen requirement (UNR), lb/bu. This is the nitrogen (N) fertilizer requirement/unit of yield, if no N were supplied from nonfertilizer sources [depends on (a) and (b) below].
   a. Unit N uptake by the crop (UNU), lb/bu. This is the N uptake in aboveground crop per unit of yield.
   b. Fraction of N fertilizer recovered by aboveground crop (e_f).
2. Realistic yield goal, bu/acre.
3. Credits for nonfertilizer sources of nitrogen (N_{nf}), lb/acre.

The corn-related example in Table 7.5 assumes that corn takes up 1.0 lb N in the aboveground crop/bu of grain produced (UNU = 1.0) and that 0.60 of the N fertilizer applied was recovered by the aboveground crop (e_f = 0.60), thus giving a unit N requirement (UNU/e_f) for corn of 1.67 lb N/bu of grain (1.0/0.6). This type of information for corn and several other crops is in Table 7.6 (Bock and Hergert, 1991). Typical e_f values are shown in Table 7.7 (Bock and Hergert, 1991). A yield goal of 160 bu/acre was then used in the example in Table 7.5, giving a N fertilizer requirement of 267 lb N/acre, if no N were supplied from nonfertilizer sources. However, it was observed that soil organic matter released nitrogen equivalent in effectiveness to 75 lb fertilizer N/acre, and residual inorganic nitrogen in the root zone was equivalent in effectiveness to 25 lb fertilizer N/acre, leaving a nitrogen fertilizer requirement of 167 lb N/acre. Additional information on determining credits for nonfertilizer sources of nitrogen is in Bock and Hergert (1991).

Embleton et al. (1979) described a project conducted in California to develop practical citrus nitrogen fertilizer management programs that would reduce groundwater nitrate-pollution potential without having adverse effects on fruit yield, size, and quality. Studies were conducted on the 'Washington' navel orange

Table 7.5 Worksheet for Calculating Nitrogen (N) Fertilizer Requirements for Maximum Yield

Unit N requirement (see table 7.6)		Realistic yield goal or Y_m		N fertilizer requirement before subtracting credits for non-fertilizer sources of N
1.67 lb N/bu[a]	×	160 bu/acre[b]	=	267 lb N/acre

Credits for nonfertilizer sources of N

Sources	N fertilizer equivalent
Soil organic matter[c]	–75 lb N/acre
Soil inorganic N	–25 lb N/acre
Irrigation water	
Manure	
Other organic wastes	
Previous legume crop	
N fertilizer requirement	167 lb N/acre

[a] For nongrain crops, units are lb N/1000 lb of harvested crop.
[b] For nongrain crops, units are 1000 lb of harvested crop/acre.
[c] Includes crop residues from nonleguminous crops.

From Bock, B. R. and Hergert, G. W., in *Managing Nitrogen for Groundwater Quality and Farm Profitability*, Follett, R. F., Keeney, D. R., and Cruse, R. M., Eds., Soil Science Society of America, Madison, WI, 1991, chap. 7, pp. 139–164. With permission.

Table 7.6 Typical Unit Nitrogen Requirements for Maximum or Near Maximum Yield with Relatively Efficient Nitrogen Management

Fraction of N recovered in above-ground crop (e_f)[a]	Corn (lb N/bu)	Grain sorghum (lb N/bu)	Hard red winter wheat (lb N/bu)	Cotton (lb N/1000 lb seed cotton)	Pastures (lb N/1000 lb forage)
	Unit N uptake (UNU) in above-ground crop				
	1.0	1.0	1.6	50	25
	Unit N requirement (UNR)[b]				
0.70	1.43	1.43	2.29	70	36
0.65	1.54	1.54	2.46	77	38
0.60	1.67	1.67	2.67	83	42
0.55	1.82	1.82	2.91	91	45

[a] See Table 7.7 for guidelines on estimating e_f.
[b] UNR = UNU/e_f.

From Bock, B. R. and Hergert, G. W., in *Managing Nitrogen for Groundwater Quality and Farm Profitability*, Follett, R. F., Keeney, D. R., and Cruse, R. M., Eds., Soil Science Society of America, Madison, WI, 1991, chap. 7, pp. 139–164. With permission.

and 'Limoneira 8a Lisbon' lemon. Supplying the needed nitrogen of the trees by means of urea foliar sprays resulted in a lower nitrate-pollution potential than by applying nitrogen fertilizer to the soil. For a given annual nitrogen addition, foliar-applied treatments were associated with a substantially greater amount of nitrogen removed in the harvested crop than were the soil-applied treatments. An increase in the annual soil-applied nitrogen rate was shown to increase the nitrate-pollution potential. Finally, nitrogen applications to lemons in November were associated with a higher nitrate-pollution potential than the same annual nitrogen rate applied in March, or in a split application (Embleton et al., 1979).

Table 7.7 General Guidelines for Estimating e_f When Using Nitrogen Rates for Maximum or Near Maximum Yield

Relative efficiency of N-application timing	e_f Values			
	Perennial grasses	Upland cereal grains	Shallow-rooted crops (e.g., onion and lettuce)	Flooded crops (e.g., rice)
Low[a]	0.55	0.45	0.35	0.25
Medium[b]	0.70	0.60	0.50	0.40
High[c]	O 80	0.70	0.60	0.50

Note: The e_f values assume medium-to-high nitrate loss potential as determined by soil type and moisture regime and no or negligible NH_3 volatilization losses.

[a] One N application (without nitrification inhibitor) well in advance of the growing season. When nitrate loss potential is low due to soil type or moisture regime, use e_f values for medium to high efficiency of N application timing.

[b] One N application near beginning of growing season.

[c] Multiple N applications with first application near beginning of growing season; use of nitrification inhibitor may substitute or partially substitute for splitting N applications.

From Bock, B. R. and Hergert, G. W., in *Managing Nitrogen for Groundwater Quality and Farm Profitability*, Follett, R. F., Keeney, D. R., and Cruse, R. M., Eds., Soil Science Society of America, Madison, WI, 1991, chap. 7, pp. 139–164. With permission.

Another approach for minimizing agriculturally-based nitrate pollution of groundwater is to limit fertilizer applications in the immediate recharge areas of wells. A more formalized approach involving drinking water protection zones will be described in the subsequent section. Andersson, et al. (1984) described an ad hoc program in Sweden to reduce the nitrogen contamination of the groundwater at the source. In well recharge areas, the agricultural practices were investigated and, if necessary, alterations were proposed to the farmer in order to decrease the nitrogen leaching from the soil profile. As a final product, an action plan was proposed for water supplies with dominating arable land in the recharge area.

The term "Best Management Practice" (BMP) means a practice or combination of practices that is determined by a state (or designated area-wide planning agency) after problem assessment, examination of alternative practices, and appropriate public participation, to be the most effective practicable (including technological, economic, and institutional considerations) means of preventing or reducing the amount of pollution generated by nonpoint sources to a level compatible with water quality goals. BMPs for control of nitrate pollution of groundwater can encompass both crop and soil management. The guiding principle is to minimize the amount of nitrate in the rooting zone, especially during periods when leaching is likely to occur (Keeney, 1986). This could involve multiple fertilizer applications, use of cover crops or deep-rooted crops, genetic selection to improve crop nitrogen use efficiency, chemical additives that inhibit the rate of nitrification, slow release inorganic or organic fertilizers, carefully managed irrigation to improve crop yields and nitrogen uptake, and accounting for available nitrogen from nitrate in the rooting zone, nitrogen mineralized from organic matter, and nitrogen from manure and crop residues. Considerable refinement of nitrogen fertilizer application recommendations, taking into account such factors as weather, soil types, and level of management, is needed. In order to appropriately develop BMPs, it is important to view the nitrogen cycle from an ecosystem perspective (Keeney, 1986). Nitrogen fertilizer recommendations using soil-plant mass balances should aid in reducing fertilizer use and the environmental consequences of over-fertilization.

Johnson, Perry, and Adams (1989) explored the economic effects of adopting alternative strategies that reduce agricultural-related groundwater pollution from nitrates. Based upon developed dynamic optimization and linear programming models for an irrigated Columbia Basin farm in Oregon, the following conclusions were drawn (Johnson, Perry, and Adams, 1989): (1) careful management of soil moisture is critical to the reduction of pollution rates; (2) some nitrate leachate is unavoidable in the production of irrigated crops; (3) weather events play a significant role in explaining the existence of nitrate leachate under optimal irrigation and fertilization practices; and (4) input taxes and restrictions on nitrogen application rates may not always reduce pollution rates.

Table 7.8 summarizes BMPs for controlling potential contamination of surface and groundwater from fertilizer applications (Office of Technology Assessment, 1990). A comprehensive review of agricultural BMPs for environmental protection was presented by Logan (1990). As shown in Table 7.9, the BMPs, including structural measures, were categorized according to environmental objective, target pollutant, environmental media affected, and the management approach of the specific practice (Logan, 1990). The effectiveness of BMPs can be considered in terms of their impact on pollutant loads, acceptability by farmers, cost effectiveness, and ease of implementation and maintenance. Table 7.10 summarizes the effectiveness of several BMPs related to groundwater protection regarding pesticides, nitrates, and salt (Logan, 1990).

Table 7.8 Examples of BMPs for Controlling Potential Contamination of Surface and Groundwater from Fertilizers

- Soil testing to determine soil nutrient content and appropriate fertilization and liming regimes
- Spring fertilizer applications in regions with wet soils, humid climates, and high infiltration
- Split applications may reduce potential losses by up to 30% compared to single applications
- Level terraces as a mechanism to reduce nitrate losses in runoff in areas with low vulnerability to nitrate leaching, contour farming is recommended in humid regions with high vulnerability to contamination
- Drainage control to reduce nitrate losses in wet and irrigated areas; to include wise irrigation management to prevent leaching losses
- Slow release nitrogen fertilizers
- Crop rotations, no-till and conservation tillage to reduce surface losses of nitrogen
- Soil incorporation of broadcast fertilizer
- Level terraces as a phosphorus control measure
- Rotation grazing, crop rotation, cover crops, and conservation tillage to reduce phosphorus losses as compared to continuous grazing or conventional tillage
- Sedimentation basins and flow control in irrigation systems to reduce phosphorus losses

From Office of Technology Assessment, "Beneath the Bottom Line — Agricultural Approaches to Reduce Agrichemical Contamination of Ground Water," OTA-F-418, 1990, U.S. Congress, Washington, D.C., pp. 90–100.

Table 7.9 Classification of Conservation Practices and Agricultural BMPs by Environmental Objective, Pollutant Type, and Medium Impacted

BMP	Primary environmental objective[a]	Pollutant type[b]	Medium impacted[c] Surface water	Ground-water	Air	Soil
Structural						
Terraces, hillside ditches	E	E, P	P	N/A	N	P
Grass waterways	E	E, P	P	N	N	P
Subsurface (tile) drains, water-table management	S	E, P, N, S	P/A	P	N	P
Irrigation systems	S, E	S, N	P	P	P	P
Chemigation back-siphon devices	Q	C, N	N	P	N	N
Sediment and water retention basins	L, Q	E, P, N	P	A	A	N
Surface drains	N, Q	N	A	P	N	P
Manure storage, runoff control, filter strips	W, L Q	N, P, B, O, M	P	P	P	P
Irrigation tailwater recovery systems	E	E, C, S	P	N	N	P
Cultural						
Conservation tillage	E, L, Q	E, P	P	N/A	N/A	P
Contour cropping	E, L	E	P	N/A	N	P
Stripcropping	E, L	E	P	N/A	N	P
Contour stripcropping	E, L	E	P	N/A	N	P
Cover cropping	E, L Q	E	P	P	N	P
Crop rotation	E	E	P	N/P	N	P
Subsoiling	S	S, E	P	N/A	N	P
Land grading	S, E	S, E	A	N/P	N/P	P
Critical area planting	E, L Q	E, P, N	P	P	P	P
Stream bank protection	E	E	P	N/P	N/P	P
Low-input farming	E, L Q	E, C	P	N/P	P	P
Management						
Integrated pest management	Q	C, M	P	P	P	P
Animal waste management	L, W, Q	N, P, M	P	P	P	P
Fertilizer management	L, Q	N, P	P	P	P	P
Pesticide management	Q	C, M	P	P	P	P
Irrigation management	S, L Q	S, N, P	P	P	P	P

[a] E = erosion control; L = eutrophication; W = animal waste management; Q = water quality; S = salinity; N = none, for example surface drains primarily eliminate wetness problems.
[b] E = sediment, P = phosphorus, N = nitrogen, C = pesticide, B = biological oxygen demand, S = salt, M = heavy metals, O = pathogenic organisms.
[c] P = positive impact, A = adverse impact, N = no impact.

From Logan, T. J., *Journal of Soil and Water Conservation*, March–April, 1990, pp. 201–206. With permission.

Additional examples of BMPs, which are intended to reduce soil erosion, runoff, and surface water pollution, as well as groundwater contamination, include (Fleming, undated):

Table 7.10 Effectiveness of Agricultural BMPs for Groundwater Protection

BMP	Degree of effectiveness[a]		
	Pesticides	Nitrates	Salt
Structural			
Irrigation systems	M	M	H
Chemigation back-siphon devices	H	H	N
Animal waste storage	N	M	L
Subsurface (tile) drainage	L	M	M
Point-of-use treatment	H	H	H
Cultural			
Critical area planting	H	H	N
Cover cropping	H	M	M
Low-input farming[b]	H	M/L	N
Management			
Integrated pest management	M	N	N
Animal waste management	N	H	L
Fertilizer management	N	M	L
Pesticide management	H	N	N
Irrigation management	M	H	H
Focusing on sensitive groundwater areas	H	H	H

[a] H = high, M = medium, L = low, N = none.
[b] Assumes restricted use of pesticides and various combinations of leguminous cover crops, manure and sludge, nitrate soil tests, split nitrogen fertilizer applications, and other measures to reduce excessive nitrogen build-up and losses in the soil.

From Logan, T. J., *Journal of Soil and Water Conservation*, March–April, 1990, pp. 201–206. With permission.

1. Setting realistic yield goals based on measured yields, rather than guesses; since fertilizer rates are generally chosen according to expected yield, farmers' tendency to overestimate yields leads to excessive applications.
2. Soil testing to match fertilizer application rates to the crop's needs; farmers may underestimate the nutrients supplied by the soil either from native fertility or by carryover from previous fertilizer applications.
3. Timing fertilizer applications to correspond with crop needs and to minimize leaching; the trend away from applying nitrogen in the fall is an example of improvement in this area.
4. Using crop rotations, including nitrogen-fixing cover crops to restore soil fertility and tilth.

The development and use of agricultural BMPs are the focus of groundwater pollution prevention programs in Arizona, Nebraska, Washington, and California (National Governors Association, 1991). While not all-inclusive, the programs collectively include components related to special protection area designations and fertilizer application rates. Many of the current components are voluntary and are being promoted via educational efforts.

The Arizona Environmental Quality Act of 1986 established the Regulated Agricultural Activities Program that is related, in part, to the use of nitrogen fertilizers (Munson and Russell, 1990). The goal statements of the program are

Table 7.11 General Goal Statements for the Arizona Regulated Agricultural Activities Program

Best management practices for the application of nitrogen fertilizer
Application of nitrogen fertilizer shall be limited to the amount necessary to meet projected crop plant needs
Application of nitrogen fertilizer shall be timed to coincide as closely as possible to the periods of maximum crop plant uptake
Application of nitrogen fertilizer shall be by a method designed to deliver nitrogen to the area of maximum crop plant uptake
Application of irrigation water to meet crop needs shall be managed to minimize nitrogen loss by leaching and runoff
Application of irrigation water shall be timed to minimize nitrogen loss by leaching and runoff
The operator shall use tillage practices that maximize water and nitrogen uptake by crop plants
Best management practices for animal feeding operations
Harvest, stockpile, and dispose of animal manure from concentrated animal feeding operations as economically as is feasible to minimize discharge of nitrogen pollutants by leaching and runoff
Control and dispose of nitrogen-contaminated water resulting from activities associated with a concentrated animal feeding operation, up to a 25-year, 24-hour storm event equivalent, as economically feasible, to minimize the discharge of nitrogen pollutants
Close facilities in an economically feasible manner to minimize the discharge of nitrogen pollutants

From Munson, B. E. and Russell, C., *Journal of Soil and Water Conservation*, March–April, 1990, pp. 249–252. With permission.

in Table 7.11 (Munson and Russell, 1990). The program includes public information components along with a general permit and enforcement provisions.

Hall (1992) described a study to determine the effects of reduced manure and fertilizer applications under nutrient management (an agricultural best management practice) on groundwater quality at a 55-acre farm in Lancaster County, Pennsylvania. After 2 years (1985–1986) of routine agricultural activity and groundwater nitrate data collection, nutrient application reductions were implemented at the field site from 1987 through 1990. Statistically significant (at the 95% confidence level) decreases in median nitrate concentrations in groundwater samples occurred at four of five sampled wells. The decreases in median nitrate concentrations in groundwater at these four wells were 32% (26 mg/l), 30% (16 mg/l), 12% (3 mg/l), and 8% (2 mg/l) of the pre-nutrient reduction concentrations; these percentage decreases corresponded to decreases in nitrogen application on the contributing areas upgradient of these wells of 67% (a decrease of 340 pounds per acre per year), 53% (a decrease of 305 pounds per acre per year), 39% (a decrease of 198 pounds per acre per year), and 60% (a decrease of 378 pounds per acre per year), respectively. The fifth well, which was sampled at a point approximately 66 ft below the water table (a greater depth than at the other wells), had a nitrate concentration increase of 8% (1 mg/liter). Application of nitrogen in the contributing area upgradient of this well had been reduced by about 67% (325 pounds per acre per year). Changes in the groundwater nitrate concentrations were observed to lag behind the reductions in applied nitrogen fertilizers (primarily manure) by about 4 to 19 months (Hall, 1992).

Table 7.12 Examples of Models Used in Management Decisions

AGNPS (Agricultual Non-Point Source) — Single event, cell-based model that simulates sediment and nutrient transport from agricultural watersheds.

DRASTIC — Empirical standardized system for evaluating groundwater pollution potential by using hydrogeologic settings; the seven parameters estimated to be most significant in controlling pollution potential are: (1) *D*epth to water table, (2) net *R*echarge, (3) *A*quifer material, (4) *S*oil, (5) *T*opography, (6) *I*mpact of the vadose zone, and (7) *C*onductivity of the aquifer.

EPIC (Erosion Productivity Impact Calculator) — A model to determine the relation between soil erosion and soil productivity; capable of simulating periods greater than 50 years; incorporates hydrology, weather, erosion, nutrients, plant growth, soil temperature, tillage, economics, and plant environment control.

GLEAMS (Groundwater Loading Effects of Agricultural Management Systems) — Developed to evalute the effects of agricultural management systems on the movement of agricultural chemicals in and through the root zone for field-size areas.

LEACHMN (Leaching Estimates and Chemistry Model Nitrogen) — Process-based model of water and N movement, transformations, plant uptake, and N reactions in the unsaturated zone.

NITWAT (Nitrogen and Water Management) — Developed especially for corn on sandy soils; evaluates N transformations and transport in relation to crop growth under certain weather and irrigation conditions.

NLEAP (Nitrate Leaching and Economic Analysis Package) — Computer application package developed to estimate potential nitrate leaching from agricultural areas and project impacts on associate aquifers.

NTRM (Nitrogen Tillage and Residue Management) — Model with emphasis on management of nitrogen sources at the soil surface in conventional and reduced till systems; N transformations and transport are detailed using the NCSOIL submodel with active and passive N pools.

RZWQM (Root Zone Water Quality Management) — In development; will compare alternative management practices and their potential for groundwater contamination; comprehensive model includes macropore flow and N cycle description; expert systems approach.

From Office of Technology Assessment, "Beneath the Bottom Line — Agricultural Approaches to Reduce Agrichemical Contamination of Ground Water," OTA-F-418, 1990, U.S. Congress, Washington, D.C., pp. 90–100.

Mathematical modeling can be used as a tool in determining appropriate fertilization schemes. Table 7.12 lists examples of models that can be used for predicting nitrogen contamination potential from agricultural practices (Office of Technology Assessment, 1990). Additional information on several of the listed models is presented in Chapters 4 and 6.

The CREAMS model, described in Chapter 6, has been used to simulate the long-term effects of seven different BMPs on nitrate-nitrogen loadings to a shallow, unconfined groundwater system (Shirmohammadi, Magette, and Shoemaker, 1991). The study area included two watersheds in the Coastal Plain physiographic region of Maryland. Seven different management practices commonly used or viable for this region were selected; the practices are listed in Table 7.13 and the associated fertilization plans are summarized in Table 7.14 (Shirmohammadi, Magette, and Shoemaker, 1991).

For simulations involving only the CT and NT systems listed in Tables 7.13 and 7.14, input data to the chemistry component of the model were the actual nutrient amounts, application dates, and methods used on the two study watersheds. The fertilization plans used for other simulations were similar. As shown

Table 7.13 Management Practices Used for Simulation Purposes and Their Ranking Based on Nitrate-Nitrogen Leached Below the Root Zone

Management practices	Ranking based on Mass	Ranking based on Conc.
Conventional till corn — contoured — winter cover (CT-CN-WC)	1	6
No-till corn — contoured — winter cover (NT-CN-WC)	2	4
Conventional till corn (CT)	3	5
Conventional till corn — contoured (CT-CN)	4	3
No-till corn (NT)	5	2
No-till corn — contoured (NT-CN)	6	1
Conventional till turfgrass (CT-TG)	7	7

From Shirmohammadi, A., Magette, W. L., and Shoemaker, L. L., *Ground Water Monitoring Review,* Winter, 1991, pp. 112–118. With permission.

Table 7.14 Fertilization Plan Used for Different BMPs

Watershed	Crop	Date	Method of application	N (kg/ha)	P_2O_5 (kg/ha)
CT	Corn[a]	5/4[d]	Surface	36.7	11.0
		5/9[e]	Surface	29.4	70.9
		6/18[b]	Surface	105.0	—
	Turfgrass	10/1[e]	Incorporated[c]	67.1	134.3
		10/20[f]	Incorporated[c]	67.1	33.6
		11/10	Surface	55.9	28.0
		3/15	Surface	55.9	28.0
NT	Corn[a]	5/7[d]	Surface	36.7	11.0
		5/15[e]	Surface	29.4	70.9
		6/18[b]	Surface	105.0	—

[a] Continuous corn with winter cover crop (barley) had the same fertilization.
[b] Liquid nitrogen was applied.
[c] 10 cm deep incorporation.
[d] Before planting.
[e] At planting.
[f] After emergence.

From Shirmohammadi, A., Magette, W. L., and Shoemaker, L. L., *Ground Water Monitoring Review,* Winter, 1991, pp. 112–118. With permission.

in Table 7.14, the total mass of fertilizer applied was apportioned over multiple dates, i.e., split-applied on both watersheds. The amounts, dates, and methods of application varied depending on the type of cropping system. For example, the fertilization program for turfgrass involved applying 25% of the total nutrient amount in early September, 25% in early October, 25% in mid-November, and 25% in mid-March. The fertilization plan for CT and NT corn production with and without winter cover was the same; that is, no additional fertilizer was applied for having winter cover as part of the cropping system (Shirmohammadi, Magette, and Shoemaker, 1991).

Simulation results were used to evaluate the impact of different agricultural BMPs on groundwater loadings of nitrogen for the two watersheds in the study area. Average 12-year concentrations leached below the root zone and mass loadings were used to rank the BMPs, using the rankings shown in Table 7.13

(Shirmohammadi, Magette, and Shoemaker, 1991). Turfgrass was ranked as BMP 7 based on both the mass and concentration of nitrate-nitrogen leached below the root zone. In this study, mass losses were considered to be the more important ranking criteria. Results indicate that using winter cover such as barley helps to decrease nitrate-nitrogen leachate to shallow, unconfined groundwater. Turfgrass production decreases surface runoff and associated sediment and nutrients, but increases nitrate-nitrogen leachate significantly. Finally, the results of the model simulations also indicated that 12-year average nitrate-nitrogen concentrations leaching below the root zone under all of the considered BMPs were above the public drinking water standards of 10 mg/l (Shirmohammadi, Magette, and Shoemaker, 1991).

A cost-benefit analysis for dealing with the effect of nitrogen fertilizer application on nitrate concentrations in groundwater was developed by de Haen (1982) and based on data collected in the Federal Republic of Germany. Graphs were developed depicting the relationships between costs of reducing nitrate levels in drinking water versus benefits in terms of prevention of damage to human health. From this, an acceptable nitrate concentration limit was chosen. Based upon this limit, the effectiveness of different management alternatives was examined. Two examples were regulation of crop type and taxation of fertilizer use (de Haen, 1982). Since different crop types have different nitrogen requirements, and if a crop choice is feasible based on economic yields, the crop with lower nitrogen requirements should be chosen. To illustrate another choice, crops with low nitrogen needs could be specified for geographical areas with higher susceptibility to groundwater pollution; for example, in areas with shallow alluvial aquifers. Another type of management alternative is to tax fertilizer usage, with the tax rate increasing with increasing fertilizer application rate above basic crop needs.

Schneider and Day (1976) described a policy study designed to curtail nitrate pollution of groundwater resulting from field-applied manure and fertilizer nitrogen. Six policies were tested in four farm sizes at four locations in Wisconsin: (1) prohibit all manure spreading in winter; (2) prohibit winter manure spreading on sloped land; (3) prohibit winter manure spreading close to streams, lakes, or open ditches; (4) restrict the excess of applied nitrogen over the estimated plant uptake of nitrogen; (5) combine 2, 3, and 4 above; and (6) combine 1 and 4 above. The effects of these policies on the efficiency of resources allocation and their differential impact in the glaciated and unglaciated regions of Wisconsin were explored. The marginal cost of pollution reduction was calculated for each policy.

McWilliams (1984) identified three institutional approaches for controlling nitrate pollution of groundwater: federal, state, and local or citizen action. Federal and/or state action could include legislation promoting groundwater management, or programs that would aid innovative farmers in implementing groundwater pollution prevention practices. At the local level, even if laws are passed, time is required to develop implementation and enforcement programs. The forcefulness of national, state, and/or local environmental and conservation groups can be an encouragement for more rapid development and implementation of institutional approaches.

Regulatory approaches for reducing groundwater nitrate pollution potential by using less nitrogen fertilizer include (Francis, undated): (1) an excise tax, based on the concept that increasing the cost of nitrogen fertilizers will reduce their use; (2) limited rights to buy nitrogen fertilizer; (3) per-acre restrictions on commercial nitrogen fertilizer application rates; and (4) zoning regulations that limit the type of land use and fertilizer inputs. These approaches are not mutually exclusive; a combination of two or more can be adopted. However, it should be noted that while each approach has advantages and limitations, the implementation effectiveness of each can be questioned.

Regulatory or other governmental management options that could be used to protect groundwater from nitrate pollution include (Fleming, undated):

1. Farm support programs can be restructured to eliminate the incentives encouraging farmers to increase their use of agrichemicals and to reduce the barriers limiting farmers from adopting low-chemical input production methods.
2. Taxes can be imposed on fertilizers and pesticides to offset the external environmental costs of agrichemical use, fund low-chemical input technology research, and influence farmers to reduce excessive applications of farm chemicals.
3. A greater proportion of agricultural research funds can be devoted to methods that reduce farmers' agrichemical inputs and minimize groundwater contamination.

Carefully managed irrigation can aid in minimizing nitrate pollution of groundwater. Supplemental irrigation is frequently used in water-deficient areas, along with fertilizers, to assure sufficient plant nutrients for higher potential yields. However, irrigation and fertilization practices combined with seasonal rainfall can cause increased leaching of nitrate-nitrogen below the root zone. For example, Timmons and Dylla (1979) described a corn crop-oriented study to evaluate the effects of irrigation and increased fertilizer use and to determine which corn management systems produce maximum yields and minimize nitrate-nitrogen leaching losses. Finally, Kepler, Carlson, and Pitts (1978) have developed a pollution control manual for irrigated agriculture. Traditional and modified irrigation practices are described and evaluated in terms of use, pollutant loading pathways, costs, and effectiveness.

OTHER MANAGEMENT MEASURES

Table 7.15 contains summary comments from 22 references on other management measures that can be used to prevent, or minimize, agriculturally related nitrate pollution of groundwater. Nine types of management measures are included in the 22 references, and Table 7.16 contains a categorized listing. Examples of each of the nine types of measures are briefly described.

Pleistocene aquifers in the southern and eastern parts of The Netherlands are used extensively as sources for public water supply. In The Netherlands, aquifer protection policies are based upon defining protection zones around well systems

Table 7.15 References Involving Other Management Measures for Prevention of Nitrate Pollution of Groundwater

Author(s) (year)	Comments
Adams and Foster (1992)	Discussion of groundwater protection zones in Great Britain
Allen (1977)	Control of nitrates via use of nitrification inhibitors
Allen (1984)	Use of slow-release nitrogen fertilizers to minimize soil leaching losses of nitrogen
Bremner, McCarty, and Gianello (1986)	Research study on nitrification inhibitors as a rapid test for measuring potentially available organic soil nitrogen
Conway and Pretty (1991)	Review of Nitrate Sensitive Areas Scheme in the U.K.
Eccles, Cline, and Hardt (1976)	Use of inflatable packer in a well to reduce nitrates in produced groundwater
Economic Commission for Europe (1986)	General discussion of groundwater protection in Europe
Francis (undated)	Review of four regulatory approaches for reducing the nitrate pollution potential of groundwater
Gilley et al. (1982)	Use of water management as a tool for minimizing nitrate pollution of groundwater
Harris and Skinner (1991)	Discussion of Nitrate-Sensitive Areas in the U.K.
Harryman (1989)	Use of vulnerable zones designation for groundwater protection in the member states of the European Community
Hergert (1978)	Proper water management for minimizing leaching losses of nitrogen
Hoeft (1984)	Use of nitrification inhibitors to delay the nitrification process in soils
Lauterbach and Klapper (1982)	Use of drinking-water protection zones and water management for minimizing nitrate pollution of groundwater
Mansell et al. (1977)	Effects of tillage practices and lime additions on nitrogen leaching from citrus groves
Mansell, Fiskell, and Calvert (1982)	Effects of tillage practices and lime additions on soil nitrification in a citrus grove
Peters and den Blanken (1985)	Use of protection zones around well systems to minimize groundwater pollution
Randall and Bandel (1987)	Delineation of increased nitrate leaching when conservation tillage is utilized
Schepers (1987)	Discussion of nitrate leaching with a program of no tillage practices
Schleyer, Milde, and Milde (1992)	Review of three groundwater protection zones established by the Association of Gas and Water Engineers in West Germany
Wendt, Onken, and Wilke (1976)	Effects of irrigation methods on groundwater pollution by nitrates
Zwirnmann (1982)	Proceedings of conference on nonpoint nitrate pollution of municipal water supplies

as a function of the travel times of the groundwater within the local area (Peters and den Blanken, 1985). Calculations were made with a groundwater contaminant transport model in order to define areas of limitations on land usage based on groundwater protection policies. An example of an assessment of the impact of applications of pesticides in water supply catchment areas is described by Peters and den Blanken (1985). This approach could also be used as a basis for the development of a fertilizer application policy.

Table 7.16 Examples of Other Management Measures

Measure	Ref. describing measure
Implement groundwater (or drinking water) protection zones (or nitrate-sensitive areas program)	Adams and Foster (1992) Conway and Pretty (1991) Economic Commission for Europe (1986) Harris and Skinner (1991) Harryman (1989) Lauterbach and Klapper (1982) Peters and den Blanken (1985) Schleyer, Milde, and Milde (1992)
Consider soil organic N in determining fertilizer requirements	Bremner, McCarty, and Gianello (1986)
Implement appropriate water management practices	Gilley et al. (1982) Hergert (1978) Zwirnmann (1982)
Choose irrigation method to minimize nitrogen losses	Wendt, Onken, and Wilke (1976)
Choose tillage practice and soil conditioning to minimize nitrogen losses	Mansell et al. (1977) Mansell, Fiskell, and Calvert (1982) Randall and Bandel (1987) Schepers (1987)
Use nitrification inhibitors	Allen (1977) Bremner, McCarty, and Gianello (1986) Hoeft (1984)
Use slow release nitrogen fertilizers	Allen (1984)
Implement regulatory approaches for nitrate management	Francis (undated)
Use well packer for selective zoning of groundwater withdrawal	Eccles, Cline, and Hardt (1976)

Lauterbach and Klapper (1982) discussed policy decisions relative to planning and implementing a groundwater protection zone program. A classification scheme for protection zones is shown in Figure 7.1 (Lauterbach and Klapper, 1982). The protected zone I is the zone of direct water obtainment; within this zone, direct pollution would be possible. Adjacent to this zone is zone II, i.e., the closer protected zone, followed by protected zone III; that is, the widest protected zone, which comprises the zone of contribution area. Some possible policy choices relative to protection against groundwater pollution are shown in Table 7.17, while Table 7.18 delineates various allowable land usages within the protection zones, with special relevance to the minimization of groundwater nitrate pollution (Lauterbach and Klapper, 1982).

According to Adams and Foster (1992), there are two key elements in a groundwater protection policy: (1) it should strike a balance between the protection of resources and sources used as potable drinking water, and (2) it should address the problems of point and nonpoint pollution sources. Land-surface zoning is a procedure used for the general framework for a groundwater protection policy (Adams and Foster, 1992). There are two main components to this procedure: (1) areas related to the catchment of individual groundwater sources (protection zones) are identified, and (2) the entire land surface is divided based on the vulnerability of the aquifers to pollution (Adams and Foster, 1992).

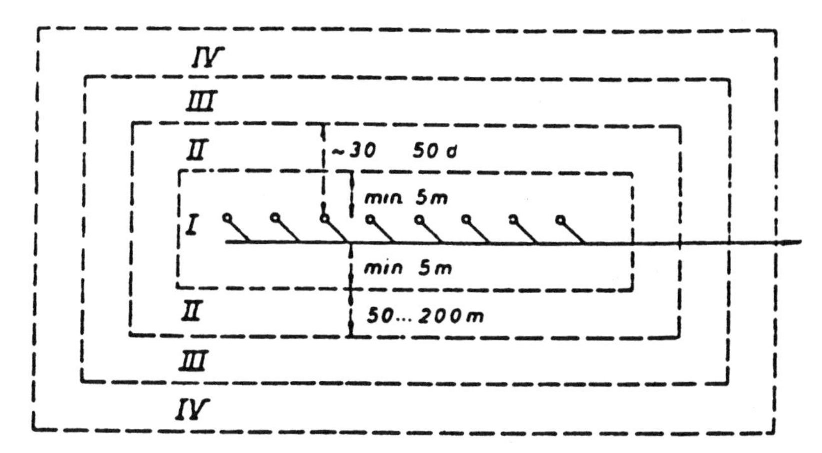

Figure 7.1 Classification scheme for groundwater protection zones. (From Lauterbach, D. and Klapper, H., *Proceedings of IIASA Task Force Meeting in Nonpoint Nitrate Pollution of Municipal Water Supply Sources*, Series No. CP-82-54, 1982, International Institute of Applied Systems Analysis, Laxenburg, Austria, pp. 135–161. With permission.)

Table 7.17 Policy Choices for Drinking Water Protection Zones

Zone division	Distances	Retention time	Protection against pollution
Zone I	Min. 5 m from all sides around the well		No direct pollution
Zone II	Favorable underground 50 to 100 m	30 to 50 days up to obtaining water in the caption installation	No microbial or biological degradable contaminants
	Medium underground conditions 100 to 300 m		
	Unfavorable underground		
	Special stipulations		
Zone III			No pollution by chemical substances, e.g., nitrate, difficult to eliminate
Zone IV			No pollution by substances which cannot be eliminated

From Lauterbach, D. and Klapper, H., *Proceedings of IIASA Task Force Meeting in Nonpoint Nitrate Pollution of Municipal Water Supply Sources*, Series No. CP-82-54, 1982, International Institute of Applied Systems Analysis, Laxenburg, Austria, pp. 135–161. With permission.

In Great Britain, the zones have been delineated relative to four areas of protection: recharge capture area, operational courtyard, inner protection zone, and outer protection zone. The outermost protection area for a catchment is called the "recharge capture zone." In this area, all aquifer recharge will be captured, whether it stems from precipitation or surface water (Adams and Foster, 1992). All activities that are potentially polluting would be prohibited or controlled to the proper degree required by the capture zone. Because of this, some divisions

Table 7.18 Land Usages in Groundwater Protection Zones

Type of utilization	Protected zone[a]			
	I	II	III	IV
Industry				
Sewage discharge	p	p	l	—
Sewage infiltration, subsoil irrigation	p	p	p	l
Waste product disposal	p	p	l	—
Camping grounds, bathing	p	p	—	—
Agriculture and forestry				
Used as farm land	p	—	—	—
Spray irrigation of agricultural acreage	p	l	l	—
Permanent pasturage	p	l	—	—
Use of solid inorganic fertilizer	l	l	l	—
Use of liquid inorganic fertilizer	p	p	l	—
Use of solid organic fertilizer	p	l	—	—
Use of liquid organic fertilizer	p	p	l	—
Storage of solid organic and inorganic fertilizer	p	p	l	—
Storage and transport of liquid organic fertilizer	p	l	—	—
Individual livestock breeding	p	l	—	—
Industrialized animal production plants (newly built)	p	p	l	—

[a] p denotes prohibited; l denotes limited.

From Lauterbach, D. and Klapper, H., *Proceedings of IIASA Task Force Meeting in Nonpoint Nitrate Pollution of Municipal Water Supply Sources*, Series No. CP-82-54, 1982, International Institute of Applied Systems Analysis, Laxenburg, Austria, pp. 135–161. With permission.

of the recharge capture zones may be necessary, with the most strict regulations being applied closest to the source (Adams and Foster, 1992). The divisions may be based on a variety of conditions; examples include horizontal distance, horizontal flow time, proportion of the recharge area, saturated zone dilution, and the attenuation capacity (Adams and Foster, 1992).

The operational courtyard is the closest protection area around the source in the Great Britain system. It is highly desired that this area be owned and controlled by the groundwater abstractor. A radius of 30 m from the source is used in the U.K. (Adams and Foster, 1992). The inner protection zone is related to the pathogenic control of point or nonpoint sources of pollution. It is based on a specific horizontal flow time. These flow times are highly varied; for example, 400 days by the Thames Water Authority and 100 days by the Yorkshire Water Authority. Comparable data include 60 days in The Netherlands and Belgium; 50 days by the Southern Water Authority and the Anglian Water Authority in the U.K., Germany, and Austria; and 10 days in Switzerland (Adams and Foster, 1992). Finally, the outer protection zone may be necessary to allow differential control of point-source or diffuse-source pollution remaining in the area.

The European community has also established a directive that requires the member states to designate "nitrate vulnerable zones"; these zones are areas where the land drains directly into, or that contains more than the limit of nitrates, or that without protection will exceed the level of allowable nitrates (Harryman, 1989). Within these vulnerable zones, the member states would be required to:

(1) introduce controls on the application of fertilizer to land, (2) establish rules regulating application of manure, (3) guarantee that sewage plants discharging into vulnerable zones would serve no more than 5000 people, and (4) apply additional agricultural methods to reduce nitrate pollution (Harryman, 1989).

Several European countries use the protection guidelines established by the Association of Gas and Water Engineers in West Germany (Schleyer, Milde, and Milde, 1992). These guidelines divide protected areas into three zones: Zone I, Zone II, and Zone IIIA and Zone IIIB. Zone I should provide protection against pollution immediately around the well. All activity is banned. The zone should be approximately 10 m beyond the well in the direction the groundwater flows. Zone II should also provide protection against pollution, but in a larger area. This zone includes an area where the water takes 50 to 60 days to reach the well. The use of organic fertilizers is permitted in Zone II. Zone III is the area beyond the first two zones that includes the entire intake area. If the intake area of a polluted source exceeds the distance of 2 km, Zone III is divided into two subzones, A and B, with B denoting the zone that is the farthest from the well.

A Nitrate Sensitive Area Scheme has been developed in the U.K. under the auspices of the 1989 Water Act (Conway and Pretty, 1991). A total of ten areas totaling 15,000 ha has been designated. Farmers in the areas are compensated at a rate of £55–95/ha for complying with a comprehensive range of practices aimed at reducing nitrate concentrations in groundwater; the practices are summarized in Table 7.19 (Conway and Pretty, 1991).

According to Harris and Skinner (1991), the 1989 Water Act was to be the first legislation in the U.K. that dealt with the control of diffuse pollution. Specifically, Section 111 provides for the establishment of "Water Source Protection Areas" to control certain other prescribed land management areas. Section 111 also allows the National Rivers Authority (NRA) to request the government to designate areas of land where land-use practices may be controlled to prevent the pollution of water. Section 112 relates solely to nitrate pollution and is therefore a special case of the water source protection area. Following recommendations from the NRA, the Ministry of Agriculture and the Department of the Environment may designate "nitrate-sensitive areas" (NSAs). Farmers within NSAs have been encouraged to enter into agreements to follow a defined scheme of agricultural management practices. In recognition of the potential loss of income through reduced yields, compensation payments can be made as mentioned above.

One way of reducing applied nitrogen fertilizer requirements is to rely on a better means of determining the nitrogen already present in the soil, and using this information in relation to planning for crop nitrogen needs through a combination of applied and available nitrogen. As discussed in Chapter 3, reductions in fertilizer application rates typically lead to reductions in nitrate concentrations in groundwater. To facilitate this approach, Bremner, McCarty and Gianello (1986) have developed a simple and rapid chemical method basic for the determination of an index of potentially available organic soil nitrogen.

Satisfactory laboratory methods basic to an index of the amount of nitrogen likely to be made available for crop growth by mineralization of soil organic

Table 7.19 Pilot Nitrate-Sensitive Areas Scheme in the U.K.: Conditions for Basic Rate Payment Scheme

- Inorganic fertilizer use
  - Do not exceed economic optimum N fertilizer amounts for each year including full allowance for organic manures used
  - Apply less than the economic optimum amount of N fertilizer for yield on the following crops as shown:
    - Winter wheat 25 kg/ha N below optimum
    - Winter barley 25 kg/ha N below optimum
    - Winter oilseed rape 50 kg/ha N below optimum
  - Do not apply inorganic N fertilizer in autumn or winter
  - Applications should not be made between Sept 1 and the following Feb 1 to grass fields
  - Applications should not be made between Aug 15 and the following Feb 15 to field not in grass
  - Do not apply extra total N fertilizer to bread-making wheat varieties more than specified above
  - Do not apply individual applications of more than 120 kg/ha of inorganic N fertilizer
- Crop cover
  - Drill autumn-sown cereals by Oct 15 if previous crop is harvested by that date
  - If the previous crop is removed before Oct 15 and the next crop will not be sown before the next Jan 1, sow a cover crop
  - Where required, approved cover crop should be established soon after harvest of previous crop; do not apply inorganic N fertilizer for the cover crop
  - The cover crop should not be removed by cultivation, herbicide, or grazing before Feb 1 on sandy soils or Dec 1 on other soils, unless next crop is sown within 4 weeks of removal
- Organic manure
  - Organic manure should be limited to an annual application of 175 kg/ha total N
  - No slurry, poultry manure or liquid sewage sludge to be applied between Sept 1 and Nov 1 to grass fields and between July 1 and Nov 1 to fields not in grass
  - Manure stored in the NSA must not be a substantial point source of nitrate leaching
- Ploughing up grassland
  - Where slurry, poultry manure or liquid sewage sludge have been applied to grass between July 1 and Sept 1 in any year, the field may not be cultivated or the grass killed off within 4 weeks of application
  - Where grass is cultivated or reseeded, the following crop must be sown as soon as soil conditions allow and in any event no later than Oct 1 in the same year; grass may not be cultivated between Oct 1 and Feb 1 in the following year
- Irrigation
  - Where irrigation is practiced, show evidence that a scheduling system is used that best uses water and avoids excessive applications
- General
  - Do not convert grass to arable unless the grass can be shown to be a ley in arable rotation
  - Do not increase application of organic manures to the NSA land on the farm by more than 25% of current application unless the manure is produced within the NSA
  - Do not remove hedgerows or woodland unless replaced by an equivalent area
  - Keep records of applications or organic and inorganic nitrogen fertilizer in terms of quantity, timing, and areas of application
- Pig/poultry plan
  - An individually agreed manure plan will be required for the unit; this will show that the unit has appropriate storage, handling transport, and land to meet the following criteria
  - Organic manure should be limited to an annual application of 175 kg/ha total N when applied to NSA land
  - No slurry or poultry manure to be applied between Sept 1 and Nov 1 to grass fields, and between July 1 and Nov 1 to fields not in grass

Table 7.19 Pilot Nitrate-Sensitive Areas Scheme in the U.K.: Conditions for Basic Rate Payment Scheme (Continued)

Alternatively, the manure may be disposed of in an approved way other than to NSA land; manure applied to vulnerable land outside the NSA must meet the NSA criteria
Manure stored in the NSA must not be a substantial point source of nitrate leaching

Note: NSA = nitrate-sensitive area.

From Conway, G. R. and Pretty, J. N., *Unwelcome Harvest — Agriculture and Pollution*, Earthscan Publications, London, 1991, pp. 540–544. With permission.

matter during the growing season have long been needed, and numerous biological and chemical methods have been developed. Potentially reliable methods include those involving determination of the inorganic nitrogen produced by incubation of the soil sample under aerobic or anaerobic conditions for various times; however, these biological methods are complicated and time-consuming. Therefore, Bremner, McCarty, and Gianello (1986) developed two rapid chemical methods for obtaining an index of potentially available organic soil nitrogen that can be used in soil testing laboratories. One method involves determination of the ammonia-nitrogen liberated from organic soil nitrogen when the soil sample is steam distilled with pH 11.2 phosphate-borate buffer solution for 8 min. The other involves determination of the ammonium-nitrogen produced from organic soil nitrogen by heating the soil sample with 2 *M* KCl solution in a stoppered tube at 100°C for 4 hours. Both methods are simple and precise, and their results are not significantly affected by air-drying or air-dry storage of soil samples before analysis.

Nitrate losses to groundwater systems can be minimized through careful water management practices in irrigated areas. If excess irrigation water is used, the amount of applied fertilizer nitrogen leached to the subsurface environment can be undesirable. For example, Hergert (1978) described a study to examine nitrogen fertilization and irrigation practices on sandy soils with the goal of quantifying losses of nitrogen from sprinkler-applied fertilizer. Leaching losses measured in 1976 were 26 and 187 kg N/ha for low and excess irrigation treatments, respectively; losses in 1977 were 101 and 127 kg N/ha for low and excess treatments, respectively. No appreciable volatilization of nitrogen from the sprinkler-applied fertilizer occurred.

The type of irrigation method should also be considered in a comprehensive program to minimize nitrogen losses to the subsurface environment. For example, Wendt, Onken, and Wilke (1976) evaluated the use of sprinkler, furrow, and manual and automatic subirrigation systems in a loamy fine sand soil overlying a shallow aquifer in Knox County, Texas. In addition to evaluating the irrigation method, several methods of fertilizer application were also studied, including banded below the level of the water furrow, banded above the level of the water furrow, and applied through the irrigation water. By irrigating on the basis of potential evapotranspiration, it was possible to maintain nitrogen fertilizers in the root zone during the growing season (Wendt, Onken, and Wilke, 1976). The ranking of irrigation systems relative to their water requirement was: furrow >

sprinkler ≅ subirrigation > automatic subirrigation. In general, the automated subirrigation system required a water application of only 50 to 60% of the sprinkler and manual subirrigation systems, and 30 to 40% of the furrow system. During the study, only 10 to 14 cm/year supplemental water was applied through the automated system to produce the sweet corn crop, compared to 18 to 38 cm/year applied through the other studied systems.

Comparison of sprinkler, furrow, and manual subirrigation systems showed that nitrate-nitrogen in soil samples and porous bulb extracts was much less for samples from below the root zone from subirrigation systems than from the sprinkler or furrow systems (Wendt, Onken, and Wilke, 1976). Through analyses of ^{15}N in soil and plant materials, it was possible to account for 92.6, 86.1, and 50.5%, respectively, of the fertilizer nitrogen applied to the sprinkler, furrow, and subirrigation systems over a 2-year period. Soil-water potential, bromide tracer, porous bulb extract, and soil data supported the concept that denitrification occurs with subirrigation systems. In any case, less nitrate-nitrogen was available to be leached in the subirrigated plots.

Movement of nitrate during and between growing seasons was also noted by Wendt, Onken, and Wilke (1976), especially with the furrow and sprinkler irrigation plots. Data from ^{15}N and bromide tracer studies showed unique patterns of movement for the different systems. Banded fertilizer moved vertically as discrete bands under the sprinkler irrigation system. In the furrow irrigation system, the fertilizer bands merged in the bed and then moved vertically. Fertilizer bands moved up and away from the buried subirrigation system. In addition to being influenced by the irrigation system, the downward movement was influenced by soil texture and the rainfall received between growing seasons.

A good water management program could help to minimize nitrate pollution of groundwater. One of the best ways of improving water management is through the use of irrigation scheduling. Application of the correct amounts of water at the right time may lessen the negative effect of irrigated agriculture on groundwater quality. Effective water management will depend on how much water a particular crop removes from the soil. Annually, monthly, and even daily crop water requirements must be accurately determined. Several irrigation techniques can be adopted to improve irrigation scheduling; they include soil moisture measurement and computer simulation to determine the interaction of various soil types and rainfall characteristics (Gilley et al., 1982).

Another way to reduce pollution from irrigated agriculture is to encourage a change from a type of irrigation system that may cause deep percolation under a given set of circumstances to one that would improve water management. Three general methods can be used for irrigation; they include surface, sprinkler, and drip or trickle irrigation systems. The potential pollution from each of these systems can be evaluated as a function of soil type and slope. Thus, the specific pollution potential of irrigated land can be evaluated for individual sites. Evaluation of the potential pollution from the three irrigation systems as functions of soil types are shown in Tables 7.20 and 7.21, respectively (Gilley et al., 1982). It should be noted that these tables are primarily oriented toward surface water pollution.

Table 7.20 Potential Pollution Rating of Sprinkler Irrigation Systems

Types of sprinkler irrigation systems	Coarse-textured soil (sand to loamy sand) Land slope (%)					Medium-textured soil (loam to silt loam) Land slope (%)					Fine-textured soil (silty clay to clay) Land slope (%)				
	0–2	2–4	4–6	6–10	>10	0–2	2–4	4–6	6–10	>10	0–2	2–4	4–6	6–10	>10
Sprinkler															
Solid set	L[a]	L	L	M	M	L	L	L	M	M	L	L	L	L	M
Center-pivot	L	L	L	M	M	L	L	M	H	H	M	M	H	H	H
Moved	M	M	M	H	H	L	L	M	H	H	L	L	M	H	H
Moving	L	L	M	H	H	L	L	M	H	H	M	M	H	H	H

[a] Rating scale (relative): H, high potential pollution hazard; M, moderate potential pollution hazard; L, low potential pollution hazard.

From Gilley, J. R., et al., "Strategies for Reducing Pollutants from Irrigated Lands in the Great Plains," EPA R-8052 49, 1982, U.S. Environmental Protection Agency, Ada, OK.

Table 7.21 Potential Pollution Rating of Surface and Trickle Irrigation Systems

Types of surface and trickle irrigation systems	Coarse-textured soil (sand to loamy sand) Land slope (%)				Medium-textured soil (loam to silt loam) Land slope (%)				Fine-textured soil (silty clay to clay) Land slope (%)			
	0–2	2–4	4–6	>6	0–2	2–4	4–6	>6	0–2	2–4	4–6	>6
Surface												
Contour ditch	H[a]	H	H	H	H	H	H	H	M	H	H	H
Graded borders	H	H	H	H	M	H	H	H	L	M	M	M
Furrows without reuse	H	H	H	H	M	H	H	H	M	H	H	H
Furrows with reuse	M	H	H	H	L	M	M	M	L	L	M	M
Automated gated pipe with reuse	M	M	M	H	L	L	L	M	L	L	L	M
Trickle												
Trickle/drip	L	L	L	L	L	L	L	L	L	L	L	L

[a] Rating scale (relative): H, high potential pollution hazard; M, moderate potential pollution hazard; L, low potential pollution hazard.

From Gilley, J. R., et al., "Strategies for Reducing Pollutants from Irrigated Lands in the Great Plains," EPA R-8052 49, 1982, U.S. Environmental Protection Agency, Ada, OK.

Tillage and soil conditioning practices also influence nitrogen losses to the subsurface environment. For example, Mansell et al., (1977) determined the movement of nitrate and several pesticides in surface and subsurface drainage waters from a citrus grove located in an acid, sandy flatwood soil in southern Florida. The influence of fertilizer and pesticide upon water quality was examined for citrus growing in three soil management treatments: ST (shallow-tilled plowed to 15 cm); DT (deep-tilled and soil mixed within the top 105 cm); and DTL (deep-tilled to 105 cm and 56 metric tons/ha of dolomitic limestone mixed with the soil). Average annual losses of nitrate-nitrogen in both surface and subsurface drainage from ST, DT, and DTL plots were equivalent to 22.1, 3.1, and 5.4% of the total nitrogen applied as fertilizer. Deep tillage was observed to greatly decrease leaching losses of nitrogen and phosphorus nutrients. The losses of nutrients in surface runoff was very small for all three plots.

The use of nitrification inhibitors was also found to aid in the minimization of nitrates in groundwater. Allen (1977) noted that a nitrification inhibitor called N-Serve (nitrapyrin) was effective in reducing the amount of nitrate leached from soil. A comprehensive study of nitrification inhibitors was conducted by Bremner, McCarty, and Gianello (1986). Based upon the suggestion that acetylene, a gaseous compound, is a potent inhibitor of nitrification in soils, the effects of 15 monosubstituted and 6 disubstituted acetylenes on nitrification in soils treated with ammonium sulfate was studied; the compounds tested are listed in Table 7.22 (Bremner, McCarty, and Gianello, 1986). The two most effective compounds were 2-ethynylpyridine and phenylacetylene.

To further evaluate 2-ethynylpyridine and phenylacetylene as soil nitrification inhibitors, the abilities of these compounds to inhibit nitrification in soil were compared with those of nitrapyrin (N-Serve), etridiazole (Dwell), and 13 other compounds that have been patented or proposed as fertilizer amendments for inhibition of nitrification in soil (Bremner, McCarty, and Gianello, 1986). This comparison was performed by determining the effects of different amounts of the test compounds on production of nitrite and nitrate in soils incubated at 25°C after treatment with ammonium sulfate. The soils used were selected to represent a range in pH (5.9–8.1), texture (11–57% sand, 20–40% clay), and organic carbon content (1.2–4.2%). The results shown in Table 7.23 indicated that 2-ethynylpyridine and phenylacetylene compared favorably with nitrapyrin and etridiazole for inhibition of nitrification in soil and were considerably more effective than the other compounds tested (Bremner, McCarty, and Gianello, 1986). Table 7.24 summarizes a comparison of the effects of different rates of application of 2-ethynylpyridine, nitrapyrin, and etridiazole on nitrification in soils; 2-ethynylpyridine was more effective than nitrapyrin or etridiazole at all rates tested (Bremner, McCarty, and Gianello, 1986).

Inhibitors incorporated with nitrogen fertilizer have been developed to delay the nitrification process (oxidation of NH_4^+ to NO_3^-). Nitrification inhibitors minimize the activity of *Nitrosomonas* bacteria. Some environmental factors have also been shown to influence the length of time for which a nitrification inhibitor effectively maintains NH_4^+ in the soil; they include soil temperature, soil organic

Table 7.22 Effectiveness of Acetylenic Compounds as Nitrification Inhibitors

Name	Formula	Effectiveness as nitrification inhibitor[a]
Monosubstituted acetylenes		
2-Ethynylpyridine	$(C_5H_4N)C⋮CH$	H
Phenylacetylene	$C_6H_5C⋮CH$	H
E-Butyn-2-one	$HC⋮CCOCH_3$	H
2-Phenyl-2-propyn-1-ol	$HC⋮CCH(OH)C_6H_5$	M
1-Pentyne	$CH_3CH_2CH_2C⋮CH$	M
4-Phenyl-1-butyne	$C_6H_5CH_2CH_2C⋮C$	M
5-Phenyl-1-pentyne	$C_6H_5CH_2CH_2C⋮CH$	M
1-Ethynylcyclohexylamine	$CH_2(CH_2)_4C(NH_2)C⋮CH$	M
2-Propyn-1-ol	$HC⋮CCH_2OH$	M
2-Propyn-1-amine	$HC⋮CCH_2NH_2$	M
1-Hexyne	$CH_3CH_2CH_2CH_2C⋮CH$	M
3-Butyn-1-ol	$HC⋮CCH_2CH_2OH$	M
1-Heptyne	$CH_3CH_2CH_2CH_2CH_2C⋮CH$	M
1-Octyne	$CH_3CH_2CH_2CH_2CH_2CH_2C⋮CH$	M
Acetylene monocarboxylic acid	$HC⋮CCOOH$	I
Disubstituted acetylenes		
3-Phenyl-2-propyn-1-ol	$C_6H_5C⋮CCH_2OH$	M
Diphenylacetylene	$H_5C_6C⋮CC_6H_5$	M
2-Butyn-1-ol	$CH_3C⋮CCH_2OH$	M
2-Butyn-1,4-diol	$HOCH_2C⋮CCH_2OH$	M
Acetylene dicarboxylic acid	$HOOCC⋮CCOOH$	I
2-Butynoic acid	$CH_3C⋮CCOOH$	I

[a] H, highly effective; M, moderately effective; and I, ineffective.

From Bremner, J. M., McCarty, G. W., and Gianello, C., *Proceedings of the Conference on Agricultural Impacts on Ground Water*, 1986, National Water Well Association, Dublin, OH, pp. 467–481. With permission.

matter, soil texture, time of application, rate of application, soil moisture, and agricultural practice (Hoeft, 1984).

Slow-release fertilizers provide nitrogen to crops in a time-release fashion; they function in one of four general ways (Office of Technology Assessment, 1990): (1) employing a physical barrier to control the escape of water-soluble materials containing ammonia or nitrate into soil; (2) possessing reduced water-solubility properties and containing plant-usable nitrogen (e.g., metal ammonium phosphates); (3) possessing low water solubility and releasing plant-available nitrogen during chemical or biological decomposition (e.g., ureaforms and oxamides); and (4) having high water solubility but a chemical structure that allows materials to decompose gradually and release plant-available nitrogen (e.g., guanylurea salts).

Slow-release nitrogen (SRN) fertilizers may be used to minimize nitrogen losses from soils subject to leaching. Conventional nitrogen fertilizers contain nitrate, urea, or ammonium ions, with the latter two sources being readily converted to nitrate by soil microorganisms. SRN fertilizers also convert urea and ammonium components to nitrate, but their conversion rate is low; thus, less

Table 7.23 Effectiveness of Various Compounds for Inhibition of Nitrification in Soils Treated with Ammonium Sulfate

Compound	Amount added (μg/g soil)	% Inhibition of nitrification[a]	
		Range	Average
2-Ethynylpyridine	5	79–100	87
Phenylacetylene	5	62–98	77
Etridiazole (Dwell)	5	61–97	76
Nitrapyrin (N-Serve)	5	45–94	65
4-Amino-1,2,4-triazole (ATC)	5	41–90	62
2,4-Diamino-6-trichloromethyl triazine	5	24–53	41
2-Amino-4-chloro-6-methyl pyridine (AM)	5	14–63	37
Guanylthiourea (ASU)	5	9–35	25
Sulfathiazole (ST)	5	8–24	17
4-Mesylbenzotrichloride	5	2–11	7
4-Nitrobenzotrichloride	5	2–12	6
Potassium azide	5	0–5	3
N-2,5-Dichlorophenyl succinamide (DCS)	5	0–5	3
Sodium thiocarbonate	5	0	0
Thiourea (TU)	5	0	0
2-Mercaptobenzothiazole (MBT)	5	0	0
Dicyandiamide (DCD)	5	6–28	19
	10	19–35	26
	50	78–90	83
Phenolic acids[b]	10	0–1	1
	100	0–7	2
	250	0–28	5
Tannins[c]	10	0–1	0
	100	0–1	1
	250	0–8	2

Note: Soils treated with 200 μg NH_4–N/g; 25°C; 25 days.

[a] Results obtained with three soils (Harps, Webster, and Storden).

[b] *p*-Hydroxybenzoic, *p*-coumaric, vanillic, ferulic, caffeic, ellagic, gallic, and chlorogenic acids.

[c] Mangrove, quebracho, mimosa, chestnut, and sumac.

From Bremner, J. M., McCarty, G. W., and Gianello, C., *Proceedings of the Conference on Agricultural Impacts on Ground Water*, 1986, National Water Well Association, Dublin, OH, pp. 467–481. With permission.

nitrate may be present in soil when leaching events occur. Special interest is increasing for the use of some specific SRN products such as urea formaldehyde (UF) coated products, particularly surface-coated urea (SCU) and isobutylidene diurea (IBDU) (Allen, 1984). The timing of SRN application is also a key factor in preventing nitrogen leaching. For example, an SRN fertilizer is also a long-release fertilizer; therefore, if dissolution of SRN granules occurs during the noncrop uptake period, they may actually increase leaching losses.

Finally, well packers for selective zoning for groundwater withdrawal can be used to minimize the nitrate in produced water when the nitrate concentration in the groundwater varies with depth. This management measure is not directed toward minimizing the nitrate in the subsurface environment, but is directed at

Table 7.24 Effects of Different Amounts of 2-Ethynylpyridine, Nitrapyrin, and Etridiazole on Nitrification in Soils Treated with Ammonium Sulfate

Inhibitor	Amount added ($\mu g/g$ soil)	% Inhibition of nitrification[a]	
		Range	Average
2-Ethynylpyridine	0.1	0–24	11
	0.5	3–45	24
	1.0	4–61	32
	5.0	37–93	67
	10.0	63–97	82
Nitrapyrin (N-Serve)	0.1	0–3	1
	0.5	0–16	7
	1.0	0–33	15
	5.0	21–80	49
	10.0	48-95	63
Etridiazole (Dwell)	0.1	0–12	4
	0.5	0–35	13
	1.0	1–45	20
	5.0	34–73	59
	10.0	60–96	77

Note: Soils treated with 200 μg NH_4–N/g; 25°C; 42 days.

[a] Results obtained with six soils (Harps, Okoboji, Webster, Nicollet, Clarion, and Storden).

From Bremner, J. M., McCarty, G. W., and Gianello, C., *Proceedings of the Conference on Agricultural Impacts on Ground Water*, 1986, National Water Well Association, Dublin, OH, pp. 467–481. With permission.

limiting the nitrate in groundwater pumped for usage. This approach has been used in Redlands, California (Eccles, Cline, and Hardt, 1976).

In the city of Redlands public water-supply well field, wells 1 and 2 normally produce water with dissolved nitrate-nitrogen concentrations of about 18 and 30 mg/l, respectively. Well 1 is a large-capacity well capable of yielding 3700 gal/min and is the major source of water in the well field (Eccles, Cline, and Hardt, 1976). The very permeable unconfined alluvial aquifer is composed of sand, gravel, boulders, and discontinuous clayey deposits. Well 1 is 742 ft deep and is perforated throughout most of the zone of saturation. A major clayey interval from 425 to 480 ft effectively separates the aquifer into an upper and lower zones. The upper zone is more contaminated in terms of nitrates. At the well field, the static water level is typically about 180 ft below land surface. Independent tests were made on wells 1 and 2 to evaluate aquifer characteristics and to determine the sources of the high-nitrate water. Based upon this testing, and in order to reduce the concentration of dissolved nitrate in water from well 1, an inflatable packer was placed in the casing at 480 ft to coincide with the bottom of the clayey interval. The packer sealed off the upper part of the well and, as determined from a final test of well 1, reduced the dissolved nitrate-nitrogen concentrations from 20 to 4 mg/l while only reducing well yield from 3700 to 2600 gal/min.

SUMMARY

This chapter provided a review of several fertilizer management and other management measures potentially useful in minimizing nitrate pollution of groundwater resulting from agricultural practices. Examples of fertilizer-related management measures include:

1. Match timing of fertilizer application to meet crop needs
2. Match fertilizer application rate to crop needs
3. Choose fertilizer application techniques to minimize nitrogen losses
4. Limit fertilizer application in well recharge areas
5. Regulate crop types to those with minimal fertilizer requirements
6. Implement management policies intended to minimize nitrogen losses
7. Adopt legislation related to minimizing fertilizer usage
8. Implement a taxation plan focused on minimizing fertilizer usage
9. Implement Best Management Practices
10. Control the usage of irrigation

Examples of other related management measures include:

1. Implement drinking water protection zones (similar to number 4 above)
2. Consider soil organic nitrogen in determining fertilizer requirements
3. Implement appropriate water management practices (similar to number 9 above)
4. Choose an irrigation method to minimize nitrogen losses (similar to number 10 above)
5. Choose tillage practice and soil conditioning to minimize nitrogen losses (similar to number 9 above)
6. Use nitrification inhibitors
7. Use well packer for selective zoning of groundwater withdrawal.

These measures represent a compendium of approaches that can be used to control nitrate pollution of groundwater from agricultural practices. The approaches can be used either singly or in various compatible combinations.

SELECTED REFERENCES

Adams, B. and Foster, S.S.D., "Land-Surface Zoning for Groundwater Protection," *Journal of the Institution of Water and Environmental Management*, Vol. 6, No. 3, June, 1992, pp. 312–320.

Allee, D.J. and Abdalla, C.W., "Policy Education to Build Local Capacity to Manage the Risk of Groundwater Contamination," A.E. Staff Report 89-29, August, 1989, Development of Agricultural Economics, Cornell University, Ithaca, New York.

Allen, S.E., "Slow-Release Nitrogen Fertilizers," in *Nitrogen in Crop Production*, R.D. Hauck, Ed., American Society of Agronomy, Madison, Wisconsin, 1984, pp. 195–205.

Allen, R.S., "The Control of Nitrate as a Water Pollutant," EPA/600/2-77/158, August, 1977, Agricultural Experiment Station, Texas A&M University, College Station, Texas.

Andersson, R. et al., "Attempt to Reduce Nitrate Content in Ground Water Used for Municipal Water Supply by Changing Agricultural Practices," *Nordic Hydrology*, Vol. 15, No. 4–5, 1984, pp. 185–194.

Batie, S.S., Kramer, R.A., and Cox, W.E., "Economic and Legal Analysis of Strategies for Managing Agricultural Pollution of Ground Water," October, 1989, Department of Agricultural Economics, Virginia Polytechnic Institute and State University, Blacksburg, Virginia.

Bock, B.R. and Hergert, G.W., "Fertilizer Nitrogen Management," in *Managing Nitrogen for Groundwater Quality and Farm Profitability*, R.F. Follett, D.R. Keeney, and R.M. Cruse, Eds., Soil Science Society of America, Inc., Madison, Wisconsin, 1991, chap. 7, pp. 139–164.

Bouldin, D.R. and Selleck, G.W., "Management of Fertilizer Nitrogen for Potatoes Consistent with Optimum Profit and Maintenance of Ground Water Quality," *Agricultural Waste Management*, Ann Arbor Science Publishers, Ann Arbor, Michigan, 1977, pp. 271–278.

Bourg, C.W., "Producers' Guide to Nitrogen Management," *Nitrogen and Irrigation Management: Hall County Water Quality Special Project*, University of Nebraska, Cooperative Extension Service, Lincoln, Nebraska, February, 1984, pp. C1–C10.

Bremner, J.M., McCarty, G.W., and Gianello, C., "Reduction of Nitrate Pollution of Ground Water by Nitrogen Fertilizers," *Proceedings of the Conference on Agricultural Impacts on Ground Water*, August, 1986, National Water Well Association, Dublin, Ohio, pp. 467–481.

Conrad, J., "Nitrates in Ground and Drinking Water: An Analysis of Policies and Regulations," *The Science of the Total Environment*, Vol. 51, 1986, pp. 209–225.

Conway, G.R. and Pretty, J.N., *Unwelcome Harvest – Agriculture and Pollution*, Earthscan Publications, Ltd., London, England, 1991, pp. 540–544.

de Haen, H., "Impact of Nitrogen Fertilizer Application on the Nitrate Concentrations in Ground Water: Cost-Benefit Analysis Considerations," *Proceedings of IIASA Task Force Meeting on Nonpoint Nitrate Pollution of Municipal Water Supply Sources*, Series No. CP-82-S4, February, 1982, International Institute of Applied Systems Analysis, Laxenburg, Austria, pp.57–80.

Eccles, L.A., Cline, J.M., and Hardt, W.F., "Abatement of Nitrate Pollution in a Public Supply Well by Analysis of Hydrologic Characteristics," *Ground Water*, Vol. 14, No. 6, November–December, 1976, pp. 449–453.

Economic Commission for Europe, "Ground Water Legislation in the ECE Region," ECE/WATER/44, 1986, United Nations, Geneva, Switzerland, pp. 20–24.

Embleton, T.W. et al., "Nitrogen Fertilizer Management of Vigorous Lemons and Nitrate-Pollution Potential of Groundwater," Report No. 182, December, 1979, California Water Resources Center, University of California at Davis, Davis, California.

Fleming, M.H., "Agricultural Chemicals in Ground Water: Preventing Contamination by Removing Barriers Against Low-Input Farm Management," *American Journal of Alternative Agriculture*, Vol. II, No. 3, undated, pp. 124–130.

Francis, D.D., "An Analysis of Control Mechanisms to Reduce Nitrogen Fertilizer Applications," undated, U.S. Environmental Protection Agency, Office of Groundwater Protection, Region 7, Kansas City, Missouri.

Fricker, W., "Origin of Nitrates in Groundwater of the Bunz Valley," *Wasser, Energie, Luft*, Vol. 75, No. 3, 1983, pp. 75–77.

Gilley, J.R. et al., "Strategies for Reducing Pollutants from Irrigated Lands in the Great Plains," EPA R-8052 49, 1982, U.S. Environmental Protection Agency, Ada, Oklahoma.

Hall, D.W., "Effects of Nutrient Management on Nitrate Levels in Ground Water Near Ephrata, Pennsylvania," *Ground Water*, Vol. 30, No. 5, September-October, 1992, pp. 720–730.

Harris, R.C. and Skinner, A.C., "Controlling Diffuse Pollution of Groundwater from Agriculture and Industry," *Annual Symposium 1991: Groundwater Pollution and Aquifer Protection in Europe*, Institution of Water and Environmental Management, 1991, pp. 3-1 to 3-14.

Harryman, M.B.M., "Water Source Protection and Protection Zones," *Journal of the Institution of Water and Environmental Management*, Vol. 3, No. 6, December, 1989, pp. 548–550.

Hergert, G.W., "Nitrogen Losses from Sprinkler-Applied Nitrogen Fertilizer," OWRT-A-045-NEB(1), September, 1978, University of Nebraska, North Platte, Nebraska.

Hills, F.J., Broadbent, F.E., and Fried, M., "Timing and Rate of Fertilizer Nitrogen for Sugarbeets Related to Nitrogen Uptake and Pollution Potential," *Journal of Environmental Quality*, Vol. 7, No. 3, July–September, 1978, pp. 368–372.

Hoeft, R.G., "Current Status of Nitrification Inhibitor Use in U.S. Agriculture," *Nitrogen in Crop Production*, R.D. Houck, Ed., American Society of Agronomy, Madison, Wisconsin, 1984, pp. 561–577.

Johnson, S.L., Perry, G.M., and Adams, R.M., "Managing Groundwater Pollution from Agriculture Related Sources: An Economic Analysis," December, 1989, Department of Agricultural and Resource Economics, Oregon State University, Corvallis, Oregon, pp. 147–155.

Kaap, J.D., "Implementing Best Management Practices to Reduce Nitrate Levels in Northeast Iowa Ground Water," *Proceedings of the Conference on Agricultural Impacts on Ground Water*, August, 1986, National Water Well Association, Dublin, Ohio, pp. 412–427.

Keeney, D.R., "Nitrate in Ground Water—Agricultural Contribution and Control," *Proceedings of the Conference on Agricultural Impacts on Ground Water*, August, 1986, National Water Well Association, Dublin, Ohio, pp. 329–351.

Kepler, K., Carlson, D., and Pitts, W.T., "Pollution Control Manual for Irrigated Agriculture," EPA/908/3-78/002, August, 1978, U.S. Environmental Protection Agency, Denver, Colorado.

Lahl, U., Zeschmar, B., Gabel, B., Kozicki, R., and Podbielski, A., "Ground Water Pollution by Nitrate," *Proceedings of a Symposium on Ground Water in Water Resources Planning*, IAHS Publication No. 142, Vol. II, 1983, pp. 1159–1170.

Lauterbach, D. and Klapper, H., "Possibilities of Water Management for Protecting and Treating Drinking Water Resources in Case of Nitrate Pollution," *Proceedings of IIASA Task Force Meeting in Nonpoint Nitrate Pollution of Municipal Water Supply Sources*, Series No. CP-82-54, February, 1982, International Institute of Applied Systems Analysis, Laxenburg, Austria, pp. 135–161.

Logan, T.J., "Agricultural Best Management Practices and Groundwater Protection," *Journal of Soil and Water Conservation*, March–April, 1990, pp. 201–206.

Mansell, R.S. et al., "Fertilizer and Pesticide Movement from Citrus Groves in Florida Flatwood Soils," EPA/600/2-77/177, August, 1977, U.S. Environmental Protection Agency, Athens, Georgia.

Mansell, R.S., Fiskell, J.G., and Calvert, D.V., "Nitrification, Denitrification, and Sorption-Desorption of NH_4–N in Sands During Water Movement to Subsurface Drains," Pub. No. 64, 1982, Water Resources Research Center, University of Florida, Gainesville, Florida.

McWilliams, L., "Bumper Crop Yields Growing Problems," *Environment*, Vol. 26, No. 4, May, 1984, pp. 25–33.

Munson, B.E. and Russell, C., "Environmental Regulation of Agriculture in Arizona," *Journal of Soil and Water Conservation*, March–April, 1990, pp. 249–252.

National Governors Association, "Along with Corn, Beans, and Berries, State Nitrate Control Plans Are Cropping Up," *Ground Water Bulletin*, Vol. 2, No. 3, Summer, 1991, pp. 1, 2, and 4.

Office of Technology Assessment, "Beneath the Bottom Line — Agricultural Approaches to Reduce Agrichemical Contamination of Ground Water," OTA-F-418, November, 1990, U.S. Congress, Washington, D.C., pp. 90–100.

Pallares, C., "Nitrogen Fertilizer Management of a Lemon Orchard as Related to Nitrate Pollution Potential of Ground Water," Ph.D. Dissertation, January, 1978, University of California at Riverside, Riverside, California.

Peters, J.H. and den Blanken, M.G., "Risk of Application of Biocides in Water Supply Catchment Areas," *Water Supply*, Vol. 3, No. 2, 1985, pp. 179–185.

Peterson, G.A. and Frye, W.W., "Fertilizer Nitrogen Management," in *Nitrogen Management and Ground Water Protection*, R.F. Follett, Ed., Elsevier Science Publishers, Amsterdam, The Netherlands, 1989, chap. 7, pp. 183–219.

Randall, G.W. and Bandel, V.A., "Overview of Nitrogen Management for Conservation Tillage Systems: An Overview," in *Effects of Conservation Tillage on Groundwater Quality: Nitrates and Pesticides*, Lewis Publishers/CRC Press, Boca Raton, Florida, 1987, pp. 39–63.

Schepers, J.S., "Effect of Conservation Tillage on Processes Affecting Nitrogen Management," in *Effects of Conservation Tillage on Groundwater Quality: Nitrates and Pesticides*, Lewis Publishers/CRC Press, Boca Raton, Florida, 1987, pp. 241–250.

Schleyer, R., Milde, G., and Milde, K., "Wellhead Protection Zones in Germany: Delineation, Research and Management," *Journal of the Institution of Water and Environmental Management*, Vol. 6, No. 3, June, 1992, pp. 303–311.

Schneider, R.R. and Day, R.H., "Diffuse Agricultural Pollution: The Economic Analysis of Alternative Controls," OWRT-A-03-WIS(2), 1976, Wisconsin Water Resources Center, University of Wisconsin, Madison, Wisconsin.

Shirmohammadi, A., Magette, W.L., and Shoemaker, L.L., "Reduction of Nitrate Loadings to Ground Water," *Ground Water Monitoring Review*, Winter, 1991, pp. 112–118.

Singh, B. and Sekhon, G.S., "Nitrate Pollution of Groundwater from Farm Use of Nitrogen Fertilizers — A Review," *Agriculture and Environment*, Vol. 4, No. 3, March, 1979, pp. 207–225.

Swoboda, A.R., "Nitrate Movement in Clay Soils and Methods of Pollution Control," *Proceedings of National Conference on Irrigation Return Flow Quality Management*, May, 1977, Colorado State University, Fort Collins, Colorado, pp. 19–25.

Timmons, D.R. and Dylla, A.S., "Nitrate Leaching on an Irrigated Soil in a Subhumid Climate," *Proceedings of Conference on Water, Water Everywhere — But Can We Use It*, February, 1979, Irrigation Association, Silver Spring, Maryland, pp. 95–102.

Wendt, C.W., Onken, A.B., and Wilke, O.C., "Effects of Irrigation Methods on Ground Water Pollution by Nitrates and Other Solutes," EPA-600/2-76-291, December, 1976, U.S. Environmental Protection Agency, Ada, Oklahoma.

Zaporozec, A., "Nitrate Concentrations Under Irrigated Agriculture," *Environmental Geology*, Vol. 5, No. 1, 1983, pp. 35–38.

Zwirnmann, K.H., "Nonpoint Nitrate Pollution of Municipal Water Supply Sources: Issues of Analysis and Control," Report No. CP-82-S04, May, 1982, International Institute for Applied Systems Analysis, Laxenburg, Austria.

8 TREATMENT MEASURES FOR NITRATES IN GROUNDWATER

INTRODUCTION

Several scientifically based measures are available for treating groundwater with excessive nitrate concentrations. This chapter provides a summary of measures that can be used following the pumpage of groundwater to an above-ground treatment system. The measures are categorized into those providing treatment by: physico-chemical means such as ion exchange, reverse osmosis, and electrodialysis; biological means such as denitrification; and by miscellaneous other approaches. Information is also included on *in situ* biological denitrification. Sections are included on each of these categories of measures; a summary section concludes this chapter.

Table 8.1 contains summary comments on six references that contain comparative information on nitrate (or nutrient) removals from surface or groundwater supplies or wastewater. Nitrogen removal from municipal wastewater has been an issue of concern for over two decades; as a result, a considerable body of knowledge has been accumulated on the comparative features of different treatment technologies. This information provides a basis for considering technologies for removing nitrates from groundwater. Table 8.2 summarizes the status of nitrogen control alternatives based on municipal wastewater treatment practice in the U.S. (U.S. Environmental Protection Agency, 1994). Biological and physico-chemical treatment represent the fundamental categories of the utilized technologies.

TREATMENT BY ION EXCHANGE, REVERSE OSMOSIS, OR ELECTRODIALYSIS

Table 8.3 contains summary comments on 18 references describing ion exchange technologies for removing nitrates from groundwater. In addition, summary comments from six references on reverse osmosis and three references on electrodialysis technologies are listed in Table 8.3. The following subsections provide information on these three categories of technologies.

Table 8.1 References Providing Background or Overview Information on Nitrate Removal from Groundwater

Author(s) (year)	Comments
Dahab (1987)	Review of the technical feasibility and economics of anion exchange, biological denitrification, and reverse osmosis for removing nitrates from groundwater supplies
Dahab and Bogardi (1990)	Discussion of evaluation of nitrate treatment alternatives using a multi-criteria decision-making technique
Horan (1994)	Conference proceedings containing over 40 papers related to conventional and innovative processes for nutrient removal from wastewaters
U.S. Environmental Protection Agency (1987)	Cost information related to removal of nitrates from potable water supplies
U.S. Environmental Protection Agency (1994)	Comprehensive review of treatment technologies for nitrogen removal from wastewaters
World Health Organization (1985)	Discussion of the use of reverse osmosis, ion exchange, electrodialysis, and biological denitrification for nitrate removal

Ion Exchange

Ion exchange involves the exchange of ions in solution (i.e., contaminated groundwater) with chemically equivalent numbers of ions associated with the exchange material (i.e., the resin). The mechanism involved in the removal of nitrate ions from groundwater is typically the replacement of these ions with chloride ions when the groundwater is passed through the resin. The ion-exchange process for the replacement of nitrate utilizes either a strongly basic or weakly basic anion exchanger. Anion exchangers or resins containing functional groups made up of weak amine bases, which are derivatives of ammonium, are called weakly basic. Those derived from quaternary ammonium compounds are referred to as strongly basic (Dahab and Bogardi, 1990). Exchange resins typically exhibit a preference or selectivity for various ions depending upon the concentration of ions in solution; therefore, resins with a selectivity toward nitrate ions would obviously be preferred.

Ion exchange is an attractive process for nitrate removal from groundwater because it offers process control, is easily automated, and is not affected by temperature in typical operating ranges. Table 8.4 contains a 1978 U.S. EPA ranking of resins for nitrate removal using the considerations of high nitrate/chloride selectivity, high capacity, and moderate sulfate/nitrate selectivity (Clifford and Weber, 1978). Organic extractables were not considered in the ranking. The resins are ranked in preference order, but the differences are not large.

Ion exchange is typically chosen over reverse osmosis and electrodialysis based on systematic cost comparisons. For example, Sorg (1980) presented cost comparisons of these three nitrate removal methods along with cost information for other treatment techniques such as lime softening, ion exchange softening, and activated alumina treatment for fluoride removal. Additional cost information is given by the U.S. Environmental Protection Agency (1987).

Table 8.2 1990 Status of Nitrogen Control Technologies in Municipal Wastewater Treatment Applications

Technology	Knowledge[a]			
	High			Low
	Well demonstrated	Limited application[b]	Found lacking[c]	Emerging[d]
Biological treatment				
Higher technology, mechanical plant approach				
Suspended growth				
Single sludge				
Multiphased				
Aerator and/or aeration basin cycling	O,R			
Sequential batch reactor	O			R
Multistaged (e.g., serial application of processes)	O,R			
Multizone (e.g., ditches)	O,R			
Two sludge	O,R			
Three sludge	O,R			
Attached growth, single- or multiphased, and/or staged applications				
Submerged media				
Fluidized bed		O,R		
Packed bed				
Downflow	R	R		O
Upflow		O,R		
Nonsubmerged media				
Stationary (e.g., trickling filter)	O	O		R
Rotating (e.g., rotating biological contactor)	O	O		R
Combination processes				
Any of the above in serial application	O,R	O,R		O,R
Submerged stationary media (vertical plates or media)				O,R
Nonsubmerged				
Stationary media with solids recycle (e.g., activated biofilter)		O		O,R

Table 8.2 1990 Status of Nitrogen Control Technologies in Municipal Wastewater Treatment Applications (continued)

Technology	Knowledge[a]			
	High			Low
	Well demonstrated	Limited application[b]	Found lacking[c]	Emerging[d]
Rotating media in solids suspension		O,R		O,R
Specific surface additives to suspended growth system				
Concurrent additive management (e.g., powdered activated carbon)		O		O,R
Separate additive separation, processing, and return (e.g., Linpor, Captor)				O,R
Lower technology, transitional and natural systems approach				
Transitional				
Aerated lagoons (suspended growth)	O	O		R
Intermittent and/or recirculating sand filtration (attached growth)	O	O,R		O,R
Aquatic-based				
Lagoons (suspended growth)		O,R		O,R
Facultative (N stripping)				R
Algae harvesting (N removal by stripping and synthesis)[e]		R		
Natural and constructed wetlands (attached growth with N removal by synthesis)[e]				
Surface flow (floating and rooted aquatic plants)	R		O,R[e]	O
Subsurface flow (with rooted aquatic plants)	R		O,R[e]	
Land-based (attached growth treatment with and without N removal by synthesis)[e]				
Slow rate infiltration	O,R			
Rapid rate infiltration	O,R			
Overland flow	O	R		
Subsurface infiltration	O	R		
Physical/chemical treatment				
Ion specific				
Ammonia stripping (NH_3–N)		R		
Ion exchange (NH_4^+–N) (NO_3^-–N)		R		
Breakpoint chlorination (NH_4^+–N)		R		

Table 8.2 1990 Status of Nitrogen Control Technologies in Municipal Wastewater Treatment Applications (continued)

Technology	Knowledge[a]			
	High			Low
	Well demonstrated	Limited application[b]	Found lacking[c]	Emerging[d]
Non-ion specific				
Reverse osmosis		R		
Avoidance of nitrogen control technologies through beneficial reuse				
Irrigation	R[e]			
Selected industrial reuse (e.g., cooling water)		R		

[a] O = nitrogen oxidation; R = nitrogen removal by biological denitrification unless otherwise noted. Classification can vary depending on particular application.
[b] Knowledge of performance capabilities is high but process has been used only on a limited basis.
[c] Knowledge of performance capabilities is high but process capabilities or economics have been found to be poor, based on limited application.
[d] Knowledge of process performance capabilities is low because of infrequent or recently emerging application.
[e] All systems that rely on synthesis for nitrogen removal ultimately must plan for harvest and disposal of the resultant biomass (10–20 times the synthesized nitrogen is a likely rule of thumb).

From U.S. Environmental Protection Agency, *Nitrogen Control*, Technomic Publishing, Lancaster, PA, 1994, pp. 35–36.

Table 8.3 References Dealing with Groundwater Treatment via Ion Exchange, Reverse Osmosis, or Electrodialysis

Author(s) (year)	Technology[a]	Comments
Buelow et al. (1975)	IX	Laboratory study of nitrate selective anion exchange resins
Clifford and Weber (1978)	IX	Comparative study of 32 anion resins for removal of nitrates from water supplies; and discussion of single-bed and two-bed ion exchange processes for nitrate removal from public water supplies
Dahab and Bogardi (1990)	IX	Illustrations of the use of ion exchange for nitrate removal
Gauntlett (1975)	IX	Discussion of methods and results of nitrate removal from water with commercially available ion exchange resins
Goodrich, Lykins, and Clark (1991)	IX	Use of ion exchange for nitrate removal by individual homeowners
Guter (1981)	IX	Use of ion exchange column tests for design of nitrate removal systems
Guter (1987a)	IX	Design and performance evaluation of a 1.0 mgd plant
Guter (1987b)	IX	Analysis of operation and maintenance costs and plant performance over a 25-month period at a 1.0 mgd plant at McFarland, California
Hoell and Feuerstein (1985)	IX	Hardness and nitrate removal from groundwater using a cation exchanger followed by an anion exchanger
Horng (1985)	IX	Mathematical model for two-stage and single-bed ion exchange resins
Lockridge (1990)	IX	Review of anion exchange resins which exhibit nitrate selectivity
Musterman et al. (1983)	IX	Effects of nitrate and sulfate loading on the regeneration efficiency of strong base ion exchange resins
Richard (1988)	IX	Two case studies from France involving ion exchange for nitrate removal from water supplies
Sheinker and Coduluto (1977)	IX	Continuous countercurrent ion exchange system for nitrate removal
Sorg (1980)	IX	Cost comparisons of nitrate removal by ion exchange, reverse osmosis, and electrodialysis
Urbanisto and Bays (1985)	IX	Proceedings of a conference with several papers on nitrate removal from groundwater
U.S. Water News (undated)	IX	Ion exchange plant in Des Moines, Iowa
U.S. Water News (undated)	IX	Ion exchange plant in Pomona, California
Bilidt (1985)	RO	Use of reverse osmosis for removal of nitrates from groundwater
Dahab and Bogardi (1990)	RO	Review of possible use of reverse osmosis for nitrate removal from groundwater
Huxstep (1981)	RO	Use of high-pressure and low-pressure reverse osmosis systems for nitrate removal from groundwater
Huxstep and Sorg (1988)	RO	Pilot study of reverse osmosis using 5 different membrane elements

Table 8.3 References Dealing with Groundwater Treatment via Ion Exchange, Reverse Osmosis, or Electrodialysis (continued)

Author(s) (year)	Technology[a]	Comments
Rautenbach and Henne (1983)	RO	Use of a pilot plant for development of design factors for a full-scale reverse osmosis system
Sorg and Love (1984)	RO	Review of laboratory and field studies for removal of inorganics by reverse osmosis
Dahab and Bogardi (1990)	ED	Principles of electrodialysis in relation to nitrate removal from groundwater
Montgomery (1985)	ED	Review of fundamental aspects of the electrodialysis process
Rautenbach et al. (1985)	ED	Review of pilot plant experiments using electrodialysis for nitrate removal from groundwater

[a] IX = ion exchange; RO = reverse osmosis; ED = electrodialysis.

Table 8.4 Ranking of Resins for Use in Nitrate Removal Services

Recommended
STY-DVB, Polyamine Resins
Amberlite IR-45
STY-DVB, Tertiary-amine, MR Resins
Amberlite IRA-93
Dowex MWA-1
Ionac AFP-329
Duolite ES-368
STY-DVB, Quat. (I & II) Amines, Gel, and MR Resins
Ionac ASB-100, AFP-100, A-641, ASB-1P, ASB-2
Duolite, A-101-D, A-102-D
Dowex 11, SAR, SBR-P, SBR
Amberlite IRA-400, IRA-900, IRA-402, IRA-910, IRA-410
Acrylic-Amine, Polyamine, MR Resins
Duolite ES-374
Phenol-HCHO, Polyamine, MR Resins
Duolite A-7
Duolite ES-561

From Clifford, D. A. and Weber, W. J., "Nitrate Removal from Water Supplies by Ion Exchange," EPA/600/2-78-052, 1978, U.S. Environmental Protection Agency, Cincinnati, OH.

The use of single-bed, strong-base anion-exchange units with a sodium chloride regeneration cycle is fairly common; however, in nonarid, noncoastal locations, the disposal of the regenerant brine can be a problem (Clifford and Weber, 1978). The typical approach in system design includes the conduction of laboratory studies to develop preliminary design information, followed by pilot-scale studies based upon the preliminary design information; the final step involves the design and performance evaluation of the full-scale system (Sheinker and Coduluto, 1977).

The regenerant sodium chloride and the associated disposal thereof represent one of the primary costs of removing nitrates with a single, strong-base ion-exchange column. Alternatives for regenerant brine disposal, according to Clifford and Weber (1978), are as follows:

1. Ocean outfall in coastal locations
2. Evaporation ponds in semi-arid regions
3. Sanitary sewers where permitted, but only where wastewater denitrification facilities exist and where the brine does not seriously dilute the wastewater
4. Deep-well injection where permitted (this is very costly for large volumes of brine)
5. Sale as fertilizer

Laboratory studies of weak-base to strong-base anion-exchange resins have been described by Buelow et al. (1975), Clifford and Weber (1978), Gauntlett (1975), and Musterman, et al. (1983). These studies typically involved comparisons of several commercially available anion resins, various resin regeneration schemes, and the effects of nitrate concentrations on the removal of nitrates in the ion exchange system combinations. One concern with nitrate removal by ion exchange is that sulfates in the groundwater can compete for the exchange sites on the resin. In fact, many ion-exchange resins are more selective for sulfate than for nitrate (Sorg, 1980).

There are two fundamental methods of ion exchange available for nitrate removal from drinking water supplies: a single-bed process (Figure 8.1) and a two-bed process (Figure 8.2) (Clifford and Weber, 1978). The single-bed process involves a strong base of anionic resin with sodium chloride regeneration with the following advantages (+) and disadvantages (–) (Clifford and Weber, 1978):

(+)Simple, no balancing of beds and regenerants
(+)Low-cost regeneration
(–)Very difficult and costly to dispose of regenerants in noncoastal locations where natural evaporation is impossible
(–)Iron must be removed to prevent resin fouling
(–)Continuous nitrate analysis required for process control

A two-bed process involves a strong acid resin with weak-base ammonia and hydrochloric acid regenerants. The following are advantages (+) and disadvantages (–) (Clifford and Weber, 1978):

(+)Partial softening in addition to nitrate removal
(+)No problem with iron fouling; precipitated iron is removed from the cation bed during each regeneration
(+)Regenerant wastewaters are expected to be easy to dispose of by land application as fertilizer
(–)The system is complex; bed sizes and regenerants must be balanced
(–)Degasifier for carbon dioxide removal is required
(–)Continuous pH and nitrate analyses required for process control
(–)High regenerant costs

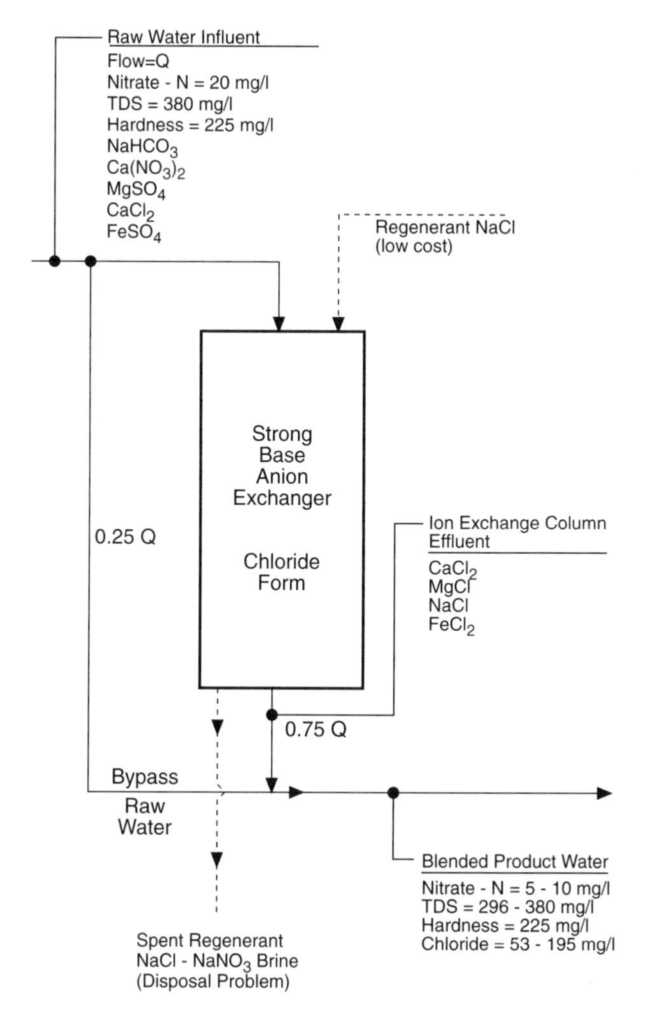

Figure 8.1 Conventional single-bed ion-exchange process. (From Clifford, D. A. and Weber, W. J., "Nitrate Removal from Water Supplies by Ion Exchange," EPA/600/2-78-052, 1978, U.S. Environmental Protection Agency, Cincinnati, OH.)

Tables 8.5 and 8.6 compare the chemical costs of regenerants for use in single- or two-bed nitrate removal processes (Clifford and Weber, 1978). In Table 8.5, sodium chloride and ammonium chloride are used for the single-bed process; and ammonia, hydrochloric acid, nitric acid, or sulfuric acid should be used for the two-bed process. The nitric acid is the most expensive regenerant to be used. Table 8.6 shows a further comparison and cost breakdown of the single-bed versus two-bed processes.

Mathematical models to provide a basis for an empirical design approach have been developed from laboratory and pilot-scale studies. Horng (1985) suggested that by assuming instantaneous equilibrium, plug flow, constant feed composition and resin capacity, constant separation factors, and uniform

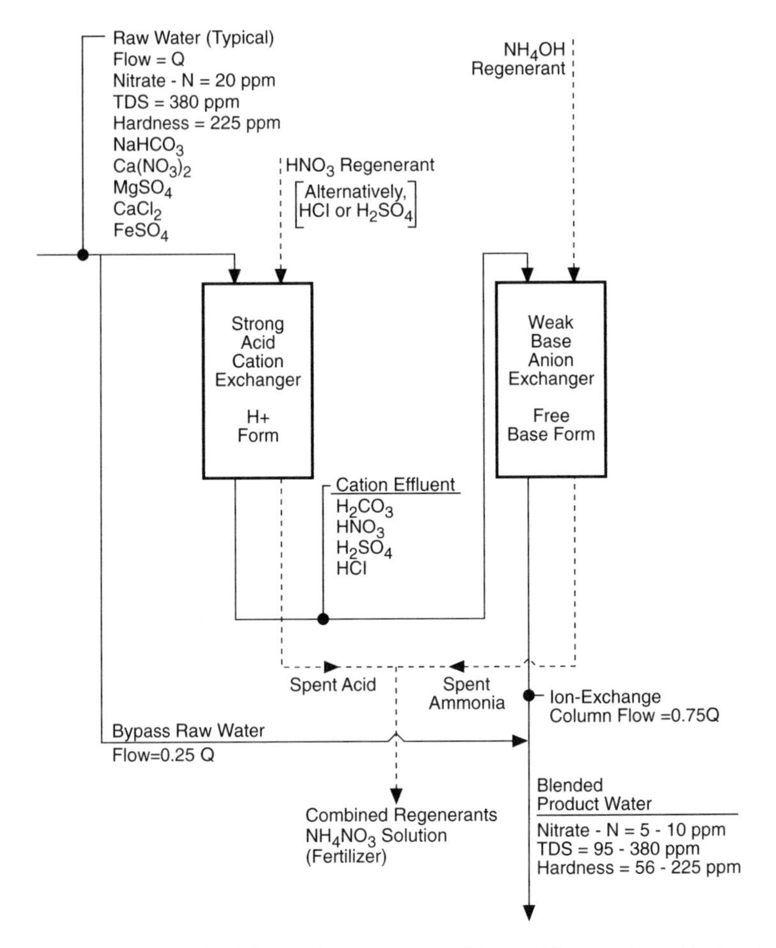

Figure 8.2 Proposed two-bed, ion-exchange process. (From Clifford, D. A. and Weber, W. J., "Nitrate Removal from Water Supplies by Ion Exchange," EPA/600/2-78-052, 1978, U.S. Environmental Protection Agency, Cincinnati, OH.)

Table 8.5 Calculated Chemical Regenerant Costs

Regenerant chemical	$/lb-equivalent	¢/1000 gal treated	¢/m³ treated
H_2SO_4	1.23	8.63	2.28
NaCl	1.30	9.15	2.42
NH_3	1.53	10.7	2.78
HCl	2.43	17.1	4.52
NH_4Cl	5.64	39.6	10.5
HNO_3	6.62	46.5	12.3

Note: 1 lb-equivalent = 14 lb of nitrogen removed.

From Clifford, D. A. and Weber, W. J., "Nitrate Removal from Water Supplies by Ion Exchange," EPA/600/2-78-052, 1978, U.S. Environmental Protection Agency, Cincinnati, OH.

Table 8.6 Economic and Regenerant Wastewater Comparisons Between the Single-Bed and Two-Bed Processes

Item	Single-bed process	Two-bed process
Regenerant chemical costs, ¢/1000 gal H_2O supplied	9.2	27.80
Regenerant chemical costs, ¢/m^3 H_2O supplied	2.43	7.34
Regenerant disposal costs, ¢/1000 gal H_2O supplied	9.2	Nil
Regenerant disposal costs, ¢/m^3 H_2O supplied	2.43	Nil
Regenerant plus disposal costs, ¢/1000 gal H_2O supplied	18.4	27.8
Regenerant plus disposal costs, ¢/m^3 H_2O supplied	4.86	7.34
Regenerant volume, %Total water supplied	0.78	1.73
Regenerant composition:		
Total concentration, N	1.08	0.650
Total dissolved solids, ppm	69,200	37,400
Undesirable cation (Na^+), ppm	24,900	0
Calcium ion, ppm	0	2,480
Magnesium ion, ppm	0	465
Ammonium ion, ppm	0	8,780
Sulfate ion, ppm	7,870	3,540
Nitrate ion, ppm	8,960	4,030
Chloride ion, ppm	27,500	18,100
Nitrogen, ppm	2,020	7,740
Nitrogen fertilizer produced, lb N/1000 gal H_2O supplied	0	1.12
Nitrogen fertilizer produced, kg/m^3 H_2O supplied	0	0.134
Per-capita fertilizer production, lb N/capita·year	0	40.9

From Clifford, D. A. and Weber, W. J., "Nitrate Removal from Water Supplies by Ion Exchange," EPA/600/2-78-052, 1978, U.S. Environmental Protection Agency, Cincinnati, OH.

presaturation, the equilibrium multicomponent chromatography theory can be used to simulate multicomponent ion-exchange column runs. Based upon this theory, Horng (1985) developed a mathematical model for calculating effluent concentration histories and compositions of exhausted ion-exchange beds. This model can be applied to both two-stage and single-bed ion-exchange operations, including both frontal and nonfrontal analyses. The theory was verified by experiments with single- and two-component homogeneously presaturated ion-exchange column runs. In addition to the model by Horng (1985), Musterman, et al. (1983) developed an empirical model to predict regeneration requirements for strong base anion-exchange resins used for nitrate removal. Extensive laboratory tests with various exchange resins were used to verify the model. A second model was developed to predict the length of time for exhaustion of the resin as a function of loading.

A summary of actual users of ion-exchange nitrate removal processes is presented in Table 8.7 (Dahab and Bogardi, 1990). Additional ion-exchange plants have been placed in operation in Des Moines, Iowa, and Pomona, California (U.S. Water News, undated a and undated b). Information from several case studies listed in Table 8.7 will now be presented.

Guter (1981) described a pilot-scale ion-exchange study conducted at McFarland, California. The influent nitrate-nitrogen concentrations ranged from 16 to 23 mg/l, and sulfate concentrations were greater than 300 mg/l. Following extensive laboratory tests, a 20-in. diameter pilot anion-exchange column containing 4.36 cu ft resin was designed and operated for over 1 year. Data from this column operation were used to verify estimates of pilot column performance and to project the cost for equipment and regenerant for an installation to treat up to 1.0 mgd nitrate-contaminated groundwater. The following conclusions were drawn from the pilot-scale studies (Guter, 1981).

1. The study indicated that automatic ion-exchange equipment, which is commonly used by the water softening industry, can be adapted for nitrate removal from groundwater. The equipment can be installed at a well site for direct treatment of groundwater and operated on demand without storage.
2. The selected resin was effective for nitrate removal at loading rates above 48.9 m/h (200 gpm/ft^2) of bed area (1.38 bed volumes/min). This rate was the upper limit of the test equipment used.
3. Capital equipment costs for an ion-exchange system to treat half of a 3800 m^3/day (1 mgd) production well were estimated to be less than $90,000 installed. This estimate was based on moderate nitrate-nitrogen levels of less than 14 mg/l in well water, sulfate levels of less than 200 mg/l, and blending of treated and raw water to produce a product containing less than 10 mg/l nitrate-nitrogen. The corresponding equipment cost estimate for a system to treat all water from a 1-mgd production well was less than $150,000. This cost estimate was based on high nitrate levels in raw water (about 23 mg/l as nitrate-nitrogen), high sulfate levels (about 300 mg/l), and ion-exchange treatment to reduce nitrate-nitrogen to less than 10 mg/l without blending.
4. A significant operating cost for the process was related to the use of sodium chloride as a resin regenerant. Because anion-exchange resins can be selective for sulfate ions, the presence of sulfate in the raw water decreased the efficiency of the resin in removing nitrate. In this study, however, sulfate was easily removed from the spent resin by the sodium chloride regenerant in nearly stoichiometric proportions; whereas excess regenerant is required for nitrate removal. Nonetheless, the overall effect of sulfate is to increase the sodium chloride required to remove nitrate per unit quantity of water treated. This study also confirmed that large quantities of regenerant (320 kg/m^3 or 20 lb/ft^3 of resin) are required to remove most of the nitrate from the spent resin. Not all nitrate needs to be removed, however, to reduce nitrate-nitrogen levels in treated water to less than 10 mg/l.
5. For the McFarland wells, the salt (sodium chloride) costs for lowering nitrate-nitrogen levels to 7 to 10 mg/l ranged from an estimated 1.9¢/1000 gal of

Table 8.7 Summary of Ion-Exchange Nitrate Removal Processes

Process	Location	Concentration		Resin type
		NO_3	SO_4	
Combined ion-exchange and biological denitrification closed-circuit process, use of nitrate selective resin and low conc. regenerant; P.S.	Wageningen, Netherlands	80	30	Macroporous Amberlite IRA 966
Combined ion-exchange and biological denitrification under different process conditions, P.S. Low and high sulfate conc.	Wageningen, Netherlands	88± 87± 100±	31.6± 31.1± 181±	Doulite A165, Amberlite IRA996
Nitrate removal from groundwater; P.S.	Wageningen, Netherlands	88.6	30	Doulite A165
Use of nitrate-selective resin in combined ion exchange/biological denitrification; P.S.	Netherlands	87	31.1	Amberlite IRA996
Azurion process, diluted to make final concentration of 25 mg/l, flow rate = 600 m³/hr; F.S.	Plouenan, France	80		Dowex SBRP, 1.4 eq/l (Cl form)
Ion exchange by use of special resin that only fixes small amount of bicarbonate ions too; flow rate = 27 m³/hr; F.S.; diluted concentration of 25 mg/l.	Ormes-sur-Voulzie, France	60		
The CARIX process, removes nitrate, sulfate and hardness, Q = 3 m³/hr; P.S.	Kulmbach, France	82–96	120–150	1661 Amberlite IRA 50 & 5001 Lewatit M 600
The CARIX process, combined use of acidic and an anion exchanger, Q = 170 m³/hr; F.S.	Bad Rappenau, France			Same as above
Ion exchange; effect on chemical quality of effluent water, use of strong base ion-exchange resin; P.S.	Paris, France			IRA 400
Pilot-scale studies before McFarland plant construction	McFarland, California	90	320	Doulite A-101D
Ion exchange plant for well water; final conc. of blended water is 30 mg/l; built in 1983; F.S.	McFarland, California	70	120	Doulite A-101D
Second plant built in 1987 identical to first one in process; F.S.	McFarland, California	50–70		Doulite A-101D
Ion exchange, flow rate = 160 m³/hr; mixed with untreated water for dilution; F.S.	Binic (cotes du nord)			
Use of ion exchange for groundwater supply; flow rate = 10 m³/hr; eluate to dilution; F.S.	Craon (Mayenne)			
Ion exchange and denitrification plant for groundwater; flow rate = 150 m³/hr; blending with untreated water; F.S.	Reignac (Indre et Loire)			
Evaluation of four types of resins in laboratory experiment, use of different types of resins; B.S.	Lincoln, Nebraska			Dowex SBR, Amberlite IRA94 400 & 410

Table 8.7 Summary of Ion-Exchange Nitrate Removal Processes (continued)

Process	Location	Concentration		Resin type
		NO_3	SO_4	
Ion exchange for nitrate removal for public water use, Q = 50 m³/hr; F.S.	Plouvenez-Lochrist (finistere)			

Note: B.S. = bench-scale; F.S. = full-scale; P.S. = pilot-scale.

From Dahab, M. F. and Bogardi, I., "Risk Management for Nitrate-Contaminated Groundwater Supplies," 1990, U.S. Geological Survey, Reston, VA, pp. 76–103.

blended water (or $6.10/acre-ft) for well No. 2 to 10¢/1000 gal of treated water (or $32.50/acre-ft) for well No. 3. Water from the latter well represents a particularly difficult water to treat as nitrate-nitrogen concentrations were near 23 mg/l and sulfate levels were above 300 mg/l. Nitrate-nitrogen concentrations in well No. 2 were near 14 mg/l, and sulfate levels were near 200 mg/l.

6. During the regeneration cycle, wastewater produced was rich in sodium sulfate, chloride, and nitrate. Continuous operation of well No. 2 would produce more than 45.4 m^3 (12,000 gal) wastewater/day. Continuous operation of well No. 3 would produce an average of 146.4 m^3 (38,700 gal) wastewater/day.

Following the pilot-scale studies, a 1.0-mgd nitrate removal plant was designed and constructed at McFarland, California (Guter, 1987a). The plant uses the ion-exchange process with commercially available resins. Table 8.8 presents data for the water quality of four wells with high nitrate levels in the McFarland, California, groundwater supply system (Guter, 1987a). The plant was located at well No. 2, which is owned and operated by the McFarland Mutual Water Company. A flow diagram for the plant is shown in Figure 8.3 (Guter, 1987a). Feed water was supplied directly from the well pump into two of the ion-exchange vessels in the service cycle. Vessel 1 is 50% exhausted when Vessel 2 starts its service period. When Vessels 1 and 2 are in service, Vessel 3 is in regeneration or standby mode. After No. 1 is exhausted, Vessels 2 and 3 are in service; thus the system is operated on a cyclical basis.

Based upon the initial 6-month operating period at the McFarland plant, the following conclusions were drawn (Guter, 1987a).

1. The plant was operated automatically and exhibited the following performance characteristics averaged over the operating period:
   a. Nitrate leakages averaged 5.2 mg/l NO_3–N (23.2 mg/l NO_3) in a blend of treated and untreated water; the blend consisted of 76.1% treated water and 23.9% untreated water.
   b. Brine dosages for regeneration were 6.36 lb salt per cubic foot resin, or 2.49 lb/1000 gal blended water.
   c. Brine efficiencies averaged 10.3 equivalents chloride per equivalent nitrate removed, and varied from a low of 8.3 to a high of 11.8.
   d. Water recovery was 96.7% of the water pumped. The remaining 3.3% was discarded as waste brine and wastewater.
   e. Wastewater per 1000 gal blended water consisted of 0.92 gal saturated brine (4.9 gal dilute brine), 17.6 gal rinse water, and 11.4 gal backwash water.

Table 8.8 Composition of McFarland, California, groundwater (ppm) in 1980

Item	Well number 1	2	3	4
Date	5-8-80	4-9-80	5-1-80	4-16-80
Calcium	28	88	156	78
Sodium	50	65	100	72
Bicarbonate	88	102	121	95
Chloride	28	86	94	51
Sulfate	51	105[a]	310	182
Nitrate-N	6.8	15.2	22.1	10.6
TDS	235	446	827	485
pH	7.7	7.2	7.3	7.7

[a] Analyses on 5/31/78 showed sulfate levels of 261 ppm and nitrate levels of 78 ppm.

From Guter, G. A., "Nitrate Removal from Contaminated Water Supplies, Vol. I," EPA/600/S2-86/115, 1987a, U.S. Environmental Protection Agency, Cincinnati, OH.

2. Maximum automation was successfully used to satisfy the minimal manpower requirements of a small water system operator. The plant was designed and demonstrated based primarily on the needs of small communities where wells and distribution systems are already in place. The plant operates at a well site rather than as a central treatment plant.
3. Raw water composition varied during this period of operation. Nitrates varied from 11.1 to 16.0 mg/l nitrate-nitrogen. This provided the opportunity to measure the effect of changing water composition on plant performance.
4. Resin beds were operated at 76% capacity during the initial adjustment period to prevent overruns that could occur because of operational problems. The effect of operating at less than 100% bed capacity was estimated to be a decrease of brine efficiency of approximately 18%.
5. Brine efficiency, nitrate leakage, and bed volumes to nitrate breakthrough can be accurately predicted from ion-exchange theory. Computer-based programs can simulate effluent histories and were comparable to those obtained from the plant. They also can be used to predict chromatographic distributions of ions within spent beds.
6. A 3-ft resin bed depth was used during this period of operation.
7. The power consumed by the plant was 244 kWh per million gallons blended water. This amount was 10% of the total power required for pumping at the well site.
8. Capital costs were estimated at $311,000 for a plant with a 3-ft-deep resin bed, and $356,000 for a plant with a 5-ft bed. The total costs were $0.245 per 1000 gal blended water for the 5-ft bed plant (1983 costs). The overall cost to the McFarland community for nitrate removal during this period was $0.162 per 1000 gal water consumed (1983 costs).
9. The plant was totally automatic in operation with automatic nitrate analysis for monitoring and automatic shutdown if nitrate exceeded the maximum contaminant level in the product water. Computer printouts of operating data were obtained on a daily basis and if alarms occurred. Operator tasks were reduced to approximately 1 hr per day and included routine inspection, maintenance, and record keeping.

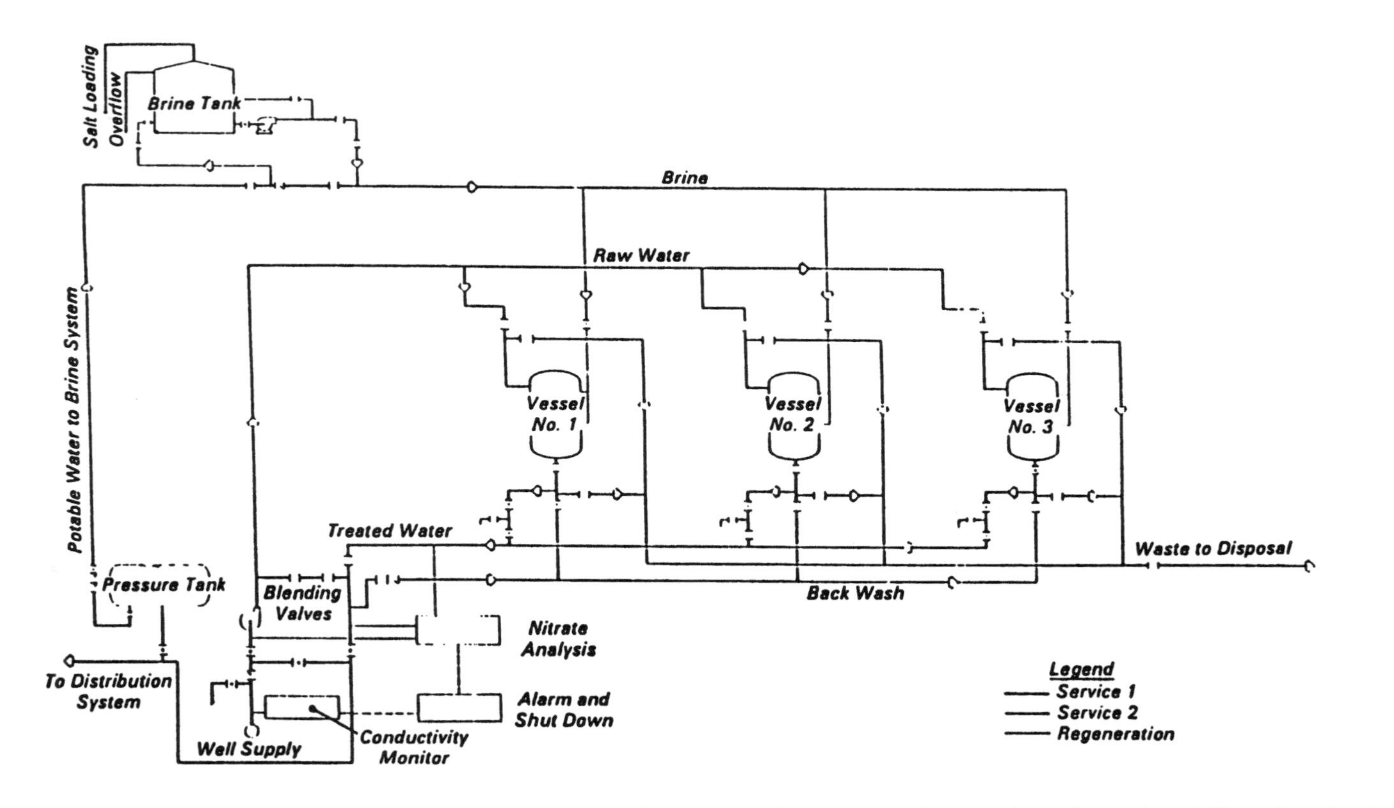

Figure 8.3 Flow diagram for McFarland, California, ion-exchange plant. (From Guter, G. A., "Nitrate Removal from Contaminated Water Supplies, Vol. I," EPA/600/S2-86/115, 1987a, U.S. Environmental Protection Agency, Cincinnati, OH.)

10. Nitrate removal was economically and technically feasible by the ion-exchange process. The most undesirable feature was the production and disposal of waste brine. At McFarland during the initial 6-month operating period, approximately 1300 lb waste salts were disposed of daily in the plant wastewater by discharging to the municipal wastewater system. If the plant were operated 24 hr per day, the daily salt discharge would be 2500 lb in 33,000 gal wastewater.
11. Although nitrate removal by the ion-exchange process is largely being considered as a process adaptable for small communities, it is the latter who will find the waste disposal problems the most difficult to solve. Improvements in the process are still required to reduce quantities of waste salts. This can probably be accomplished using highly selective nitrate resins, brine recirculation, recovery and separation of sodium nitrate and sodium chloride, and close adjustment of plant operation to changes in raw water composition.

Operation and maintenance costs and plant performance data for the 1.0-mgd plant at McFarland, California, have also been reported for the 25-month period from December 1, 1984, through January 1, 1987 (Guter, 1987b). Table 8.9 contains example operating data for the plant for a 1-month period (Guter, 1987b). Approximately 69% of the water to the system had been treated in the ion-exchange plant; this water was then blended with untreated groundwater.

The conclusions of the 25-month study were as follows (Guter, 1987b).

1. The plant was successfully operated during 1985 and 1986 to provide the community of McFarland with 343 million gallons (over the 2 years) of drinking water meeting the nitrate standard. This amount of water met 57% of the total demand for the community. On a yearly basis, the plant produced 197.4 million gallons (65.8% of the water) in 1985 and 145.58 million gallons (48.5% of the water) in 1986.
2. The capital cost for the McFarland plant was $356,000 (1983). Total annual cost for capital amortized over 20 yr at 8% interest was $36,200. The average annual operation and maintenance cost over the 2-year (1985–1986) period was $30,700. Based on the design capacity of 1.0 mgd, the capital cost was 9.9¢/1000 gal and the operation and maintenance cost was 8.5¢/1000 gal for a total cost of 18.4¢/1000 gal.
3. The plant actually processed 343 million gal water over the 1985–1986 period for an average of 0.47 mgd. Based on the actual flow of 0.47 mgd, which is 47% of the design capacity, the capital cost was 21.1¢/1000 gal (amortized over 20 yr at 8% interest) and operation and maintenance costs were 13.1¢/1000 gal, for a total of 34.2¢/1000 gal. These costs do not include the costs of disposing wastes from the plant.
4. Over the 2-yr period of operation, 98.2% of the water pumped from the well was distributed to the system after nitrate reduction to approximately 6.8 mg/l nitrate-nitrogen. The 1.8% not distributed was discharged as wastewater.
5. The amount of wastewater produced per 1000 gal distributed water consisted of 1.4 gal brine, 6.6 gal rinse water, and 10.3 gal backwash water.
6. Experimental work was continued on the development of resins with nitrate-to-sulfate selectivity. One resin, a tributyl amine strong-base resin, showed unusually high selectivity. A U.S. patent was issued for the use of this resin in nitrate removal as a result of this work.

Table 8.9 Operating Data for McFarland, California Ion-Exchange Plant for August, 1985

	Gallons of water				Average nitrate mg/l in blended water		Gallons of wastewater			
Date	To system	Treated by IX	By-passed	Gallons saturated brine	As NO_3	As N	Dilute brine	Slow rinse	Backwash	Total gallons wastewater
1	854,000	608,200	245,800	718	27.0	6.11	1,440	11,560	7,990	20,990
2	907,300	636,600	270,700	658	25.6	5.80	1,460	14,173	10,646	26,279
3	907,300	636,600	270,700	658	—	—	1,460	14,173	10,646	26,279
4	907,300	636,600	270,700	658	—	—	1,460	14,173	10,646	26,279
5	867,000	604,700	262,300	719	28.0	6.34	1.540	14,420	10,480	26,440
6	933,000	640,400	292,600	718	27.6	6.25	1,540	12,310	8,000	21,850
7	900,000	636,800	263,200	547	29.0	6.57	1,160	12,930	10,030	24,920
8	957,000	657,200	299,800	711	31.0	7.02	1,470	14,420	10,670	26,560
9	879,700	605,200	274,500	697	30.0	6.79	1,590	14,286	9,727	25,603
10	879,700	605,200	274,500	697	—	—	1,590	14,286	9,727	25,603
11	879,700	605,200	274,500	697	—	—	1,590	14,286	9.727	25,603
12	905,000	627,100	277,900	855	29.0	6.57	6,930	18,030	13,310	38,270
13	902,000	618,900	283,100	719	27.0	6.11	1,390	14,420	10,610	26,420
14	925,000	649,100	275,900	718	29.0	6.57	1,620	14,420	9,960	26,000
15	953,000	708,300	244,700	665	28.4	6.43	1,640	14,420	11,470	27,530
16	926,000	829,600	96,400	539	29.4	6.66	1,250	10,820	8,110	20,180
17	906,000	623,300	282,700	719	30.4	6.88	1,120	11,920	10,465	23,505
18	906,000	623,300	282,700	719	—	—	1,120	11,920	10,465	23,505
19	912,000	626,800	285,200	719	27.4	6.20	1,670	19,420	10,520	31,610
20	948,000	650,700	297,300	718	25.6	5.80	1,650	12,790	7,190	21,630
21	955,000	650,600	304,400	712	27.4	6.20	1,610	13,830	10,520	25,960
22	274,000	189,000	85,000	180	31.0	7.02	340	5,830	5,380	11,550
23	931,300	434,200	497,100	760	29.8	6.75	1,696	13,526	9,743	24,965
24	931,300	434,200	497,100	760	—	—	1,696	13,526	9,743	24,965
25	931,300	434,200	497,100	760	—	—	1,696	13,526	9,743	24,965

Table 8.9 Operating Data for McFarland, California Ion-Exchange Plant for August, 1985 (continued)

	Gallons of water				Average nitrate mg/l in blended water		Gallons of wastewater			
Date	To system	Treated by IX	By-passed	Gallons saturated brine	As NO_3	As N	Dilute brine	Slow rinse	Backwash	Total gallons wastewater
26	1,250,100	1,250,100	—	539	29.0	6.57	1,240	14,170	10,210	25,620
27	473,500	322,400	151,100	359	26.6	6.02	825	7,215	5,240	13,280
28	473,500	322,400	151,100	359	—	—	825	7,215	5,240	13,280
29	921,500	621,700	299,800	643	28.4	6.43	1,490	12,620	9,080	23,190
30	921,500	621,700	299,800	643	—	—	1,490	12,620	9,080	23,190
31	954,500	627,100	327,400	705	30.2	6.84	1,680	14,420	12,960	29,060
Total	27,172,500	18,737,400	8,435,100	20,269	28.5[a]	6.45[a]	49,278	407,675	298,128	755,081

[a] Average value.

From Guter, G. A., "Nitrate Removal from Contaminated Water Supplies, Vol. II," EPA/600/52-87/034, 1987b, U.S. Environmental Protection Agency, Cincinnati, OH.

7. During the 1985–1986 period, over 250 tons salt were consumed in the nitrate removal process. The water containing these waste salts was disposed of to the McFarland wastewater collection system where it was blended with raw municipal wastewater, treated in aeration ponds, and disposed of to 120 acres of irrigated cotton crops. The disposal of this large quantity of waste salt to the environment poses serious questions about the fate of these materials and their impact on the local environment.

One method developed to provide simultaneous removal of nitrate and sulfate, and also removal of hardness, is called the CARIX process. The CARIX process is based on the combined use of a weakly acidic cation-exchanger in the hydrogen ion form and a strongly basic anion-exchanger in the bicarbonate (HCO_3^-) form. Both resins are regenerated simultaneously with carbon dioxide in a nonpolluting fashion. In pilot-plant studies in different waterworks in West Germany, Hoell and Feuerstein (1985) have shown that the CARIX process allows a more efficient and economical partial demineralization of water than other possible options. Urbanisto and Bays (1985) include a paper on combined hardness and nitrate/sulfate removal from groundwater by the CARIX ion-exchange process.

Finally, Richard (1988) reported on a case study involving nitrate removal from groundwater in France. The village of Ormes-Sur-Voulzie is supplied with groundwater with a flow rate of 0.17 mgd. The nitrate concentration in the groundwater is 60 mg/l (as nitrate), with an alkalinity of 16 mg/l (as calcium carbonate). The Ormes-Sur-Voulzie plant treats only part of the flow and mixes the nontreated water with treated water to yield a nitrate concentration of equal to or less than 25 mg/l (as nitrate). The plant services a small distribution system and uses ultraviolet rays as the disinfectant. The Ormes-Sur-Voulzie plant accomplishes regeneration by an upward flow of regenerant, as shown in Figure 8.4 (Richard, 1988). The resin is held in place by a mechanical blocking system. Rinsing and regeneration takes place in the fixing column. The effluent from the Ormes-Sur-Voulzie plant regeneration cycle is stored and then pumped into the local sewer system at a constant flow.

Reverse Osmosis

Treatment process comparisons for removal of nitrate from groundwater have typically included reverse osmosis and electrodialysis along with ion-exchange. Reverse osmosis (RO) refers to a process whereby ionic species (e.g., nitrates in groundwater) present in water are removed by forcing the water to be transported across a semipermeable membrane and effectively leaving the nitrates behind. This process is accomplished by subjecting the water supply in the RO cell to pressures exceeding its corresponding osmotic pressure. The RO process was developed over 25 years ago as a desalination technology for seawater. Such pressures can easily reach 300 to 400 psig when treating brackish water and up to 1000 psig when used in desalinating seawater (Dahab and Bogardi, 1990).

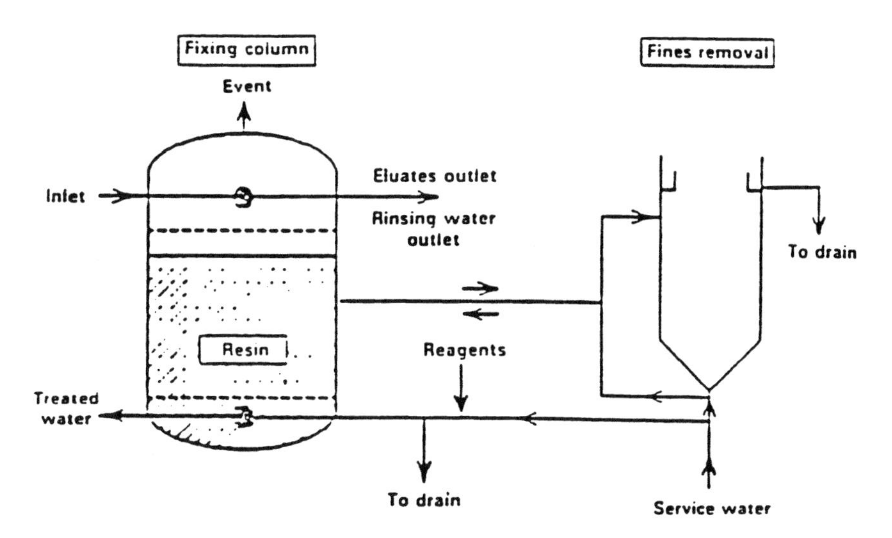

Figure 8.4 Counter-current regeneration in upward flow denitrification (UFD) process. (From Richard, Y. R., *Journal of the Institution of Water and Environment Management*, April, 1988, pp. 154–167.)

Membranes in RO units are often comprised of cellulose acetate and similar polymeric materials (i.e., polyamides); they must be able to withstand system pressures. The membranes generally do not exhibit high selectivity for any given ion, although the degree of salt rejection seems to be directly related to the valency of ions present in the water. As a consequence, the RO process generally results in removal of many ionic species, including nitrates. Potential problems associated with RO membranes include fouling, compaction, hydrolytic deterioration, and concentration polarization. There are currently no RO plants in the U.S. that are specifically designed to remove nitrates from water supplies. However, there are some full-scale RO plants in the U.S. that were built for total dissolved solids (TDS) reduction. It should be noted that it might be possible to economically justify RO for nitrate removal by allocating a significant portion of the process capital and operating costs to the reduction of dissolved solids or hardness if they are present in the groundwater and need to be reduced (Dahab and Bogardi, 1990).

Five RO membrane elements have been tested in a pilot study by the Charlotte Harbor Water Association in Harbor Heights, Florida (Huxstep and Sorg, 1988). The pilot test involved a 5000-gallons-per-day system that was modified to accept the five different RO membranes. The flow diagram for the pilot system is shown in Figure 8.5 (Huxstep and Sorg, 1988). The groundwater was spiked with a variety of contaminants, and test runs of 1 to 13 days were made for all the spiked contaminants and several naturally occurring substances (Huxstep and Sorg, 1988). The five membranes tested were: Toray® SC 3100, Filmtec® BW 30-4021, DOW® 5K, Dupont® B-9, and Hydranautics® P/N 4040 (SY-IFCI). The spiked contaminants included fluoride, cadmium, mercury, chromium, arsenic, selenium, nitrate, and lead. Limited tests were conducted on removal of nitrate, molybdenum, copper, uranium, and radium. During the test runs, data were also collected

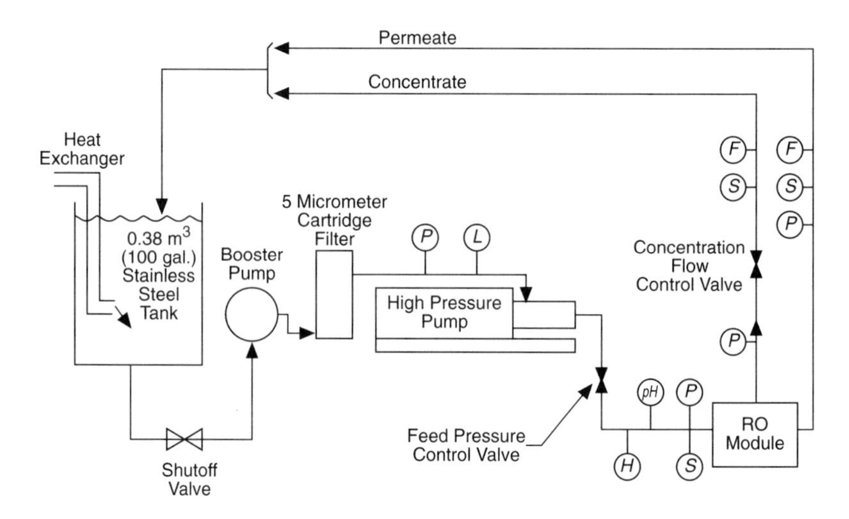

Figure 8.5 Flow diagram of RO research units. L, low pressure shutoff switch; H, high pressure shutoff switch; P, pressure valve; S, sample valve; F, flowmeter; pH, pH meter. (From Huxstep, M. R. and Sorg, T. J., "Reverse Osmosis Treatment to Remove Inorganic Contaminants from Drinking Water," EPA/S2-87/109, 1988, U.S. Environmental Protection Agency, Cincinnati, OH.)

on hardness, chloride, sulfate, TDS, calcium, and sodium (Huxstep and Sorg, 1988).

Removal of the spiked and natural contaminants was generally in agreement for all five RO membranes, with removal above 95% exhibited for arsenic V, calcium, cadmium, chromium III and VI, copper, lead, molybdenum, sodium, radium, selenium IV and VI, uranium, hardness, and TDS. Moderate removals of 85 to 94% were obtained for nitrate, nitrite, fluoride, and chloride. The lowest removals (below 85%) were obtained for arsenic III and inorganic mercury (Huxstep and Sorg, 1988). Wide variations of the removal of arsenic III, mercury, fluoride, and nitrate occurred. Variability in the removal of these contaminants was found to be a function of the chemistry of the contaminants, test waters, and test conditions (Huxstep and Sorg, 1988).

Electrodialysis

Electrodialysis (ED) refers to an electrically driven unit operation in which ions are selectively transported through semipermeable membranes from one solution to another under the influence of a direct current electric field. Electrodialysis membranes are typically used in cell pairs, the pairs (or stages) are alternately placed as cation (such as sodium, calcium, and magnesium) and anion (such as chloride, sulfate, nitrate, and bicarbonate) transfer membranes. When a direct current is passed through the membrane array, ions are selectively transferred through the membranes (i.e., anion or cation). As a result, alternate compartments are formed in which the ionic concentration is greater or less than the concentration in the original water. By manifolding the appropriate compartments, treated water (low electrolyte concentration) and brine (high electrolyte concentration) can be collected (Dahab and Bogardi, 1990).

The ED process was developed over 25 years ago for desalination of brackish or seawater. Desalination by ED is based on the following principles (Montgomery, 1985).

1. Most of the salts dissolved in water are ionic and thus can be removed under the influence of an electromagnetic field.
2. The ions, both positively and negatively charged, are attracted by an opposite electrical charge, i.e., positive ions (cations) are attracted to a negative electrode (cathode) and negative ions (anions) are attracted to a positive electrode (anode).
3. Membranes can be devised that will be selective in the type of charged ion they will pass or reject; that is, membranes can be made that allow negative ions (anions) to pass but reject cations, and vice versa.

There are three fundamental components in an ED system: (1) a supply of pressurized water, (2) a membrane stack consisting of several stages, and (3) a direct current power supply. Flow through an ED system varies with the pressure drop across the membrane stack. Current designs have a maximum pressure of about 60 psi; pressures higher than this are likely to cause excessive leakage through the membranes. A maximum pressure of 60 psi will limit the number of ED stages to 8 to 10, due to pressure drops across the stages.

ED membranes can be susceptible to fouling by calcium carbonate, barium, calcium, and strontium sulfates, iron and manganese oxides, colloids, microorganisms, and organic chemicals. These problems may be reduced via pretreatment (e.g., coagulation, settling, filtration, and activated carbon adsorption) and/or by addition of a small amount of acid to the feed stream. Organic fouling can be reduced by periodically cleaning the membranes with an enzyme detergent solution (Dahab and Bogardi, 1990).

The development of nitrate-selective membranes could prove the ED process a viable technology for nitrate removal from groundwater; however, the energy demands of ED systems cause them to be more expensive when compared to ion-exchange or biological denitrification.

TREATMENT BY BIOLOGICAL DENITRIFICATION

Biological processes are often used to biochemically convert constituents of concern to acceptable end-products. For domestic wastewater, this involves biologically converting the organics in the wastewater to acceptable end-products (water and carbon dioxide). The microorganisms that are active in treating domestic wastewater utilize the carbon and nitrogen in the wastewater as nutrients; and free oxygen (O_2) is added to serve as an electron acceptor for the microbes (thus, the microorganisms are referred to as aerobic, utilizing free oxygen).

In the absence of free oxygen as an electron acceptor, certain microorganisms can utilize other compounds as electron acceptors. For example, some microorganisms are able to utilize nitrate as an electron acceptor (this situation can occur under anoxic conditions — no free oxygen). When nitrate is utilized as an electron acceptor, it is reduced to nitrogen gas. This process is referred to as biological denitrification. Table 8.10 contains summary comments from 16 references dealing with denitrification measures for groundwater treatment. The basic concept of denitrification was described in Chapter 2.

Table 8.10 References Describing Denitrification Measures for Groundwater Treatment

Author(s) (year)	Comments
Anderson (1977)	Patented process for biological denitrification of nitrate-containing waters
Battelle Pacific Northwest Laboratories (1989)	Laboratory studies to determine the applicability of biological denitrification for removal of nitrates in Hanford groundwaters
Bockle, Rohmann, and Wertz (1986)	Description of process involving heterotrophic denitrification in an activated carbon filter and aerobic post-treatment in an underground system
Bourbigot (1985)	Removal of nitrogen from river water via injection into an alluvial aquifer
Dahab and Bogardi (1990)	Review of fundamentals of biological denitrification process; summary of case studies of usage for water treatment
Dahab and Lee (1988)	Study of biodenitrification in static-bed upflow reactors
Davidson and Cormack (1993)	Studies of sulfur-mediated biological denitrification of nitrate-contaminated groundwater
Gauntlett and Zabel (1982)	Use of fluidized sand beds or expanded sand beds for biological denitrification
Gayle, Boardman, Sherrard, and Benoit (1989)	Review of European literature on biological denitrification of water
Ginocchio (1984)	Carbon sources for promotion of biological denitrification
Lance (1985)	Promotion of denitrification via the use of groundwater recharge for wastewater renovation
Leprince and Richard (1982)	Pilot plant testing of denitrification using biological reactions, aeration, and filtration
Putzien (1984)	Discussion of technical problems in achieving biological denitrification
Richard (1988)	Results from the usage of the Nitrazur process at two water plants in France
Van der Hoek and Klapwuk (1987)	Use of a combination of ion exchange and biological denitrification for nitrate removal from groundwater
World Health Organization (1985)	Description of biological denitrification using sulfur substrate and heterotrophic bacteria

Microorganisms that can utilize nitrate as an electron acceptor are often facultative; they can utilize free oxygen if it is present, but will switch to nitrate in the absence of free oxygen. Facultative microbes also require organic carbon and nitrogen as nutrients. For a wastewater or groundwater that has nitrate as a contaminant, in the absence (or at low levels) of free oxygen and in the presence of adequate organic carbon (carbon:nitrate ratios of approximately 3:1), biological denitrification can convert the nitrate to nitrogen gas (an acceptable environmental constituent). Frequently, it is necessary to add organic carbon (possibly in the form of methanol) to the wastewater, thereby allowing biological denitrification to occur. Thus, for biological denitrification, the electron acceptor is present in the wastewater or groundwater (nitrate) and an organic carbon source is typically added. This is in contrast to conventional domestic wastewater treatment for the removal of organics where the organic carbon source is present in the wastewater and the electron acceptor (free oxygen) must be added.

Biological denitrification of groundwater involves contacting facultative microorganisms with groundwater containing nitrates and an added carbon source in an anoxic environment. Under these conditions, the bacteria utilize nitrates as a terminal electron acceptor in lieu of molecular oxygen. In the process, nitrates are reduced to nitrogen gas. The carbon source is necessary since it supplies the energy required by the microorganisms for respiration and synthesis. Most denitrification studies for groundwater have used methanol (CH_3OH) as the carbon source. The stoichiometric relationships for this process are as follows (Dahab and Bogardi, 1990).

Bacterial energy reaction, step 1:

$$6\ NO_3^- + 2\ CH_3OH \rightarrow 6\ NO_2^- + 2\ CO_2 + 4\ H_2O$$

Bacterial energy reaction, step 2:

$$6\ NO_2^- + 3\ CH_3OH \rightarrow 3\ N_2 + 3\ CO_2 + 3\ H_2O + 6\ OH^-$$

Overall energy reaction:

$$6\ NO_3^- + 5\ CH_3OH \rightarrow 3\ N_2 + 5\ CO_2 + 7\ H_2O + 6\ OH^-$$

Bacterial synthesis reaction:

$$3\ NO_3^- + 14\ CH_3OH + CO_2 + 3\ H^+ \rightarrow 3\ C_5H_7O_2N + H_2O$$

In the bacterial synthesis reaction, nitrate is converted to cell tissue, and $C_5H_7O_2N$ is a representation of the tissue formed. It is estimated that 25 to 30% of the added methanol is used for bacterial synthesis. On the basis of experimental laboratory studies, the following empirical equation was developed to describe the overall nitrate-removal reaction (Dahab and Bogardi, 1990):

$$NO_3^- + 1.8\ CH_3OH + H^+ \rightarrow 0.065\ C_5H_7O_2N + 0.47\ N_2 + 0.76\ CO_2 + 2.44\ H_2O$$

If only nitrate is present and organic carbon is limited, as in most nitrate-contaminated groundwater, the overall reaction can be used to determine the methanol requirement. If some nitrite and dissolved oxygen are present in the groundwater, the methanol requirement would be correspondingly higher (Dahab and Bogardi, 1990).

Bacteria involved in the denitrification process are generally facultative anaerobes that can use nitrite and nitrate as electron acceptors. The process occurs in four distinct steps as follows (Dahab and Bogardi, 1990).

$$NO_3^- \rightarrow NO_2^- \rightarrow NO^- \rightarrow N_2O \rightarrow N_2$$

When an organic compound serves as the electron donor, the process is heterotrophic and the bacteria are known as heterotrophs. Bacteria using hydrogen and reduced sulphur as electron donors are known as autotrophs. Bacterial genera known to contain denitrifying species include *Achromobacter, Alcaligenes, Bacillus, Chromobacter, Corynebacterium, Halobacterium, Methanonas, Moraxella, Paracoccus, Propionibacterium, Pseudomonas, Spirillium, Thiobacillus,* and *Xanthonas* (Dahab and Bogardi, 1990).

One of the basic requisites for biological denitrification is the presence of a carbon source; the source in the subsurface environment, wherein *in situ* biological denitrification occurs, could be either dissolved organic carbon or soil organic carbon (Trudell, Gillham, and Cherry, 1986). For designed processes treating pumped groundwater, the carbon source could be methanol (Andreoli, et al., 1979; Gauntlett and Zabel, 1982), cellulosic material (Anderson, 1977), acetic acid (Gauntlett and Zabel, 1982; Ginocchio, 1984), ethanol (Ginocchio, 1984), glucose (Ginocchio, 1984; Trudell, Gillham, and Cherry, 1986), or molasses (Ginocchio, 1984). When glucose is used as the carbon source, the process of denitrification can be represented by the following reaction (Trudell, Gillham, and Cherry, 1986):

$$4\ NO_3^- + 5/6\ C_6H_{12}O_6 + 4\ H^+ \rightarrow 2\ N_2 + 5\ CO_2 + 7\ H_2O$$

The biological denitrification process for nitrate removal from abstracted groundwater can occur in either suspended or attached growth systems. In suspended growth systems, the bacterial culture is suspended within the contents of the reactor vessel by mixing. In these systems, sedimentation is required as a follow-on process to settle the bacterial biomass so it can be returned to the reactor vessel. In attached growth systems, the bacterial biomass is physically attached to a solid matrix. Attached growth systems can utilize static media or an expanded-bed system. In static media systems, the solid matrix is typically comprised of synthetic modules stacked in the reactor vessel. Static media attached growth systems can be operated in either a downflow or upflow pattern. Upflow systems are fairly common due to the reduced chance of plugging associated with their operation and the fact that the bacterial biomass is constantly submerged (Dahab and Bogardi, 1990).

Expanded-bed systems are operated in an upflow manner so that the bacterial growth matrix bed is expanded hydraulically. In expanded-bed systems, the support media are generally of the granular type (both natural and synthetic) to facilitate bed expansion. As the bed is expanded, the entire surface of the granular material becomes available for bacterial support (Dahab and Bogardi, 1990).

The heterotrophic denitrification process can be considered in four stages as shown in Figure 8.6 (World Health Organization, 1985). The functions of the four stages are summarized in Table 8.11 (World Health Organization, 1985).

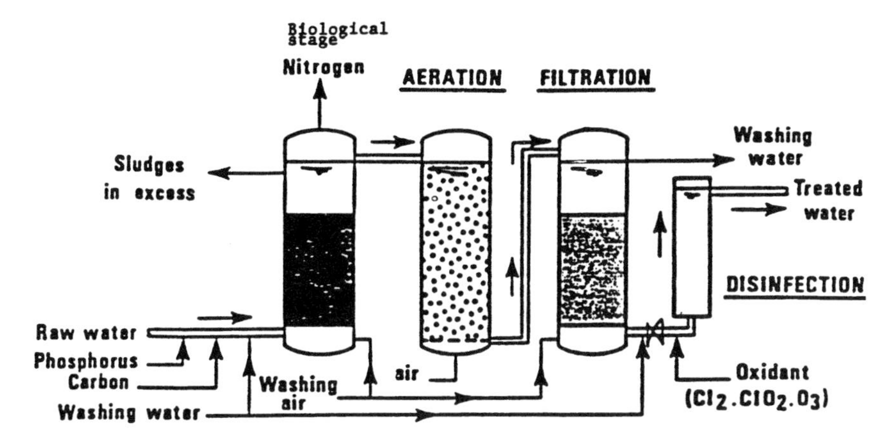

Figure 8.6 Four stages in the heterotrophic denitrification process. (From World Health Organization, "Health Hazards from Nitrates in Drinking Water," 1985, Regional Office for Europe, Copenhagen, Denmark, pp. 73–94. With permission.)

Table 8.11 Functions of the Four Stages in a Heterotrophic Denitrification Process

	Stage			
	Biological	Aeration	Filtration	Disinfection
Nitrate	++	0	+	0
Total organic carbon	–	0	+	0
Dissolved oxygen	–	++	0	0
Turbidity	–	0	++	0
Bacteria	–	0	+	++

Note: ++ = Large positive effect relative to removal of listed constituent; + = Positive effect relative to removal of listed constituent; 0 = No effect on listed constituent; – = Negative effect relative to removal of listed constituent.

From World Health Organization, "Health Hazards from Nitrates in Drinking Water," 1985, Regional Office for Europe, Copenhagen, Denmark, pp. 73–94. With permission.

Table 8.12 contains a summary of biological denitrification processes used for nitrate removal from groundwater (or possibly surface water) (Dahab and Bogardi, 1990). The processes include both suspended and attached growth systems. A detailed review of European practices related to biological denitrification of drinking water is given in Gayle, et al. (1989). Comparative advantages and disadvantages of the biological denitrification process for nitrate removal from groundwater are shown in Table 8.13 (Dahab and Bogardi, 1990).

Several illustrations can be cited of denitrification as an engineered process for pumped groundwater containing high nitrate levels. For example, Anderson (1977) proposed a process for biological denitrification of nitrate-containing waters, particularly agricultural drainage waters, wastewater effluents, and

Table 8.12 Summary of Biological Denitrification Processes for Nitrate Removal from Water Supplies

Process description	Location	Influent conc NO_3/l	Removal %
Sulfur limestone filtration, (P.S.) Q = 35 m^3/h	Monterland, Netherlands	70–75	95–100
Fixed-film reactor biological filtration, (F.S.) Q = 400 m^3/h	Guernes, France	40–65	—
Biological filters, (F.S.), downflow	Eragny, France	60	50
Upflow static bed reactor, (P.S.)	Lincoln, Nebraska	100	95–100
"Nitrazur," upflow, (F.S.) Q = 50 m^3/h	Chateau Landon, France	80–90	72
"Nitrazur," upflow, (F.S.) Q = 70 m^3/h	Champfleur, France	72	65
"Biodenit," downflow, (F.S.) Q = 80 m^3/h	Eragny, France	64	100
Slow sulfur limestone filtration, (P.S.)	The Netherlands	65–70	>90
Biological filters, (F.S.)	Eragny, France	160–170	84
"Denitropur" process, comprises four fixed-film upflow, (F.S.)	Monchengiabach Rassein, FRG	75	95–100
"Denitropur" process, fixed-film, packed with floating styroper spheres, (F.S.)	Langenfield & Monheim, Germany	65	95
Downflow filters, (F.S.)	Neuss, Germany		92
Fluidized-bed, (P.S.), experiment with two flows	Toulese cedex, France	150	90
Fixed-film granular bed, demonstration plant, sugar substrate nitrate, nitrite	Italy	40–50	80–90
Biological upflow, bed filters, fixed culture, (P.S.)	Great Britain	N.R.	78–100
Fluidized bed with varying methanol dose, (P.S.)	Buklesham, Great Britain	62	100
Fluidized bed, (F.S.)	Stevenage, Great Britain	67	63
Autotrophic denitrification, (P.S.)	Eragny, France	109–168	78–85
Autotrophic denitrification, (P.S.)	France	80	90–100
Fixed-bed rotating biological contactor, (F.S.)	Lawrence, California	60–80	91–93
Soil aquifer dune infiltration, (F.S.)	Castricum, The Netherlands	100	72
Submerged upflow type anaerobic filters, (P.S.)	France	100–150	N.R.
Post-treatment using $FeCl_3$, following fixed-film process to reduce effuent nitrite conc.	Germany	18 NO_2/l	100
Packed bed with polystyrene beads, (P.S.)	France	55	95
Columns packed with immobilized Pseudomonas cells, (P.S.)	Europe	104	—
Cost evaluation of substrates in a fluidized-bed, (P.S.)		—	—
Three-stage process of oxygen removal, denitrification, and reaeration, (P.S.)	Germany	N.A.	—
Autotrophic, using hydrogen in fluidized bed reactor, (P.S.)	Switzerland	N.A.	—
Autotrophic, columns packed with elemental sulfur and activated carbon, (P.S.)	West Germany	35	—
Autotrophic, calcium alginate beads suspended in incomplete-mixed reactor		120	—

Table 8.12 Summary of Biological Denitrification Processes for Nitrate Removal from Water Supplies (continued)

Process description	Location	Influent conc NO_3/l	Removal %
Combined ion-exchange bio-denitrification, (B.S.)	The Netherlands	87–88	40–50
Use of hydrogen-otrophic denitrifiers in a polyurethane carrier reactor, (P.S.)	Belgium	200–220	80
Bank filtration, soil aquifer, (F.S.)	Rhine River, Germany	N.R.	75

Note: P.S. = pilot scale; F.S. = full scale; B.S. = bench scale; N.R. = no report; N.A. = not available.

From Dahab, M. F. and Bogardi, I., "Risk Management for Nitrate-Contaminated Groundwater Supplies," 1990, U.S. Geological Survey, Reston, VA, pp. 76–103.

Table 8.13 Advantages and Disadvantages (Limitations) of Biological Denitrification for Nitrate Removal from Abstracted Groundwater

Advantages

1. The process is cheaper to install with comparable operation and maintenance cost to other treatment alternatives.
2. The excess biological growth produced as waste is much easier and less expensive to dispose of than waste salts and brines from physical-chemical methods. Furthermore, these solids can safely be discharged to local wastewater treatment systems without causing any long-term problems.
3. The process is extremely effective in reducing nitrates to virtually near-zero concentration in the treated water regardless of the influent water concentration.
4. Process stability is excellent, particularly when using static-media reactor systems (i.e., biofilm systems).
5. The process does not impart excess undesirable chemicals such as chlorides to the treated water.
6. Biological treatment, in general, is probably better suited to the removal of various toxic and hazardous micro-pollutants (if such pollutants are in groundwater) than most physical-chemical systems.

Disadvantages

1. Extreme variations in groundwater characteristics, such as dissolved oxygen, total organic carbon (TOC), nutrient concentrations, pH, temperature and the presence of inhibitors can contribute to performance variability, particularly in suspended growth systems. Because groundwater supplies are generally uniform in quality and temperature, this disadvantage may not be consequential to biological denitrification.
2. The carbon source level is critical in denitrification operations. Insufficient supply may result in high levels of nitrates or nitrites, while overdosing will probably result in high concentrations of residual carbon in the treated water.
3. Products of microbial activity, such as endotoxins, soluble microbial products, and incompletely degraded organic compounds, may be imparted to the treated water.
4. Start-up periods for biological processes are generally long and proper maintenance and monitoring of biomass accumulation and its composition is required. However, excellent performance of biological treatment units can be achieved by proper application of process principles, regular laboratory testing, and frequent monitoring and assessment.

From Dahab, M. and Bogardi, I., "Risk Management for Nitrate-Contaminated Groundwater Supplies," 1990, U.S. Geological Survey, Reston, VA, pp. 76–103.

groundwaters. The process focused on waters containing from 10 to about 250 mg/l nitrate-nitrogen and an insufficient quantity of organic material to support the necessary anaerobic bacterial action for effecting reduction of substantial quantities of the nitrate content. The water was contacted with a solid substrate, usually a cellulosic material comprised chiefly of paper (or paper pulp) in the presence of an anaerobic bacterial culture with sufficient time and favorable growth conditions to permit the anaerobic bacteria to undergo an active growth phase and effect reduction of the nitrate content of the water.

Gauntlett and Zabel (1982) described a fluidized sand bed for removing up to 10 mg/l nitrates with a bed depth of 4 to 5 m, temperatures as low as 2°C, and an upflow rate of 20 m/hr. Methanol was an effective source of carbon for the denitrifying bacteria; in contrast, acetic acid produced excessive slime and occasional high nitrate levels in treated water. Sand cleaning and aeration were necessary parts of the fluidized bed process. An expanded sand bed, an alternative to the fluidized bed, was found to remove 10 mg/l nitrogen at an upflow rate of 14 m/hr, temperatures of 3 to 5°C, and a bed depth of 1.5 m (Gauntlett and Zabel, 1982).

Dahab and Bogardi (1990) reported on laboratory-scale experiments using biological denitrification for nitrate reduction in groundwater supplies in Nebraska. Results from several studies indicate that the process can be expected to reduce the nitrate concentration in the influent water supply from as high as 100 mg/l (as N) to levels within the 1.0-mg/l (as N) range. While high nitrate removal efficiencies have been observed, it should be noted that some residual soluble as well as insoluble organic matter should be expected in the denitrified water supply.

A laboratory study of nitrate removal from simulated groundwater using static-bed upflow reactors has been described by Dahab and Lee (1988). Two laboratory reactors were constructed of plexiglass tubes with an inner diameter of 125 mm and a height of 1.20 m. The reactors were packed with different synthetic media as the bacterial support matrices; one medium was made of spherical modules about 25 mm in diameter, and the other of 16-mm cylindrical Pall rings. Sampling ports were installed along each reactor at 150-mm intervals.

A feed solution was metered to these reactors using positive displacement tubing pumps (Dahab and Lee, 1988). The solution simulated groundwater with a high nitrate concentration, and was fortified with trace elements and a buffer. Potassium nitrate was used as the nitrate source, and acetic acid was added as a carbon source. A small quantity of sodium sulfite was added, along with a cobalt chloride catalyst, to eliminate dissolved oxygen and ensure anoxic conditions in the systems. The reactors were housed in a constant temperature room at 25°C. Table 8.14 displays the operational variables used during the study (Dahab and Lee, 1988).

Based on the experimental conditions, it was found that both reactors removed nearly 100% of the nitrate if the influent chemical oxygen demand (COD) was at the minimal stoichiometric requirement. A carbon:nitrogen ratio of 1.5 minimized the effluent COD. Reducing the carbon:nitrogen ratio below 1.5 caused the breakthrough of nitrates. Also, it was noted that nitrate was present in increasing amounts past the initial 0.3-m increment of reactor height as the hydraulic retention time (HRT) was reduced from 36 to 9 hours under the given

Table 8.14 Reactor Operational Sequence for Laboratory Study of Biological Denitrification

Study phase	HRT[a] (hours)	NO_3^-–N (mg/l)	COD (mg/l)	C:N	Days of operation
Start-up	72	100	534	2.0	72
1	36	100	534	2.0	40
2	18	100	534	2.0	37
3	9	100	534	2.0	25
4	36	100	400	1.5	36
5	18	100	400	1.5	29
6	9	100	400	1.5	33
7	36	100	267	1.0	33

[a] HRT = hydraulic retention time.

From Dahab, M. F. and Lee, Y. W., *Journal of the Water Pollution Control Federation*, Vol. 9, No. 9, 1988, pp. 1670–1674. With permission.

operating conditions. The following empirical formulation was suggested for denitrification when using acetic acid as the carbon source (Dahab and Lee, 1988):

$$NO_3^- + 0.85\ CH_3COOH + H^+ \rightarrow 0.10\ C_5H_7O_2N + 0.45\ N_2 + 1.2\ CO_2 + 2.1\ H_2O$$

Two biological denitrification processes have been developed in France; they are referred to as (Richard, 1988): (1) the heterotrophic Nitrazur process, which uses acetic acid and ethanol as a carbon source; and (2) the Biodenit process, also heterotrophic, which uses ethanol as a carbon source. Figure 8.7 displays the components of a typical Nitrazur plant (Richard, 1988). The nitrogen gas released by the denitrification process is entrained by the co-current upflow to the surface of the reactor without increasing the loss of head. The plant then operates under gravity flow. Figure 8.8 is a schematic of the Biodenit process (Richard, 1988). The Biodenit process is operated under pressure because of the head loss in the filter after accumulation of nitrogen gas. Operational results from two Nitrazur plants in France have provided encouraging performance data and cost information (Richard, 1988).

The Orange County Water District in California has conducted laboratory bioreactor studies involving autotrophic denitrification (Davidson and Cormack, 1993). Elemental sulfur was used as an energy source for biological denitrification catalyzed by the autotrophic bacterium, *Thiobacillus denitrificans*. The bacterium, in the absence of free oxygen, reduces nitrate to nitrogen gas by oxidizing sulfur to sulfate. The carbon required for biosynthesis was derived from carbon dioxide/bicarbonates present in the groundwater. A fluidized bed bioreactor and an agitated (stirred) tank reactor was studied. Both reactor types were capable of maintaining prolonged periods of stable operation (effluent containing less than 0.3 mg/l nitrate from influent streams containing from 45 to 100 mg/l nitrate). The process resulted in a pH drop from approximately 8.2 (influent) to 7.0 (effluent). It was found that 1.64 mg sulfate needed to be added to the nitrate-laden groundwater for every 1.0 mg nitrate removed.

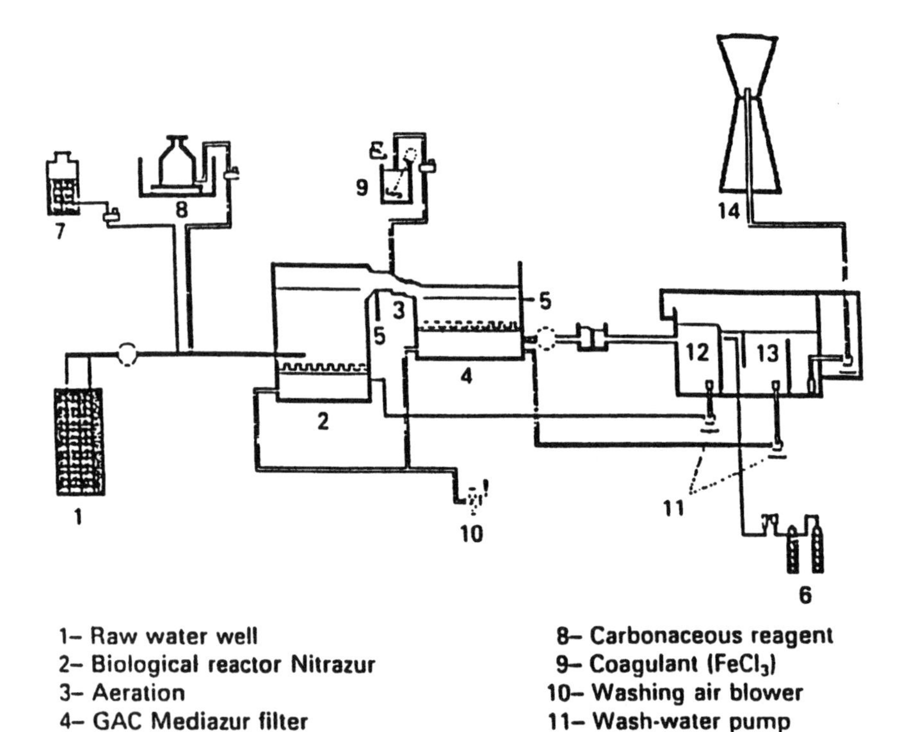

Figure 8.7 General layout of Nitrazur process. (From Richard, Y. R., *Journal of the Institution of Water and Environment Management,* April, 1988, pp. 154–167.)

Van der Hoek and Klapwuk (1987) have suggested a combined ion-exchange and denitrification process as a means of minimizing operational problems associated with each separate technique. A schematic of the combined process is in Figure 8.9 (Van der Hoek and Klapwuk, 1987). One ion-exchange column (column 1) is used for production of potable water, while another ion-exchange column (column 2) is regenerated via the combined use of the center denitrification column (reactor). When ion-exchange column 1 is exhausted and ion-exchange column 2 is regenerated, the denitrification reactor is then connected to the exhausted ion-exchange column 1 and the regenerated resin (column 2) is used for potable water production.

The regeneration process for the combined system is depicted in Figures 8.10 and 8.11 (Van der Hoek and Klapwuk, 1987). Regeneration can be accomplished with either a sodium chloride solution or a sodium bicarbonate solution. For example, from Figure 8.11 where $NaHCO_3$ is the regenerant, the solution is passed over the ion-exchange column to exchange nitrate ions for bicarbonate ions. After passage, the nitrate-rich regenerant passes through the denitrification

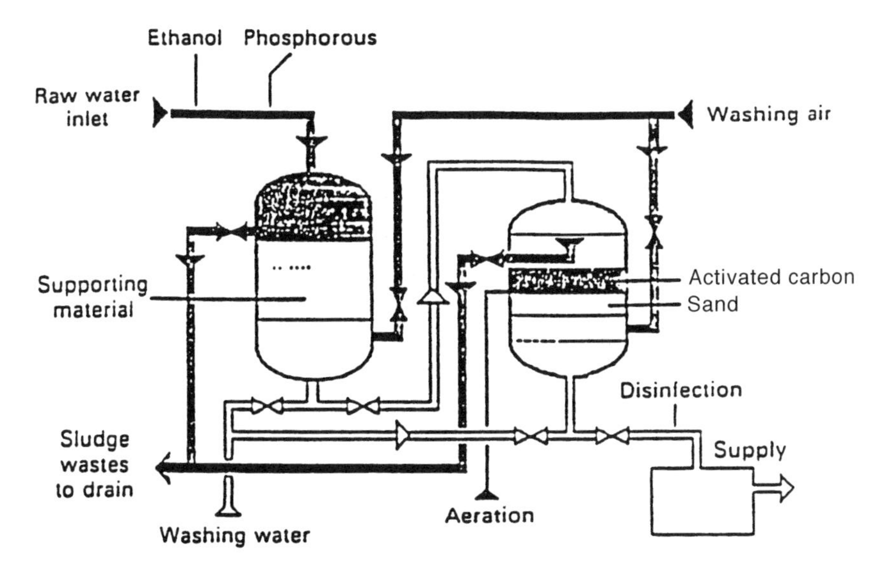

Figure 8.8 General layout of Biodenit process. (From Richard, Y. R., *Journal of the Institution of Water and Environment Management*, April, 1988, pp. 154–167.

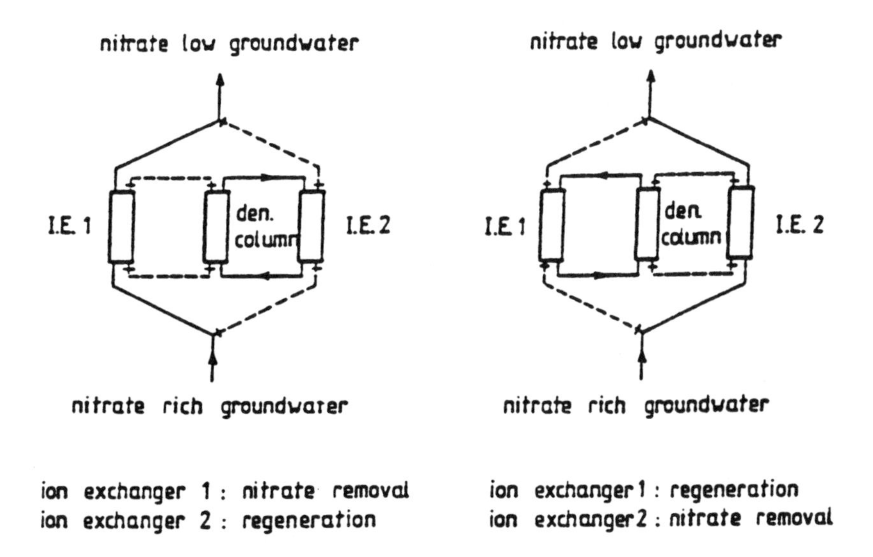

Figure 8.9 Combination of ion exchange and biological denitrification: biological/physical-chemical nitrate removal from groundwater. (From Van der Hoek, J. P. and Klapwuk, A., *Water Research*, Vol. 21, No. 8, 1987, pp. 989–997. With permission of Elsevier Science, Kidlington, U.K.)

reactor where denitrifying bacteria convert nitrate to nitrogen gas. The added organic source (methanol) is converted into bicarbonate, carbonate, and water. The regenerant continues to be recirculated through the ion-exchange column

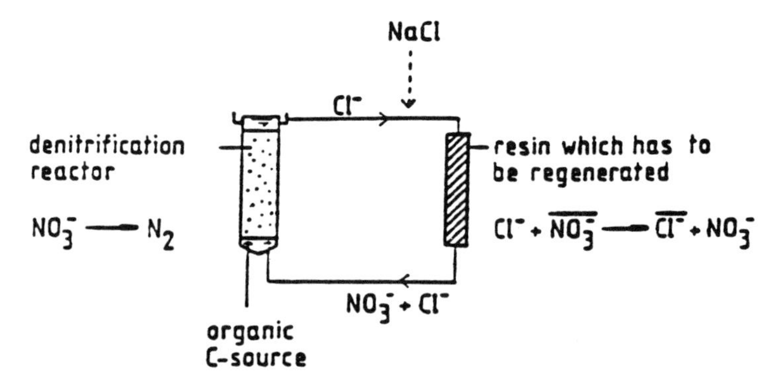

Figure 8.10 Regeneration of a nitrate-loaded resin into the chloride form with a denitrification reactor. (From Van der Hoek, J. P. and Klapwuk, A., *Water Research*, Vol. 21, No. 8, 1987, pp. 989–997. With permission of Elsevier Science, Kidlington, U.K.)

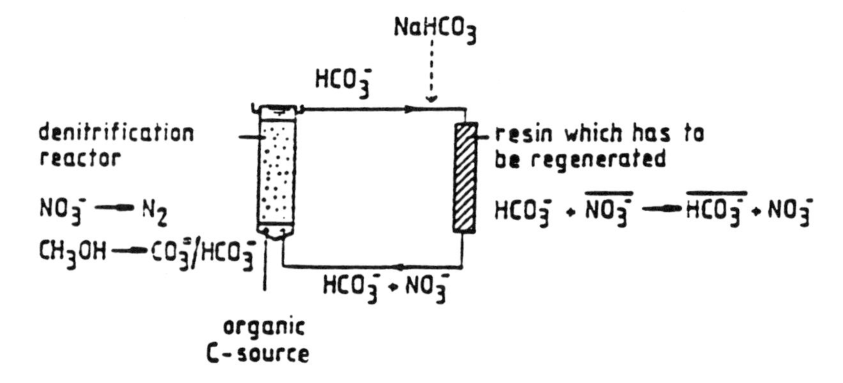

Figure 8.11 Regeneration of a nitrate-loaded resin into the bicarbonate form with a denitrification reactor. (From Van der Hoek, J. P. and Klapwuk, A., *Water Research*, Vol. 21, No. 8, 1987, pp. 989–997. With permission of Elsevier Science, U.K.)

and the denitrification reactor, until the ion-exchanger has reached a sufficient bicarbonate loading (Van der Hoek and Klapwuk, 1987).

The basic design criteria for the combined process, along with initial pilot plant results, were presented by Van der Hoek and Klapwuk (1987). The identified advantages of the combined process include (Van der Hoek and Klapwuk, 1987):

1. The regeneration is conducted in a closed system in which the production of a voluminous brine can be avoided and the salt requirements are minimized. The use of $NaHCO_3$ as a regenerant has the advantage that the system itself produces the salt necessary for regeneration because bicarbonate is an end-product of biological denitrification. When NaCl is used as the regenerant, only the stoichiometric required amount has to be dosed.
2. As the biological process does not take place in direct contact with the groundwater, there is no risk that nitrite production will affect the water quality.
3. There is no direct contact of bacteria and the carbon source with the groundwater, thus minimizing concomitant contamination. However, pollution of the

resin by carryover of suspended material from the denitrification reactor to the ion-exchange column is possible. Conversely, measures against this can be taken in the regeneration circuit itself, so there is no need for extensive post-treatment.

TREATMENT BY *IN SITU* DENITRIFICATION

Table 8.15 contains summary comments from 12 references dealing with *in situ* biological denitrification. *In situ* biological denitrification is similar to the conventional denitrification process except that it is conducted in the subsurface environment without having to pump the groundwater from the aquifer, and without having to provide the process equipment, tanks, and related physical plant appurtenances, instruments, and supplies (Dahab and Bogardi, 1990). The potential advantages of *in situ* denitrification are summarized in Table 8.16 (Dahab and Bogardi, 1990).

Table 8.15 References Addressing *in situ* Biological Denitrification

Author(s) (year)	Comments
Andreoli et al. (1979)	Use of methanol as a carbon source for promoting denitrification beneath a subsurface wastewater disposal system
Braester and Martinell (1988a)	Description of Nitredox process for *in situ* denitrification
Braester and Martinell (1988b)	Mathematical modeling of fluid flow and transport processes in Nitredox subsurface treatment plants
Chalupa (1988)	Laboratory, pilot plant, and full-scale studies of *in situ* heterotrophic denitrification in Czechoslovakia
Dahab and Bogardi (1990)	Summary of advantages of *in situ* denitrification; and description of four alternative methods for accomplishment
Heinonen-Tanski and Airaksinen (1988)	Laboratory studies of *in situ* denitrification in Finland
Janda, Rudovsky, Wanner, and Marha (1988)	Use of ethyl alcohol as an organic carbon source in pilot plant and full-scale studies of *in situ* denitrification
Kristiansen (1981)	Use of sand-filter trenches for denitrification of septic tank effluents
Mercado, Libhaber, and Soares (1988)	Pilot-plant studies of two alternative *in situ* denitrification methods in Israel
Sikora and Keeney (1976)	Column studies to evaluate a sulfur-Thiobacillus denitrificans nitrate removal system for septic tank effluents
Soares, Belkin, and Abeliovich (1988)	Laboratory denitrification studies using a sand column and sucrose as a carbon source
Trudell, Gillham, and Cherry (1986)	Field study of the occurrence and rate of denitrification in a shallow unconfined sand aquifer

Four fundamental designs for *in situ* denitrification include (Dahab and Bogardi, 1990):

1. Same well for recharge and pumping: In this design one well is used for recharge of nitrate-polluted groundwater with substrate. At a later stage, nitrate-free water is pumped from the same well after a period of time has allowed

Table 8.16 Potential Advantages of *in situ* Biological Denitrification

1. Cost savings could be realized in plant capital investment for site development, equipment, recirculation pumping, piping tankage, control instrumentation, and related physical facilities.
2. Savings in physical plant operating costs by reducing the energy used to operate treatment units. Also, savings would be realized by reducing the need for water pretreatment to avoid membrane fouling such as pre-chlorination and pH adjustment.
3. Savings in plant maintenance and replacement costs. It is appropriate to point out that frequent membrane or resin replacement is one of the major costs associated with physical-chemical nitrate removal systems.
4. Savings in plant residuals disposal costs. These residuals include brines and regenerants from ion exchange, reverse osmosis, and electrodialysis systems and excess sludge from bio-denitrification plants. These residuals usually must be treated and disposed of in an environmentally acceptable method at considerable cost.
5. The need for highly trained operations and maintenance personnel is significantly reduced. This would result in additional treatment cost savings.
6. Subsurface bio-denitrification appear to be more environmentally sound than other nitrogen removal systems since no waste products are generated and no ecological changes should occur if indigenous bacterial populations are utilized.

From Dehab, M. and Bogardi, I., "Risk Management for Nitrate-Contaminated Ground Water Supplies," 1990, U.S. Geological Survey, Reston, VA, pp. 76–103.

for sufficient biodenitrification to take place and the creation of a nitrate-free zone in the aquifer in the immediate area of the well. Generally, one additional well is needed to provide recharge water. Because this system is dependent on intermittent operation, several recharge and discharge wells are required to provide continuous operation.

2. Horizontal doublet system: This system utilizes two separate wells for injection and pumping at nominal distances ranging from 10 to 20 m. This system can supply nitrate-free water of constant quality on a continuous basis, without the need of any surface treatment excluding disinfection. In this system, additional wells are also required to supply water for recirculation.
3. Vertical doublet system: This system consists of two wells of different depths, located in close proximity to each other or two wells constructed in one large-diameter borehole. In this system, one well is used for injection of nitrate-contaminated water, while the other is for the recovery of denitrified water. The basic difference between the horizontal doublet scheme and the vertical scheme is that in the latter, the biological reaction and the filtration zones are located vertically between the two wells. As in the previous case, an additional well is also needed to recirculate water through the system to accelerate the reaction rate.
4. Daisy system: The Daisy system consists of a production well surrounded by smaller diameter substrate injection wells. Organic substrate and nutrients are introduced into the aquifer through the injection wells by dilution using either treated water from the production well or from other nearby wells.

One field study of denitrification in a shallow groundwater system was conducted by Trudell, Gillham, and Cherry (1986). Information from this study can be used in the development of a design approach for *in situ* denitrification. To provide direct evidence of denitrification within the saturated zone, and to determine the rate of denitrification, an *in situ* injection experiment was conducted using a specially designed injection-withdrawal-sampling drive point. Nitrate and a conservative tracer (bromide) were added to natural groundwater and injected at a 3-m depth into a shallow, unconfined sand aquifer. After 356 hr, the concentration of nitrate-nitrogen in the injected water declined from the initial 13 mg/l to less than 0.1 mg/l. The decrease in nitrate concentration was much greater than the corresponding decrease in the concentration of bromide, thus confirming a preferential loss of nitrate. The loss of nitrate was preceded by a decline in dissolved oxygen concentration to less than 0.1 mg/l, and coincided with an increase in bicarbonate concentration of 142 mg/l. The production of bicarbonate observed in the injection experiment, 2.59 mmole HCO_3^- per mmole nitrate denitrified, agreed with calculations associated with an equilibrium geochemical model of the denitrification process. An increase in the population of denitrifying organisms (from 1 to 23 organisms per gram of soil) was detected in core samples collected at the depth of injection (169 hours after the start of the experiment). The measured rate of denitrification ranged from 0.0078 to 0.13 mg/l nitrate-nitrogen per hour, and is in reasonable agreement with published rates for saturated soils. The organic carbon source required for denitrification was either dissolved organic carbon or soil organic carbon. Soil organic carbon, at 0.08 to 0.16% by weight, was adequate to denitrify nitrate in this field test.

Several field studies have been conducted to examine the possibility of using *in situ* denitrification for septic tank system effluents (Andreoli, et al., 1979; Kristiansen, 1981; Sikora and Keeney, 1976). Although conducted for septic tank system effluents, this information may be relevant for agricultural sources of nitrate pollution in that certain basic design considerations have been explored. For example, Andreoli, et al. (1979) described the design, construction, and operation of a full-scale system consisting of a conventional septic tank-leaching field wastewater disposal system combined with a subsurface system using natural soil treatment mechanisms to accomplish nitrogen removal from disposal system leachate. A pan-shaped impervious membrane was installed in the soil below the leaching field to detain the leachate. Methanol, used as a carbon source for denitrification, was injected into the pan. After more than 1 year of operation, it was found that the system was capable of removing nitrate-nitrogen to below the recommended public health limit for groundwater.

Kristiansen (1981) reported on a 17-month study of three sand-filter trenches for improving septic tank system effluent quality. Trenches A and C were loaded with septic tank effluent at a rate of 4 to 6 cm per day, while trench B received 12 to 18 cm per day. Ambient temperatures of 4 to 16°C were maintained in trenches B and C, while the temperature in A was 12 to 15°C. This study found that insignificant amounts of nitrogen were removed from the effluent passing through the filters. During the cold seasons, nitrification was significantly affected by the sand temperature. Thus, it appeared that denitrification did not occur in continuously loaded sand filters. Accordingly, some potential means of improving denitrification include intermittent loading, recycling of nitrified effluent, or the addition of a supplemental carbon source.

Finally, relative to septic tank systems, Sikora and Keeney (1976) evaluated a sulfur-*Thiobacillus denitrificans* nitrate removal system as a means of denitrifying nitrified septic tank effluent. Duplicate 10 × 64-cm columns were filled with a 1:1 mixture (by weight) of elemental sulfur and dolomite chips, and were pretreated by recycling an enrichment culture of *Thiobacillus denitrificans* ATCC 23642 through the columns for 3 days. Continuous passage of the nitrified septic tank effluent containing 40 μg nitrate/ml through the columns resulted in nearly complete nitrate removal in 3.3 hours at steady-state conditions. The denitrification kinetics appeared to be first order in the range of the nitrate concentrations used. Sulfate was the major sulfur end-product and was present at relatively high concentrations (90 μg/ml). Passage of the column effluent through subsequent 10 × 60-cm Plainfield sand columns did not significantly decrease sulfate levels.

NITREDOX is a Swedish trademark name that refers to an *in situ* denitrification method which, when coupled with naturally occurring microorganisms, can reduce nitrate to an acceptable concentration (Braester and Martinell, 1988a). A typical NITREDOX plant consists of several injection-pumping wells located on the circumference of two concentric circles and a pumping well in the center through which water, partly free of nitrates, and iron and/or manganese, is produced. Water with added carbon is injected through the wells located on the outer circle, while the wells located on the inner circle play the role of VYREDOX (another Swedish trademark name) injection wells. VYREDOX wells serve to inject degassed aerated water into the aquifer for purposes of iron and/or manganese removal. The NITREDOX process is associated with the formation of nitrogen gas, which is then removed through the wells located on the inner circle. Both processes (NITREDOX and VYREDOX) include flow phenomena, transport, chemical reactions, and bacteriological processes. These phenomena have been mathematically modeled by Braester and Martinell (1988b).

TREATMENT BY OTHER MEASURES

Table 8.17 contains summary comments from four references dealing with other miscellaneous measures for treating or blending groundwater with excessive nitrate concentrations. For example, Brown (1971) evaluated an algal system consisting of algae growth, harvesting, and disposal as a possible means of removing nitrate-nitrogen from subsurface agricultural drainage in the San Joaquin Valley, California. The study was initiated to determine optimum conditions for growth of the algal biomass, seasonal variations in assimilation rates, and methods of harvesting and disposal of the algal product. A secondary objective of the study was to obtain preliminary cost estimates and process design data. The growth studies showed that about 75 to 90% of the 20-mg/l influent nitrogen was assimilated by shallow (12-in. culture depth) algal cultures receiving 2 to 3 mg/l additional iron and phosphorus and a mixture of 5% carbon dioxide. The most economical and effective algal harvesting system tested was flocculation and sedimentation followed by filtration of the sediment. There was a market value for this product as a protein supplement.

In a comprehensive review of nitrates in surface and groundwaters in Poland, Roman (1982) identified several "nontreatment" methods for dealing with high nitrate concentrations in groundwater. Examples of such methods included elim-

Table 8.17 References Involving Other Measures for Groundwater Treatment or Blending

Author(s) (year)	Comments
Brown (1971)	Use of an algal system for removing nitrates from subsurface agricultural drainage
Murphy (1991)	Description of an aluminum addition chemical process involving nitrate reduction to ammonia, nitrogen, and nitrate
Roman (1982)	Blending of dual quality water supplies
Wang, Tian, and Yin (1984)	Use of ozonation and biological activated carbon for nitrate removal

ination or reduced utilization of groundwater sources with a high nitrate content; and a dual water supply system in which higher-quality water (in this case, low nitrate concentration) is distributed through a separate system for the human consumption and the food industry, while lower-quality water (higher nitrate content) is distributed through another network to the remaining users. From an economic point of view, the dual water supply system can be feasible when the increased costs of transportation and distribution of two types of water are compensated for by cost savings resulting from the elimination of expensive methods of nitrogen removal.

An intensified water treatment scheme has been evaluated by Wang, Tian, and Yin (1984). In order to determine effective processes for purifying polluted source waters in Harbin City (Heilongjiang Province, China), various processes consisting of ozonation (O3), sand filtration (SF), and/or granular activated carbon (GAC) filtration and adsorption, i.e., ozonation ("O3" process), ozonation/sand filtration ("O3 + SF" process), ozonation/biological activated carbon ("O3 + BAC" process), ozonation/sand filtration/biological activated carbon ("O3 + SF + BAC" process), and granular activated carbon ("GAC" process) were tested in an 8 m^3/d capacity pilot plant. In addition, a small plant of 500-l/d capacity was used to conduct comparative studies between two processes ("GAC" and "O3 + BAC"), as well as two types of carbon. Of the processes tested, both the "O3 + BAC" and "O3 + SF + BAC" processes were most effective in removing various pollutants, including turbidity, color, odor, iron, manganese, organic substances, and ammonium, nitrate, and nitrite ions. Based on the results of this study, the two treatment flow sheets were suggested for purifying polluted surface and groundwater sources.

Murphy (1991) proposed a chemical process based on the addition of aluminum (Al) and the reduction of nitrate to nitrite and then to either ammonia or nitrogen gas. The potential reactions are:

$$3\ NO_3^- + 2\ Al + 3\ H_2O \rightarrow 3\ NO_2^- + 2\ Al(OH)_3$$

and from nitrite to either ammonia

$$NO_2^- + 2\ Al + 5\ H_2O \rightarrow NH_3 + 2\ Al(OH)_3 + OH^-$$

or nitrogen gas

$$2\ NO_2^- + 2\ Al + 4\ H_2O \rightarrow N_2 + 2\ Al(OH)_3 + 2\ OH^-$$

Optimum pH conditions for the above reactions are between 9.0 and 10.5 (Murphy, 1991). If ammonia is the final product, air stripping may be required to transfer the ammonia from the water phase to the atmosphere. The process can be controlled so as to minimize the aluminum concentration in the treated effluent. Cost comparisons relative to more conventional nitrate removal options are needed.

Finally, nitrate has actually been added to subsurface environments to enhance biorestoration of fuel spills. For example, nitrate has been used as the primary electron acceptor in the clean-up of a spill of jet fuel (JP-4) in a drinking water aquifer (Hutchins, et al., 1991a). Nitrate was chosen because it is more soluble than oxygen, and less costly and less toxic than hydrogen peroxide. Laboratory tests were conducted prior to the field study to determine degradation rates with the nitrate amendments (Hutchins, et al., 1991b). While this illustration is not focused on nitrate removal, the findings have potential relevance to *in situ* denitrification.

SUMMARY

This chapter has provided a review of several treatment measures for removing excessive concentrations of nitrate from groundwater systems. The measures are primarily for application to pumped groundwater. The most cost-effective treatment process is ion-exchange involving anionic resins. Several treatability studies involving laboratory work are described. In addition, more detailed information is presented on a pilot-plant study and the design and operation of a 1.0-mgd ion-exchange plant in McFarland, California. Other potential treatment measures described in this chapter include: (1) reverse osmosis, (2) electrodialysis, (3) denitrification, (4) algal growth and harvesting and disposal, (5) dual water supply and blending, and (6) combinations of ozonation, sand filtration, and granular activated carbon filtration. Biological denitrification of pumped groundwater is a viable treatment measure in many locations.

SELECTED REFERENCES

Anderson, D.R., "Biological Denitrification of Water," *Official Gazette of the U.S. Patent Office*, Vol. 961, No. 1, August, 1977, p. 253.

Andreoli, A., et al., "Nitrogen Removal in a Subsurface Disposal System," *Journal of the Water Pollution Control Federation*, Vol. 51, No. 4, April, 1979, pp. 841–854.

Battelle Pacific Northwest Laboratories, "Development of a Biological Process for Destruction of Nitrates and Carbon Tetrachloride in Hanford Groundwater," CONF-8910193-5, October, 1989, Richland, Washington.

Bilidt, H., "Use of Reverse Osmosis for Removal of Nitrate in Drinking Water," *Desalination*, Vol. 53, No. 1–3, September, 1985, pp. 225–230.

Bockle, R., Rohmann, U., and Wertz, A., "Process for Restoring Nitrate Contaminated Groundwaters by Means of Heterotrophic Denitrification in an Activated Carbon Filter and Aerobic Post Treatment Underground," *Aqua*, No. 5, 1986, pp. 286–287.

Bourbigot, M.M., "Artificial Injection of River Water into the Alluvial Aquifer of the Garonne for the Production of Potable Water," *Water Supply*, Vol. 3, No. 2, 1985, pp. 79–87.

Braester, C. and Martinell, R., "Vyredox and Nitredox Methods of *In Situ* Treatment of Groundwater," *Water Science and Technology*, Vol. 20, No. 3, 1988a, pp. 149–163.

Braester, C. and Martinell, R., "Modelling of Flow and Transport Processes in Vyredox and Nitredox Subsurface Treatment Plants," *Water Science and Technology*, Vol. 20, No. 3, 1988b, pp. 165–188.

Brown, R.L., "Removal of Nitrate by an Algal System," DWR-174-10, April, 1971, State Department of Water Resources, Fresno, California.

Buelow, R.W., et al., "Nitrate Removal by Anion-Exchange Resins," *Journal of the American Water Works Association*, Vol. 67, No. 9, September, 1975, pp. 528–534.

Chalupa, M., "Problems in Czechoslovakia Regarding Methods of Removal of Nitrates from Drinking Water," *Water Science and Technology*, Vol. 20, No. 3, 1988, pp. 211–213.

Clifford, D.A. and Weber, W.J., "Nitrate Removal from Water Supplies by Ion Exchange," EPA-600/2-78-052, June, 1978, U.S. Environmental Protection Agency, Cincinnati, Ohio.

Dahab, M.F., "Treatment Alternatives for Nitrate Contaminated Groundwater Supplies," *Journal of Environmental Systems*, Vol. 17, No. 1, 1987, pp. 65–75.

Dahab, M. and Bogardi, I., "Risk Management for Nitrate-Contaminated Groundwater Supplies," November, 1990, U.S. Geological Survey, Reston, Virginia, pp. 76–103.

Dahab, M.F. and Lee, Y.W., "Nitrate Removal from Water Supplies Using Biological Denitrification," *Journal of the Water Pollution Control Federation*, Vol. 9, No. 9, September, 1988, pp. 1670–1674.

Davidson, M.S. and Cormack, T., "Bioreactor Treatment of Nitrate Contamination in Ground Water: Studies on the Sulfur-Mediated Biological Denitrification Process," *Symposium on Bioremediation of Hazardous Wastes: Research, Development, and Field Evaluations*, EPA/600/R-93/054, 1993, U.S. Environmental Protection Agency, Dallas, Texas, pp. 161–165.

Gauntlett, R.B., "Nitrate Removal from Water by IonExchange," *Water Treatment and Examination*, Vol. 24, Part 3, 1975, pp. 172–193.

Gauntlett, R. and Zabel, T.F., "Biological Denitrification for Potable Water Treatment," *Water Services*, Vol. 86, No. 1031, January, 1982, pp. 17–18.

Gayle, B.P., Boardman, G.D., Sherrard, J.H., and Benoit, R.E., "Biological Denitrification of Water," *Journal of Environmental Engineering*, Vol. 115, No. 5, October, 1989, pp. 930–943.

Ginocchio, J., "Nitrate Levels in Drinking Water Are Becoming Too High," *Water Services*, Vol. 88, No. 1058, April, 1984, pp. 143, 147.

Goodrich, J.A., Lykins, Jr., B.W., and Clark, R.M., "Drinking Water from Agriculturally Contaminated Groundwater," *Journal of Environmental Quality*, Vol. 20, No. 4, October-December, 1991, pp. 707–717.

Guter, G.A., "Nitrate Removal from Contaminated Water Supplies: Vol. I. Design and Initial Performance of a Nitrate Removal Plant," EPA/600/S2-86/115, April, 1987a, U.S. Environmental Protection Agency, Cincinnati, Ohio.

Guter, G.A., "Nitrate Removal from Contaminated Water Supplies: Vol. II," EPA/600/S2-87/034, August, 1987b, U.S. Environmental Protection Agency, Cincinnati, Ohio.

Guter, G.A., "Removal of Nitrate from Contaminated Water Supplies for Public Use," EPA-600/S2-81/029, April, 1981, U.S. Environmental Protection Agency, Cincinnati, Ohio.

Heinonen-Tanski, H. and Airaksinen, A.K., "The Bacterial Denitrification in Groundwater," *Water Science and Technology*, Vol. 20, No. 3, 1988, pp. 225–230.

Hoell, W. and Feuerstein, W., "Combined Hardness and Nitrate/Sulfate Removal from Groundwater by the Carix Ion Exchange Process," *Water Supply*, Vol. 3, No. 1, 1985, pp. 99–109.

Horan, N.J., Ed., *Nutrient Removal from Wastewaters*, Technomic Publishing Company, Inc., Lancaster, Pennsylvania, 1994.

Horng, L.L., "Modeling of Ion Exchange Process," *Journal of the Chinese Institute of Chemical Engineers*, Vol. 16, No. 2, April, 1985, pp. 91–101.

Hutchins, S.R., Downs, W.C., Smith, G.B., Wilson, J.T., Hendrix, D.J., Fine, D.D., Kovacs, D.A., Douglass, R.H., and Blaha, F.A., "Nitrate for Biorestoration of an Aquifer Contaminated with Jet Fuel," EPA/600/S2-91/009, April, 1991a, U.S. Environmental Protection Agency, Ada, Oklahoma.

Hutchins, S.R., Sewell, G.W., Kovacs, D.A., and Smith, G.B., "Biodegradation of Aromatic Hydrocarbons by Aquifer Microorganisms Under Denitrifying Conditions," *Environmental Science and Technology*, Vol. 25, No. 1, January, 1991b, pp. 68–76.

Huxstep, M.R., "Inorganic Contaminant Removal from Drinking Water by Reverse Osmosis," EPA-600/S 2-81-115, October, 1981, Charlotte Harbor Water Association, Inc., Harbor Heights, Florida.

Huxstep, M.R. and Sorg, T.J., "Reverse Osmosis Treatment to Remove Inorganic Contaminants from Drinking Water," EPA/S2-87/109, March, 1988, U.S. Environmental Protection Agency, Cincinnati, Ohio.

Janda, V., Rudovsky, J., Wanner, J., and Marha, K., "*In Situ* Denitrification of Drinking Water," *Water Science and Technology*, Vol. 20, No. 3, 1988, pp. 215–219.

Kristiansen, R., "Sand-Filter Trenches for Purification of Septic Tank Effluent: Part II. The Fate of Nitrogen," *Journal of Environmental Quality*, Vol. 10, No. 3, July-September, 1981, pp. 358–361.

Lance, J.C., "Denitrification During Wastewater Renovation Groundwater Recharge," *Proceedings of Second International Conference on Ground Water Quality Research*, 1985, Oklahoma State University, Stillwater, Oklahoma, pp. 22–24.

Leprince, A. and Richard, Y., "The Use of Bio-Technics in Water Treatment: Feasibility and Performance of Biological Treatment of Nitrates," *Aqua*, No. 5, 1982, pp. 455–462.

Lockridge, J.E., "New Selective Anion-exchange Resins for Nitrate Removal from Contaminated Drinking Water and Studies on Analytical Anion Exchange Chromatography," IS-T-1395, January, 1990, Ames Laboratory, U.S. Department of Energy, Ames, Iowa.

Mercado, A., Libhaber, M., and Soares, M.I., "*In Situ* Biological Groundwater Denitrification: Concepts and Preliminary Field Tests," *Water Science and Technology*, Vol. 20, No. 3, 1988, pp. 197–209.

Montgomery, J.M., Consulting Engineers, Inc., *Water Treatment, Principles and Design*, John Wiley & Sons, New York, New York, 1985.

Murphy, A.P., "Chemical Removal of Nitrate from Water," *Nature*, Vol. 350, March 21, 1991, pp. 223–225.

Musterman, J.L., et al., "Removal of Nitrate, Sulfate and Hardness from Groundwater by Ion Exchange," Report No. 112, February, 1983, Water Resources Research Institute, Iowa State University, Ames, Iowa.

Putzien, J., "Nitrates in Drinking Water—Problems of Hygiene and Treatment," *Acta Hydrochimica et Hydrobiologica*, Vol. 12, No. 6, 1984, pp. 577–593.

Rautenbach, R., et al., "Electrodialysis for Nitrate Removal from Groundwater," *Gas und Wasserfach, Wasser Abwasser*, Vol. 126, No. 7, July, 1985, pp. 349–355.

Rautenbach, R. and Henne, K.H., "Removal of Nitrates from Groundwater by Reverse Osmosis," RFP-TRANS-409, August, 1983, Rockwell International, Golden, Colorado.

Richard, Y.R., "Operating Experiences of Full-Scale Biological and Ion-Exchange Denitrification Plants in France," *Journal of the Institution of Water and Environment Management*, April, 1988, pp. 154–167.

Roman, M., "Possibilities of Controlling Nitrate Concentrations in Drinking Water," *Proceedings of IIASA Task Force Meeting on Nonpoint Nitrate Pollution of Municipal Water Supply Sources: Issues of Analysis and Control*, Series No. CP-82-54, February, 1982, International Institute of Applied Systems Analysis, Laxenberg, Austria, pp. 117–134.

Sheinker, M. and Coduluto, J.P., "Making Water Supply Nitrate Removal Practicable," *Public Works*, Vol. 108, No. 6, June, 1977, pp. 71–73.

Sikora, L.J. and Keeney, D.R., "Evaluation of a Sulfur-Thiobacillus Denitrificans Nitrate Removal System," *Journal of Environmental Quality*, Vol. 5, No. 3, July-September, 1976, pp. 298–303.

Soares, M.I., Belkin, S., and Abeliovich, A., "Biological Groundwater Denitrification: Laboratory Studies," *Water Science and Technology*, Vol. 20, No. 3, 1988, pp. 189–195.

Sorg, T.J., "Compare Nitrate Removal Methods: For Some Communities, Ion Exchange May Be the Most Economical, Practical Way of Handling Nitrate-Contaminated Groundwater," *Water and Wastes Engineering*, Vol. 17, No. 12, December, 1980, pp. 26–31.

Sorg, T.J. and Love, O.T., "Reverse Osmosis Treatment to Control Inorganic and Volatile Organic Contamination," EPA-600/D-84-198, July, 1984, U.S. Environmental Protection Agency, Cincinnati, Ohio.

Trudell, M.R., Gillham, R.W., and Cherry, J.A., "*In-situ* Study of the Occurrence and Rate of Denitrification in a Shallow Unconfined Sand Aquifer," *Journal of Hydrology*, Vol. 83, No. 3–4, March, 1986, pp. 251–268.

U.S. Environmental Protection Agency, "Cost Supplement to Technologies and Costs for the Removal of Nitrates and Nitrites from Potable Water Supplies," February, 1987, Office of Drinking Water, Washington, D.C.

U.S. Environmental Protection Agency, *Nitrogen Control*, Technomic Publishing Company, Inc., Lancaster, Pennsylvania, 1994, pp. 35–36.

U.S. Water News, "Des Moines Adds Nitrate Removal, But System May Not Be Enough," undated a.

U.S. Water News, "Largest Denitrification Plant Being Built in California," undated b.

Urbanisto, R. and Bays, L.R., Eds., *Proceedings of the Specialized Conference on Contamination of Groundwater and Groundwater Treatment*, April, 1985, International Water Supply Association, London, England.

Van der Hoek, J.P. and Klapwuk, A., "Nitrate Removal from Ground Water," *Water Research*, Vol. 21, No. 8, 1987, pp. 989–997.

Wang, B., Tian, J., and Yin, J., "Purification of Polluted Water Source Water with Ozonation and Biological Activated Carbon," *Ozone: Science and Engineering*, Vol. 6, No. 4, 1984, pp. 245–260.

World Health Organization, "Health Hazards from Nitrates in Drinking Water," 1985, Regional Office for Europe, Copenhagen, Denmark, pp. 73–94.

INDEX

H

I

J

K

L

M

N

O

P

R